Genetics and Evolution of Abiotic Stress Tolerance in Plants

Genetics and Evolution of Abiotic Stress Tolerance in Plants

Editor

Patrizia Galeffi

MDPI • Basel • Beijing • Wuhan • Barcelona • Belgrade • Manchester • Tokyo • Cluj • Tianjin

Editor
Patrizia Galeffi
BIOAG Division
ENEA
Rome
Italy

Editorial Office
MDPI
St. Alban-Anlage 66
4052 Basel, Switzerland

This is a reprint of articles from the Special Issue published online in the open access journal *Genes* (ISSN 2073-4425) (available at: www.mdpi.com/journal/genes/special_issues/Abiotic_Stress_Tolerance_Plants).

For citation purposes, cite each article independently as indicated on the article page online and as indicated below:

LastName, A.A.; LastName, B.B.; LastName, C.C. Article Title. *Journal Name* **Year**, *Volume Number*, Page Range.

ISBN 978-3-0365-7409-7 (Hbk)
ISBN 978-3-0365-7408-0 (PDF)

Cover image courtesy of Patrizia Galeffi

Contents

About the Editor

Patrizia Galeffi

Graduated in Biology *cum Lauda* at the University of Rome La Sapienza (1982). PhD Thesis in "Molecular and Cellular Basis of Human Biology" (1987, Excellent judgement) at University of Rome, "La Sapienza". The PhD experimental work was partially carried out at the Laboratory directed by Prof. G.G. Brownlee, Institute Sir William Dunn, School of Pathology, University of Oxford, UK (1985-86). Other training periods were: (a) at the International Institute of Genetic and Biophysics, IIGB, CNR Naples, Italy, directed by Prof. F. Blasi (1983); (b) at the Laboratory of Neurobiology, CNR, Roma, Italy, directed by Prof. Rita Levi Montalcini (September 1987–April 89).

Since May 1989, enrolled as scientist (permanent staff) at ENEA CR Casaccia, Italy; responsible for a research team that worked on a pioneering project of "Antibodies expressed in Plants for virus resistance" (Nature 1993). Then, she was responsible for various projects funded by Italian and International organizations; chair-woman and invited speaker at International Conferences. Since 2004 up to today, she is ENEA-Responsible for an International collaboration with CIMMYT. Inventor of three patents, member of the ENEA Patenting Committee since 2018, and awarded by the EU certification as "Technology Transfer Manager". Tutor of several graduate/PhD students from national and international universities, and since 2020, Professor at La Sapienza for Plant Functional Genomics Course; evaluator of research projects, author of several peer-reviewed papers (around 70s), reviewer for many scientific journals with IF.

Scientific topics:

- Molecular studies of genes involved in drought plant response, as *DRF1* gene, expression profile under different water stresses in cereals;
- Plant Functional Genomics;
- Plant transcription factors;
- Genotyping cereals varieties;
- Beer production from triticale;
- Antibody molecular biology and engineering;
- Bioinformatics;
- Transgenic plants for production of useful pharmaceutical proteins.

Preface to "Genetics and Evolution of Abiotic Stress Tolerance in Plants"

This reprint represents a wide vision of what and how the research on plants at molecular level (genetics, genomics, transcriptomics, proteomics, metabolomics, and so on) contribute to a good equilibrium among human needs, food security, and future strategies for mitigating the effects of global climate changes.

Now more than ever, understanding the genetics and evolution of the gene mechanisms and the networks of different molecular pathways acting on plant abiotic stress tolerance has an important role to find new solutions for modern agricultural problems.

Scientists have the relevant task of increasing knowledge in the complex area of plant genetics and genomics, the genes responsive to specific abiotic stresses (such as drought, salts, or heat) and their inducible promoters, and various gene expression control and modulation mechanisms, including alternative splicing, micro-RNA interference, post-transcriptional mRNA decay, and post-translational protein degradation. At the same time, evolution has played a key role in the establishment of the current traits, so that major insights into the genetic diversity producing different alleles, adaptation, phylogenesis, and evolution of genomes and gene families can be translated and applied as tools for developing new tolerant plant varieties capable of satisfying the needs in terms of food security, protection of the planet, and conservation and recovery of natural resources such as water and soils.

This reprint is full of technical and specialized terms and, for this reason, it is addressed to scientists and students trained in plant functional genomics, breeding, agronomy, and genetics. It seems an exciting virtual tour through plant molecular responses to various environmental stresses and new ideas and applications will be derived.

I wish to thank all colleagues and authors who contributed with their valuable work to the success of this Special Issue, submitting their articles, up to the publishing the present reprint. Furthermore, I wish to thank the MDPI Editorial group and the *Genes* editorial officers for their constant help and presence along all period dedicated to this Special Issue.

Patrizia Galeffi
Editor

genes

Editorial

Genetics and Evolution of Abiotic Stress Tolerance in Plants

Patrizia Galeffi

Italian National Agency for New Technologies, Energy and Sustainable Economic Development, ENEA, Casaccia Research Center, 00123 Rome, Italy; patrizia.galeffi@enea.it

Citation: Galeffi, P. Genetics and Evolution of Abiotic Stress Tolerance in Plants. *Genes* **2022**, *13*, 1380. https://doi.org/10.3390/genes13081380

Received: 26 May 2022
Accepted: 10 June 2022
Published: 1 August 2022

Publisher's Note: MDPI stays neutral with regard to jurisdictional claims in published maps and institutional affiliations.

Now more than ever, the understanding of the genetics and evolution of the gene mechanisms and the networks of different molecular pathways acting on plant abiotic stress tolerance has an important role in the finding of new solutions and approaches mitigating the effects of global climate changes, thus contributing to a correct equilibrium among human needs, food security and human health and wellbeing.

Science and research have a major role in this context.

Scientists have the important task of increasing the level of knowledge, which is still not sufficient, in the complex area of plant genetics and genomics, particularly in relation to genes responsive to specific abiotic stresses (such as drought, salts or heat) and their inducible promoters, and to various gene expression control and modulation mechanisms, including alternative splicing, micro-RNA interference, post-transcriptional mRNA decay, and post-translational protein degradation. At the same time, evolution has played a key role in the establishment of the current plant molecular traits, so that a better understanding of the genetic diversity producing different alleles, adaptation, phylogenesis, and evolution of genomes and gene families can be translated and applied as a tool for developing new tolerant plant varieties able to satisfy our needs, in terms of food security, protection of the planet, and conservation and recovery of natural resources, such as water and soils.

I found that reading the various articles published in the current Special Issue represented an exciting virtual tour through plant molecular responses to various environmental stresses. To put together this book, I have divided the 10 papers (9 original research manuscripts and 1 review) into three groups and identified a common thread linking one article to another to accompany the reader on this virtual tour through plant genetics and molecular responses to abiotic stresses.

The first group of papers consists of two interesting studies based on field trials: the first one written by Marco Dettori and colleagues and focused on the development of a method to predict the ability of different varieties of durum wheat to respond to drought/heat stress. The prediction is based on a statistical approach and mathematical model using a large number of field data collected over 10 years of field experiments. The second paper, written by Arun K. Joshi and his team, is based on a large quantity of field trial data collected over two seasons, using more than 3000 varieties of wheat grown in India and other Asiatic regions, where the climate produces a strong negative effect on the grain yield and other yield-related traits. Their agronomic evaluations relate to the molecular effects of the *Vrn-1* gene, the *Ppd-1* gene and their alleles on complex traits, such as flowering time and photoperiod sensitivity. Their research suggests how to avoid the damage due to heat/drought stress by using the large numbers of allele combinations involved to regulate growth habits and achieve optimal adaptation.

The second group includes two other papers: one written by Arianna Latini et al. regarding functional genomics research into the expression profiles of a drought-responsive transcription factor gene *(DRF1)* of durum wheat in fields under full- and reduced-irrigation conditions. This article reports the expression profiles of the three *TdDRF1* gene transcripts, from durum wheat genotypes during different plant growth stages. In addition, the expression profile of one putative target gene (*Wdnh13*) is investigated and some analogies are found. The results highlight significant differences in the molecular patterns and suggest that the expression profile of this stress-related transcription factor gene is genotype dependent. Furthermore, a statistical association between the expression of *TdDRF1* transcripts and agronomic traits is also revealed, with significant differences among genotypes.

The other article, written by Giuseppe E. Condorelli et al., focuses on a GWAS work which reveals 15 QTLs (Quantitative Trait Loci) for Osmotic Adjustment (OA) found to have a high capacity when positively associated with Relative Water Content and/or Chlorophyll Content, and a low capacity when negatively associated with leaf rolling, thus indicating the beneficial effect of the OA-QTLs on the global plant water status.

These two articles, moreover, succeed in the difficult task of bringing together genomics studies with extensive field experiments.

The third group of papers consists of five articles that span from transcriptomics to metabolomics and cover the molecular behavior of different genes taken as examples. The article, authored by Deheng Yao et al., reports a comparative transcriptome analysis of rosemary suspension cells carried out at different concentrations of Methyl Jasmonates (MeJA). A large number of differentially expressed genes were identified, with totals of, respectively, 7836, 6797, and 8310 genes from the lowest to the highest concentration. This study demonstrates a strong involvement of these genes in the regulation of the biosynthesis of active compounds relevant to the physiological response to stresses.

The article written by Ashraful Islam et al. analyzes the Trehalose-6-Phosphate Phosphatase (*TPP*) genes in *Triticum aestivum* L. and identifies their gene family, composed of 31 genes organized in 5 clades, also found in *Hordeum vulgare, Brachypodium distachyon,* and *Oryza sativa,* and provides evidence of an evolutionary status. These TPP genes are involved in trehalose-metabolism, which has a role in stress tolerance. Most of the results achieved in this study were obtained using bioinformatics and silico tools, which helped in the prediction and assessment of gene structure and functioning.

The article written by Joran K. Waititu et al. describes transcriptome profiling in *Zea mays* and reveals different key cold-responsive genes, transcription factors and metabolic pathways regulating cold stress tolerance, using a comparative approach on 24 cold-tolerant and 22 cold-sensitive inbred lines affected by cold stress at the seedling stage. Valuable insights arise from the identification of 2237 differentially expressed genes (DEGs), one-third of them being annotated.

The study written by Basmah M. Alharbi et al. focuses on the biochemical and molecular effects induced by Triacontanol (TRIA) in acquired drought stress tolerance in rice. After 10 days of plant exposure to drought stress, data regarding photosynthetic pigments, stomatal conductance and photochemical efficiency were collected and analyzed. The cultivar Giza 177 (*Oryza sativa* L.) was found to be sensitive to drought stress. TRIA treatment enhanced both growth and acquired plant tolerance, by increasing the content of free amino acids and sugars, improving the ability of plant tissues to retain more water under scarcity conditions and regulating Aquaporins (AQPs), which are a class of intrinsic proteins playing an important role in transmembrane water transport.

The article written by Pierre Jacob et al. focuses on the *Heat Shock Factor A2* (*HSFA2*) regulator and co-regulated genes involved in multiple environmental stress responses required for stress acclimation. They identified 43 genes strongly co-regulated with *HSFA2* during multiple stresses. A motif of the site II element (SIIE) found in the promoters of these genes was identified to be closely related to R2R3-MYB transcription factors TT2 and MYB5. The over-expression of these factors was also investigated by transient transformation to evaluate their involvement in heat stress tolerance.

Last but not least, the review written by Rahat Sharif et al. addresses the problem of the HD-ZIP gene family by trying to clarify, through a comprehensive literature analysis, the potential roles of this gene family in improving plant growth and regulating stress-responsive mechanisms in plants. There are many interesting insights and the reader will be positively impressed.

This Special Issue was "not easy to manage", as it encompasses many aspects of different genes with the role of regulating and modulating the expression profiles of themselves and other downstream genes in very complex genetic, molecular, and metabolic networks.

The common thread linking all these articles together is the challenging study of the big data coming from genomics, transcriptomics, and metabolomics with the aim of highlighting the various, still unknown, mechanisms and regulatory systems, that are underlying the abiotic stress tolerance responses in plants.

I wish to thank very much all the authors and their co-authors for their valuable contributions, their sharing of their scientific work, their experience, and their commitment to this Special Issue, which was a "difficult task" but, in the end, proved a success in terms of acquired knowledge, new ideas, and future perspectives.

Furthermore, I wish to thank all the editorial staff and the MDPI group for their tireless assistance to me and all the authors, resulting in the publication of our Special Issue in a captivating and interesting book with a certain original appeal.

Finally, good reading to all!

Funding: This research received no external funding.

Conflicts of Interest: The authors declare no conflict of interest.

genes

Article

Performance Prediction of Durum Wheat Genotypes in Response to Drought and Heat in Climate Change Conditions

Marco Dettori [1,*], Carla Cesaraccio [2], Pierpaolo Duce [2] and Valentina Mereu [1,3]

[1] Agricultural Research Agency of Sardinia, Viale Trieste 111, 09123 Cagliari, Italy; vmereu@agrisricerca.it
[2] Institute of BioEconomy (IBE), National Research Council (CNR), Traversa La Crucca 3, 07100 Sassari, Italy; carla.cesaraccio@ibe.cnr.it (C.C.); pierpaolo.duce@ibe.cnr.it (P.D.)
[3] Impacts on Agriculture, Forestry and Ecosystem Services (IAFES) Division, Euro-Mediterranean Center on Climate Changes (CMCC), Via E. de Nicola 9, 07100 Sassari, Italy
* Correspondence: mdettori@agrisricerca.it

Abstract: With an approach combining crop modelling and biotechnology to assess the performance of three durum wheat cultivars (Creso, Duilio, Simeto) in a climate change context, weather and agronomic datasets over the period 1973–2004 from two sites, Benatzu and Ussana (Southern Sardinia, Itay), were used and the model responses were interpreted considering the role of DREB genes in the genotype performance with a focus on drought conditions. The CERES-Wheat crop model was calibrated and validated for grain yield, earliness and kernel weight. Forty-eight synthetic scenarios were used: 6 scenarios with increasing maximum air temperature; 6 scenarios with decreasing rainfall; 36 scenarios combining increasing temperature and decreasing rainfall. The simulated effects on yields, anthesis and kernel weights resulted in yield reduction, increasing kernel weight, and shortened growth duration in both sites. Creso (late cultivar) was the most sensitive to simulated climate conditions. Simeto and Duilio (early cultivars) showed lower simulated yield reductions and a larger anticipation of anthesis date. Observed data showed the same responses for the three cultivars in both sites. The CERES-Wheat model proved to be effective in representing reality and can be used in crop breeding programs with a molecular approach aiming at developing molecular markers for the resistance to drought stress.

Keywords: climate change; drought tolerance; abiotic stress; crop modelling; durum wheat

Citation: Dettori, M.; Cesaraccio, C.; Duce, P.; Mereu, V. Performance Prediction of Durum Wheat Genotypes in Response to Drought and Heat in Climate Change Conditions. *Genes* **2022**, *13*, 488. https://doi.org/10.3390/genes13030488

Academic Editors: Patrizia Galeffi and Bin Yu

Received: 31 December 2021
Accepted: 8 March 2022
Published: 10 March 2022

1. Introduction

1.1. Climate Change: Overall Projected Effects

Climate change in the twenty-first century is projected to cause increasing mean air temperatures, more frequent and intense extreme events such as longer lasting heatwaves and droughts, and more variable precipitation and surface water flows, unless strong mitigation actions occur in the next decades [1]. IPCC reports increasing air temperatures ranging from 1.0–1.8 °C under SSP1-1.9, 2.1–3.5 °C under SSP2-4.5, and 3.3–5.7 °C under SSP5-8.5) where, "SSPx refers to the Shared Socio-economic Pathway describing the socio-economic trends underlying the scenario, and y refers to the approximate level of radiative forcing (in W m^{-2}) resulting from the scenario in the year 2100. SSP1-1.9 represents the low end of future emissions pathways. At the opposite end of the range, SSP5-8.5 represents the very high warming end of future emissions pathways from the literature. SSP2-4.5 represents a scenario with stronger climate change mitigation and lower GHG emissions" [1]. Climate change has already affected food security in several regions, with negative impacts especially at lower latitudes, while at high latitudes positive impacts have been recorded for some crops [2]. Recently released global projections of crop yields show an emergence of climate impacts (before 2040) on the major breadbasket regions, with larger losses for maize, soybean and rice and additional gains for wheat [3]. The projected positive effect for wheat is mainly due to the stronger CO_2 response of C3 crops with respect to C4 crops and the expected increase in wheat yields at high latitudes that are currently limited by non-optimal minimum air temperatures [3]. Notwithstanding the great uncertainty in the scientific debate about the effects of CO_2 on crop yields [4,5], due to the physiological and genetic complexity of the phenomena [6], the ability of crop models to capture the main effects of CO_2 on crop yields under various growing conditions is fully acknowledged. However, high uncertainty still remains in crop model responses under high CO_2 concentrations and further investigations are needed [6]. As an example, high temperatures may lower the beneficial effects of increased CO_2 by reducing grain number, size and quality, as shown in rice [7] and in soybean [8,9]. In addition, increased levels of CO_2 have negative impacts on food quality, by reducing the content of micronutrients such as iron and zinc [10], and on the protein content of cereals [11,12], with detrimental effects on the baking quality of wheat [13]. Importantly, the net effect of elevated CO_2, lower growing-season rainfall and high temperature will likely increase 'haying-off', thereby limiting production of rain-fed wheat in Mediterranean-type environments [14]. As a result, instability in yields owing to increasing temperatures and higher frequency of extreme events may overcome the positive effects of a slight temperature increase [15]. Focusing on cereals, the negative impact of climate change on yields is very likely due to heat stress, increased plant water demand causing a higher transpiration rate, and a shortened growing period as well as anticipated maturity [16–22].

Observations show decreases of wheat and maize yields due to climate change in many low-latitude areas, while increases are reported for high-latitudes during the recent decades [2]. Global yield projections show yield decreases for maize ranging from −6% under SSP1-2.6 to −24% under SSP5-8.5, while for wheat the global projected changes in crop yield range from +9% under SSP1-2.6 to +18% under SSP5-8.5, and for rice from +3% under SSP1-2.6 to +2% under SSP5-8.5 by the end of the century [3]. The larger yield losses are expected at lower latitudes, while at higher latitudes potential yield gains are projected, even if with high uncertainty associated to simulations with the most pessimistic scenario (SSP5-8.5) [3]. In addition, maize yields are expected to be highly affected by climate change throughout Europe, while wheat yields could even increase as a consequence of more favorable conditions projected for Northern Europe [23].

However, adaptation strategies such as cultivar choice focused on drought tolerant genotypes, changes in planting dates and/or in irrigation scheduling, may counterbalance or even outweigh the effects of climate change [22,24,25]. From this point of view, using conventional breeding as well as a molecular approach focusing on the relationship between water stress tolerance and expression of specific genes will greatly help to develop better adapted crops for projected harsh growing conditions. In this context, plant genomic research is crucial to provide information related to the possible mechanisms involved in abiotic stress tolerance where an increasing number of genes, transcripts and proteins are involved in stress response pathways [26]. Likewise, increased water-use efficiency as well as soil conservative management techniques will become crucial goals in the next decades [27].

Concerning climate projections in the Mediterranean Region, in the next decades the effect of climate change on agriculture will very likely result in increasing plant water stress, decreasing crop yields, especially in spring sown crops, and increasing yield variability basically due to abiotic stresses such as heatwaves and droughts [16,28]. The Mediterranean Region is considered a "hot-spot", with observed rates of climate change exceeding the global trends for most variables and future projections showing a temperature increase higher than 20% of the global average and decreases in precipitation especially for central and southern areas [29,30]. Consequently, in this area the impacts are likely to exceed the global average trend [29,31]. However, these impacts will be likely related to crop and cultivar characteristics, including their genetic mechanisms of water stress tolerance as well as their response to CO_2 in terms of increase of biomass and water use efficiency. Moreover, other issues linked with climate change such as limitations in available lands, soil erosion, salinization, decreasing natural rainfall and increasing population may exacerbate the predicted negative impact of global warming, especially on the southern side of the Mediterranean Region [32]. For example, Maghreb countries as well as Egypt and Libya, are bound to face water scarcity due to the average annual growth of population and the reduction of long-term freshwater resources [33]. Therefore, food safety and security in the Mediterranean Region are expected to be seriously threatened as a result of expected climate and socio-economic changes [29].

1.2. Projected Effects on Durum Wheat Production

In the Mediterranean Basin durum wheat (*Triticum turgidum* L. subs. *durum* [Desf.]) is the most widely grown crop, accounting for half of the total world production [34]. Italy is the main producer of the area, followed by Turkey, Southern France, Algeria, Morocco, Syria, Greece, Spain and Tunisia.

Concerning the effects of climate change on durum wheat production (i.e., yield and grain quality), warmer and drier climate is projected to increase the risk of yield losses, especially for temperature increases exceeding 2 °C across the whole Mediterranean area [35]. Even greater yield reductions with a decrease of 30–50% as a result of a 4 °C increase in temperature were projected in the province of Foggia (Apulia, Southern Italy) [22]. In another study [36], an increase in grain yield of about 10% in the "Anomaly_2" scenario (+1.7 °C; 10.4% rainfall mean reduction) in response to elevated CO_2 together with a decrease of about 8% under "Anomaly_5" scenario (+4.2 °C; 20.9% rainfall mean reduction) in the outstanding durum growing area of Capitanata (Apulia, Italy) were predicted. In a more recent study [37], negative impacts (−30% by 2100 under the business-as-usual scenario, RCP (Representative Concentration Pathway) 8.5, no CO_2 effect) in central and southern Italy (e.g., Apulia, Basilicata, Campania, Lazio, Sardinia and Sicily) were reported, together with increases in northern areas (e.g., Po river plains), especially if the CO_2 effect is included in the simulations.

As for grain quality, durum wheat is basically used for pasta (Italy), cous-cous (North Africa) and bread (semolina high volume heart breads and flat breads). These end-products are traditionally associated with the Mediterranean cuisine as the basis of Mediterranean diet. Hence, grain quality is crucial to meet the requirements of downstream, large-scaled processing activities such as milling and pasta-making. Grain size in barley and wheat is negatively affected by high temperatures during grain filling [38–41]. Hence, climate change might cause an increase of shrivelled grains with low yield in semolina, thereby jeopardizing the technological value of durum wheat production. Therefore, the negative effect of increasing temperatures and drought on grain production, including grain quality and yield components, is an issue of utmost importance not only for agriculture but also for the food industry of the entire Mediterranean Region. Thus, a negative impact on durum wheat caused by climate change may also result in dramatic socio-economic consequences.

1.3. Crop Modelling

Given this context, using crop models to analyze the responses of crops across different environments in order to assess the impact of climate change becomes crucial [42,43] both to study the plant responses and to find adaptation and mitigation strategies in association with the study of gene-induced stress tolerance. The combined contribution of crop modelling and genetics can play an innovative and crucial role in targeting cultivar choice in different environments and outweighing risks. To date, only a relatively low number of studies based on crop modelling focused on crops under Mediterranean conditions have been carried out [16,35–37,44–51].

Among several crop simulation models developed since the sixties [52], the Decision Support System for Agrotechnology Transfer (DSSAT) [53] has been extensively used worldwide over the last 20 years for most major crops and many different applications including: (i) simulated management options such as fertilization, irrigation, pest management; (ii) site-specific farming and study of potential impacts of climate change on agricultural production (see [36] for an exhaustive bibliography). The basis for the DSSAT cropping system model design is a modular structure of which CERES-Wheat is a component model continuously refined and modified over the years [54–56]. CERES-Wheat simulates crop growth, development and yield taking into account the effects of weather, management, genetics, soil water, carbon (C) and nitrogen (N_2). This model provided successful performance under a wide range of agro-climatic conditions [57]. In order to evaluate the performance of the model [58], using various statistical tests to analyze simulated and observed results from several studies across Asia and Australia, proved the good ability of this model to predict phenology (i.e., anthesis and maturity) and, to a slightly lesser extent, grain and biomass yields. Furthermore, they underlined good performance in favorable conditions (e.g., high-yielding and/or irrigated environments). In contrast, results were less satisfactory in low-yielding environments. Importantly, these authors emphasized the crucial role of long-term data-sets to better evaluate the effectiveness of this model in representing reality.

In the Mediterranean environments, the CERES-Wheat model has been used in a limited number of studies [36,37,45,50,51,59–62]. In particular, refs. [45,50], focusing on the use of long-term data-sets (approximately 30 years) and statistical indices to assess the model performance, found that the CERES-Wheat model provided good to fair predictions of production (i.e., grain yield), with a tendency to overestimate, and good to very good predictions of phenology (i.e., anthesis date). In contrast, predictions of grain quality (i.e., grain weight and grain number) proved to be reliable but less satisfactory.

This study applies the CERES-Wheat crop model included in DSSAT v. 4.0 to assess the adaptation of cultivar choice as well as the environmental effect of genetic mechanisms involved in water stress conditions with a special focus on Dehydration Responsive Element-Binding (*DREB*) related genes in durum wheat, in order to evaluate the simulated negative impact of increasing temperatures and decreasing rainfall on grain production and phenology of three durum wheat genotypes grown in two different Mediterranean environments in Southern Sardinia (Italy).

2. Materials and Methods

2.1. Experimental Setting

This study was carried out at Benatzu and Ussana, two sites from AGRIS (Agricultural Research Agency of Sardinia) experimental farm "S. Michele" (Lat. 39°24′ N, Long. 9°5′ E; about 20 km S from the sea) in the Campidano plain, the main durum growing area of Southern Sardinia, Italy. The climate is Mediterranean, with warm and dry summers and mild winters. Mean air temperatures range from 4.8 °C in January and 33.0 °C in August. Precipitations are concentrated in autumn, winter and early spring, with a long-term annual amount of about 450 mm. The area shows a great soil variability due to its ancient geological origin. Ussana soil (114 m a.s.l.) is a *Petrocalcic Palexeralf* [63]. It is a sandy clay loam soil, with a percentage of sand greater than 50%, characterized by alluvial conglomerate substrate, in a weak red colored clay matrix. The drainage is moderate and the stone percentage is about 20%. This soil is located in a hilly area and accounts for medium- and low-fertility durum growing areas of Sardinia and the Mediterranean. Benatzu soil (80 m a.s.l.), is a *Vertic Epiaquet* [63]. It is a clay loam soil with a soil substrate alluvial gravel, a fraction of stones of about 30% and a clay percentage of about 40%. This soil is located in a flat area and accounts for the most fertile durum growing areas of Sardinia and the Mediterranean.

2.2. Experimental Data

Daily maximum and minimum air temperatures (°C) and annual rainfall totals (mm) over the period 1973 to 2004 were recorded by an automatic weather station (SILIDATA AD2, SILIMET s.r.l., Modena, Italy) located in the experimental farm. Daily global solar radiation values (MJ m^2 d^{-1}) were estimated using the software RadEst3.00, where: (i) radiation is calculated as the product of the atmospheric transmissivity of radiation times the radiation outside the earth atmosphere and (ii) the atmospheric transmissivity of global solar radiation is estimated based on the difference between maximum and minimum temperatures [64]. Several physical and chemical characteristics of soils required as inputs by the model were determined at both sites (Table 1): content in sand (%), silt (%), clay (%), total N (%), pH in H_2O, C.E.C. (Cation Exchange Capacity in cmol kg^{-1}), organic C (%), organic matter (%), texture, color, runoff value, slope and a fertility factor. The analysis procedures are described in DM 13.09.1999, points: II.4 and II.5 for Sand, Silt, Clay and Texture; VII.1 for Organic carbon and Organic matter; III.1 for pH in H_20; XIV.1 for Total nitrogen; and XIII.2 for C.E.C.

Table 1. Physical and chemical characteristics of soil at Benatzu and Ussana experimental sites.

	Benatzu	Ussana
Sand (%)	26.2	56.4
Silt (%)	34.4	21.5
Clay (%)	39.4	22.1
Texture	Clay Loam	Sandy Clay Loam
pH in H_2O	8.5	7.9
Organic carbon (%)	1.62	0.83
Organic matter (%)	2.80	1.20
Total nitrogen (%)	0.15	0.07
C.E.C. (cmol kg^{-1})	2.9	2.3

Management and durum wheat performance data over the period 1973 to 2004 from both sites were taken from the evaluation trials of the Italian Durum Wheat Network (http://qce.entecra.it/RISULTATI.htm, (accessed on 1 December 2021)). The experimental design consisted in a triple lattice with 8-rowed plots. Each plot was 5.9 m long and 1.5 m wide with an approximate surface of 10 m^2 and rows spaced 0.18 m apart. Plant density was about 350 viable seeds m^{-2}.

2.3. Cultivar Description

Creso, Duilio and Simeto, three hallmark Italian durum wheat cultivars, well adapted to Mediterranean environmental conditions, were used to test the performances of the CERES-Wheat model. In addition to their agronomic and economic importance, these varieties were chosen for the availability of reliable long-term experimental data. Creso, released in 1973, is a medium-late, short variety with good grain quality. Despite its longstanding cultivation, it is still widespread in the high rainfall spring areas of Central Italy. Duilio, released in 1984, is an early-medium, medium-tall variety, well adapted to the durum growing areas of Southern Italy owing to its high-yielding potential, grain quality and resistance to drought. Simeto, released in 1988, is an early and short genotype with good performances both in yield and grain quality, especially in the dry areas of Southern Italy. This cultivar still ranks among the most widespread in Italy for the production of certified seed (https://www.crea.gov.it/web/difesa-e-certificazione/-/statistiche-di-certifcazione-superfici-controllate, (accessed on 1 December 2021)).

2.4. Molecular Responses to Drought Stress

These three cultivars had previously been studied in regards to their molecular responses to abiotis stresses in general and drought stress in particular. In this study, the expression of the endogenous *DREB2A*-homologous gene activation, belonging to the Dehydration-Responsive Element-Binding (*DREB*) transcription factor gene family, was considered through RT-PCR analyses obtained from time-course experiments of drought stress both in controlled greenhouse and in field conditions [65,66].

2.5. Model Simulations

The CERES-Wheat model, included in the Decision Support System for Agrotechnology Transfer (DSSAT) version 4.0 [53,67] was used to perform crop growth simulations of Creso, Duilio and Simeto. This model describes daily both phenology and growth in response to environmental factors (e.g., soil properties and weather patterns) and management. The model, which includes subroutines to simulate soil and crop water balance and nitrogen balance, can be used to simulate the effects of nitrogen deficiency and soil water deficit on photosynthesis and pathways of carbohydrate allocation in plants.

CERES-Wheat had already been calibrated and validated in the test area for these cultivars [45]. An iterative procedure for minimizing the differences between predicted and observed values to obtain the genetic coefficients values was used [68]. In particular, observed and predicted values of grain yield (kg ha^{-1}), anthesis date (days after planting, dap) and average seed weight (g) were compared and the cultivar coefficients were modified until the model responses matched the real data or fell within a defined error threshold. Table 2 exhibits the genetic coefficients found for the three cultivars in the area.

Table 2. Genetic coefficient values for Creso, Duilio, and Simeto durum wheat varieties during CERES-Wheat model calibration using data collected at two experimental sites located in Southern Sardinia, Italy [45]. P1D: Photoperiod sensitivity coefficient (% reduction/h near threshold); P1V: Vernalization sensitivity coefficient (%/d of unfulfilled vernalization); P5: Thermal time from the onset of linear filling to maturity (°C d); G1: Kernel number per unit stem + spike weight at anthesis (#/g); G2: Standard kernel size under optimum conditions (mg); G3: Standard, non-stressed dry weight (total, including grain) of a single tiller at maturity (g); PHINT: Thermal time between the appearance of leaf tips (°C d).

Genetic Coefficients	Creso	Duilio	Simeto
P1V	30.0	25.0	25.0
P1D	55.0	50.0	58.0
P5	450.0	480.0	450.0
G1	25.0	25.0	25.0
G2	55.0	55.0	55.0
G3	1.7	1.7	1.7
PHINT	100.0	90.0	90.0

Concerning the simulation runs, 1 August was set as the starting day for each year. The cropping season was between October and June of the following year. The planting date was set on the observed date of each year, depending on the amount of natural rainfall fallen from autumn until late early winter. The end of the growing season was set according to the observed harvest dates for each year. All agronomic information, such as previous crops and fertilizer management, was set in the experimental file. The following data were registered as initial conditions: previous crop, sowing depth and dates, row spacing, plant population, fertilizer applications and dates, harvest dates. The same data were subsequently set as inputs in the experimental simulation design.

The following indices based on simple and squared differences between predicted and measured values were calculated: normalized Root Square Error (nRMSE), index of agreement (D-index) and Coefficient of Residual Mass (CRM) [69]. Ideally, a model reproduces experimental data perfectly when nRMSE is 0 and D-index is 1 [58]. CRM measures the tendency of the model to over- (i.e., negative values) or under-estimate (i.e., positive values) observed data [70].

2.6. Meteorological Trends and Climate Scenarios

Mean annual air temperature data (°C), along with Standardized Anomaly Index (SAI) values, and the annual and seasonal amount of rainfall (mm) over the period 1974 to 2004 were observed in order to evaluate the real trend of air temperature and rainfall in the study area.

As for climate scenarios, a set of 48 synthetic climates based on global and regional climate model simulations predicting a substantial drying and warming over the Mediterranean Region by the end of the century, with annual precipitation decrease exceeding

−25–30% and warming exceeding +4–5 °C compared to the actual climate, was developed [71,72]. The baseline air temperature, as well as the precipitation records for the actual climate recordings over the 1973–2004 period at Benatzu and Ussana sites, were adjusted by between +1 and +6 °C at 1 °C intervals, and by between −5% and −30% at 5% intervals, respectively (Table 3). For more details concerning the pro and cons of the incremental approach followed in this study and for an exhaustive review, see [50].

Table 3. Simulated climate change scenarios. R = rainfall; T = temperature.

		Decreasing Rainfall						
		0	−5%	−10%	−15%	−20%	−25%	−30%
	0	-	R5	R10	R15	R20	R25	R30
Increasing Temperature	**+1 °C**	T1	T1_R5	T1_R10	T1_R15	T1_R20	T1_R25	T1_R30
	+2 °C	T2	T2_R5	T2_R10	T2_R15	T2_R20	T2_R25	T2_R30
	+3 °C	T3	T3_R5	T3_R10	T3_R15	T3_R20	T3_R25	T3_R30
	+4 °C	T4	T4_R5	T4_R10	T4_R15	T4_R20	T4_R25	T4_R30
	+5 °C	T5	T5_R5	T5_R10	T5_R15	T5_R20	T5_R25	T5_R30
	+6 °C	T6	T6_R5	T6_R10	T6_R15	T6_R20	T6_R25	T6_R30

This set of 48 synthetic scenarios was used in conjunction with the CERES-Wheat crop model to determine the potential effects of increasing temperatures and decreasing rainfall on crop production (i.e., grain yield and grain size), and phenology (i.e., anthesis date) of the three durum wheat cultivars Creso, Duilio and Simeto by scenario and site.

2.7. Calibration, Validation and Evaluation of CERES-Wheat Model Performances

The whole study period for Benatzu and Ussana sites, covers a time span of 30 years (1974–2004). Datasets used for calibration were: 1996–2004, 1997–2004 and 2000–2004 for Creso, Duilio and Simeto, respectively. Datasets used for validation were: 1974–1995, 1985–1996 and 1989–1999 for Creso, Duilio and Simeto, respectively. The differences in time span both in calibration and validation depend on the availability of data from the cultivar evaluation trials of the Italian durum wheat network owing to the different year of release of each cultivar. Detailed information about calibration, validation and evaluation of the CERES-Wheat model in the two experimental sites of the study area can be found in [45].

3. Results

3.1. Meteorological Trends

Trends in mean annual temperature (T_{mean}) (°C) along with Standardized Anomaly Index (SAI) values, and annual and seasonal amounts of rainfall (mm) observed in the study area over the period 1974 to 2004, are shown in Figures 1 and 2, respectively. The T_{mean} linear trend shows an increase of 0.44 ± 0.59 °C per decade (Figure 1A). Lower than average temperatures are prevalently scattered over the left-hand side of the SAI graph (approximately from 1974 to 1985), i.e., in the first years of the study period, whereas in the following years higher than average temperatures become more frequent (Figure 1B). A non-significant negative trend for annual and seasonal rainfall amounts was observed except in autumn, which showed a non-significant increasing trend (Figure 2A,B). This seasonal downward trend is clearer in winter than in spring and summer.

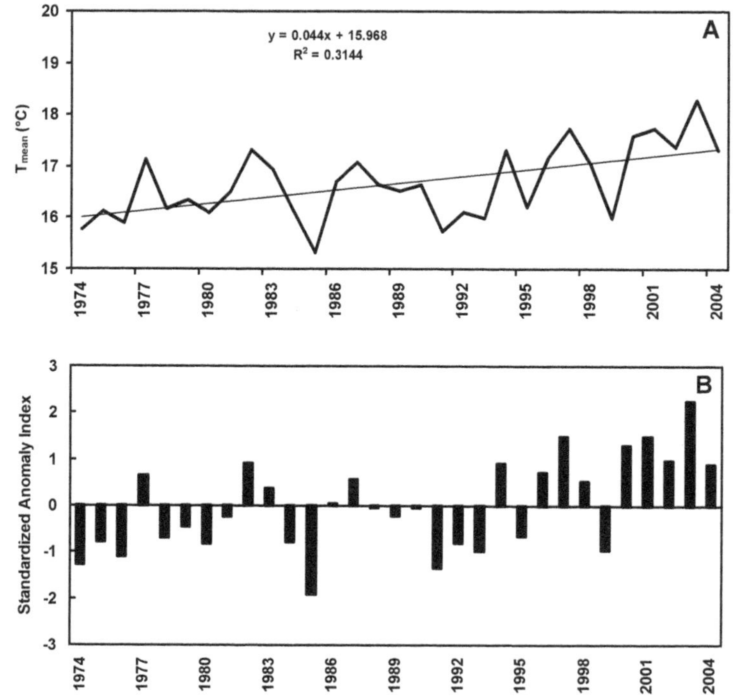

Figure 1. Trend of mean annual air temperatures (**A**) and Standardized Anomaly Index, SAI (**B**) over the period 1974–2004 at the AGRIS experimental station "S. Michele" (Southern Sardinia, Italy).

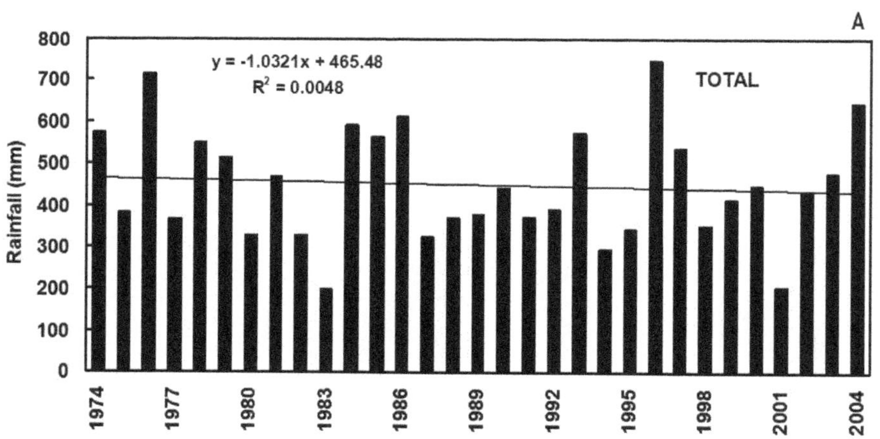

Figure 2. *Cont.*

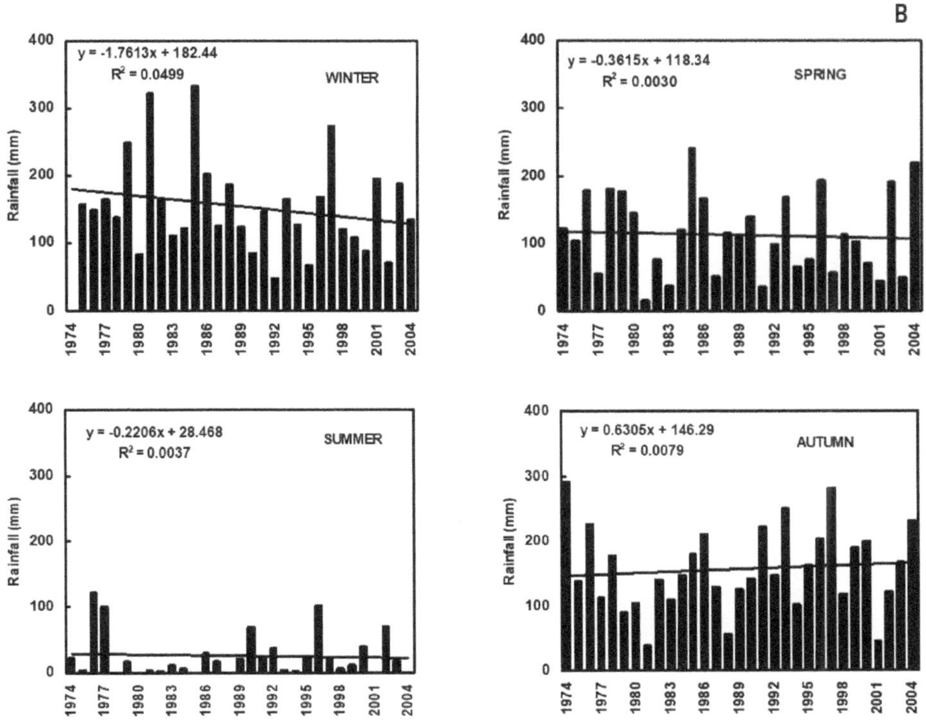

Figure 2. Annual (**A**) and seasonal (**B**) trends of rainfall over the period 1974–2004 at the AGRIS experimental station "S. Michele" (Southern Sardinia, Italy).

3.2. Calibration, Validation and Evaluation of CERES-Wheat Model Performances

Concerning grain yield predictions, the CERES-Wheat model provided good to fair performances for all three cultivars. As for phenology, all results proved the effectiveness of the CERES-Wheat model in predicting anthesis dates for these experiments. On the contrary, the model performances proved to be less effective for estimating the average seed with a tendency of the CERES-Wheat model to underestimate predictions. However, considering the results as a whole, statistical indices show that this model proves to be an effective tool to represent reality. For further details and an exhaustive presentation and discussion about results and model performances, see [45].

3.3. Climate Change Scenarios: General Responses

In order to evaluate the impacts of climate change on durum wheat production and phenology, the general analysis was performed using data sets from the whole study period (1974–2004). The experimental conditions observed during calibration and validation were left unchanged. Hence, weather was the only factor of variation.

The responses of the CERES-Wheat model to 48 simulated scenarios (Table 3) at the two experimental sites "Benatzu" and "Ussana" for the annual values of grain yield, anthesis date and average seed weight and for three durum wheat cultivars were analyzed by comparing observed and simulated values. Figure 3 shows the observed mean grain yield data in comparison with the CERES-Wheat model responses to 2 (mildest and worst scenarios, respectively) of the 48 simulated climate change scenarios for Creso (time span: 1974–2004), Duilio, (time span: 1985–2004) and Simeto (time span: 1989–2004) at Benatzu and Ussana sites, respectively. The mildest simulated scenario shows a +1 °C increase in temperature and a 5% reduction in rainfall compared to the actual mean temperatures and total rainfall amount, respectively. The worst-case scenario shows a +6 °C increase in temperature and a 30% lower annual rainfall. The detrimental effect on simulated yield determined by increasing temperatures and decreasing rainfall for all cultivars and sites cannot be questioned.

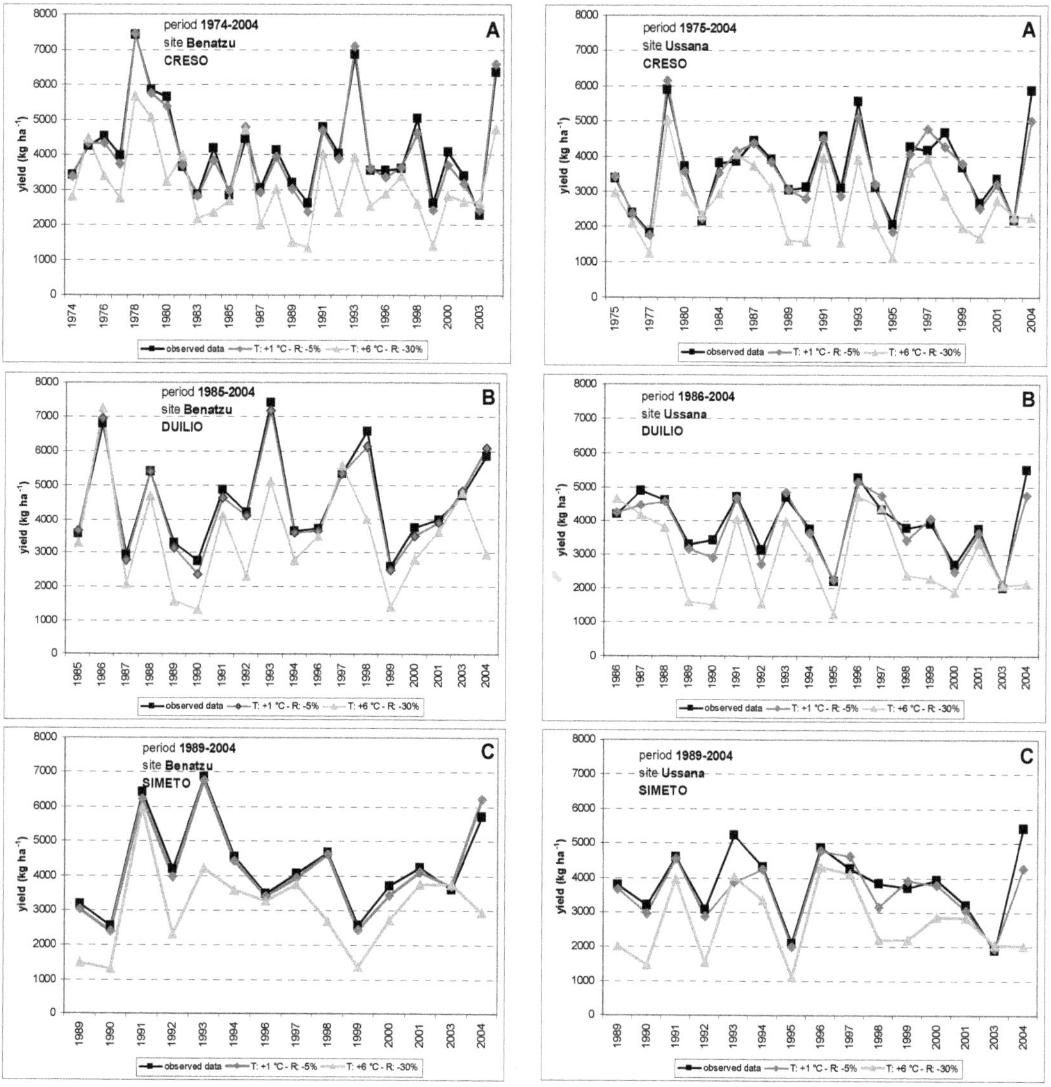

Figure 3. Effects of climate change scenarios on grain yield of durum wheat varieties Creso (**A**), Duilio (**B**), and Simeto (**C**) at Benatzu and Ussana sites. Simulation results from scenarios T1_R5 (temperature increase: +1 °C; rainfall reduction: 5%) and T6_R30 (temperature increase: +6 °C; rainfall reduction: 30%) are compared to observed yield data.

3.4. *Climate Change Scenarios: Cultivar Responses*

To compare the simulated impact of increased temperatures and decreased rainfall on each cultivar, the analysis was limited to the years when field trials, were conducted simultaneously for all cultivars (i.e., years 1990–2004 for Benatzu and 1989–2004 for Ussana).

Figures 4 and 5 exhibit the percentage reductions in grain yield of all cultivars between the mean values observed at Benatzu and Ussana, respectively and simulation results from twelve climate change scenarios with increasing temperatures (from +1 °C to +6 °C) and decreasing rainfall (6 scenarios with a 5% reduction and 6 scenarios with a 30% reduction in annual rainfall). For each cultivar, the predicted negative impact on grain yields rises steadily from the least unfavorable scenarios to the most severe ones.

Comparing the different responses of the three cultivars to simulated scenarios at Benatzu site, Creso (medium-late cultivar) proved to be the most sensitive, with the greatest yield reduction especially when temperature increases were combined with strong decreasing rainfall. For this cultivar, the reduction in grain yield from mean observed values ranged from 2.4% (scenario T1_R5) to 14.9% (scenario T6_R5) for a 5% lower annual rainfall amount (Figure 4A), and from 19.9% (scenario T1_R30) to 29.2% (scenario T6_R30) for a 30% decrease in annual rainfall (Figure 4B).

The reduction in grain yield of Duilio (early cultivar) and Simeto (early cultivar) ranged from 2.7% and 1.7% (scenario T1_R5) to 9.1% and 8.6% (scenario T6_R5), respectively, for a 5% decrease in rainfall (Figure 4A). The reduction in grain yield of Duilio and Simeto was much higher using a 30% rainfall decrease scenario, ranging from 22.4% and 21.4% (scenario T1_R30) to 25.5% and 26.4% (scenario T6_R30), respectively (Figure 4B).

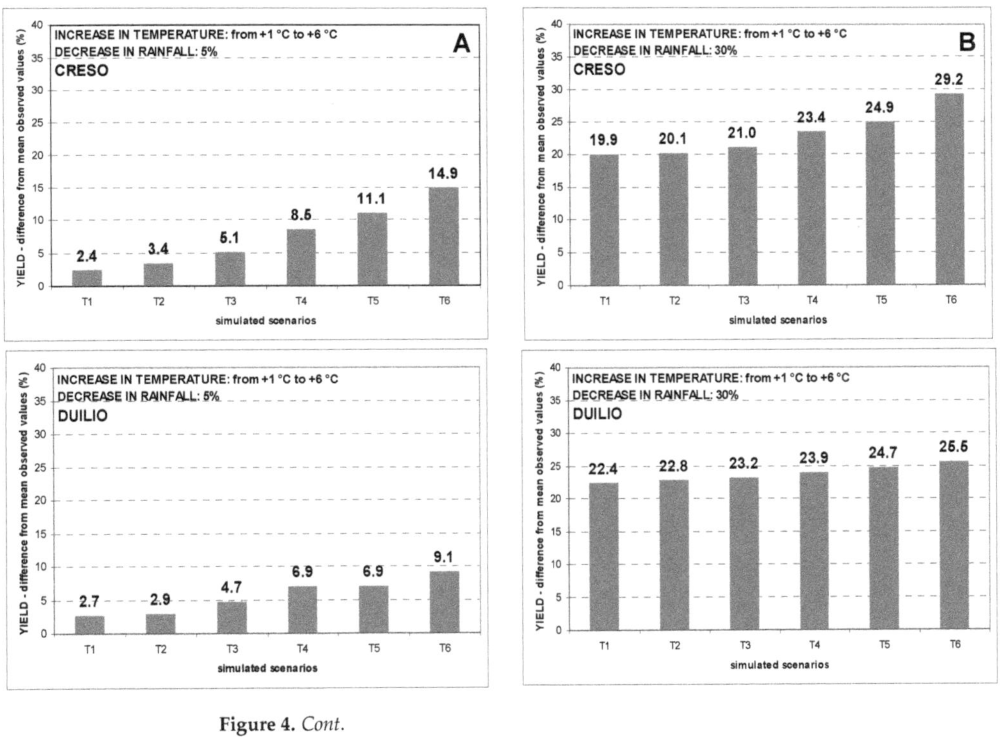

Figure 4. *Cont.*

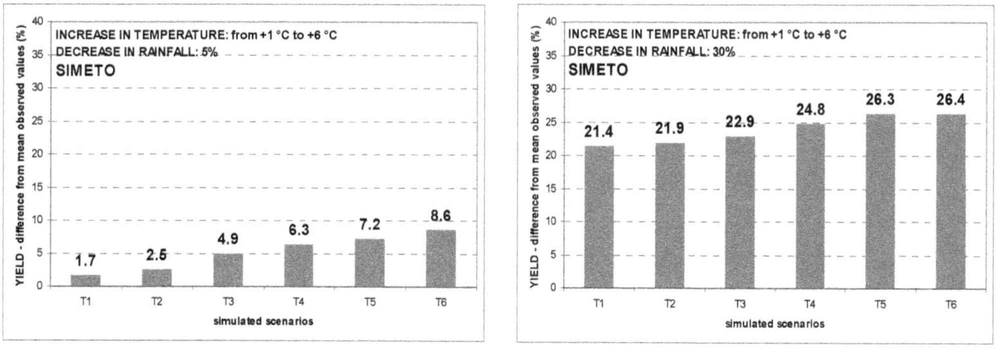

Figure 4. Percentage decline of grain yield over the period 1990–2004 at the experimental site of Benatzu for climate change scenarios characterized by increasing temperature (from +1 °C to +6 °C) and rainfall reduction by 5% (**A**) and 30% (**B**).

Creso was also confirmed to be the most sensitive cultivar at Ussana site over the period 1989 to 2004, with a grain yield reduction ranging from 4.2% (scenario T1_R5) to 15.3% (scenario T6_R5) for a 5% rainfall reduction (Figure 5A), and from 26.0% (scenario T1_R30) to 33.3% (scenario T5_R30) for a 30% rainfall decrease (Figure 5B).

Duilio showed grain yield declines ranging from 3.3% (scenario T1_R5) to 10.0% (scenario T6_R5) for a 5% rainfall reduction (Figure 5A), and from 25.6% (scenario T2_R30) to 29.3% (scenario T5_R30) for a 30% rainfall decrease (Figure 5B).

A similar trend was observed for Simeto with a decrease of grain yield ranging from 6.7% (scenario T1_R5) to 11.2% (scenario T6_R5) for a 5% rainfall reduction (Figure 5A), and from 27.8% (scenario T2_R30) to 30.1% (scenario T6_R30) for a 30% rainfall reduction (Figure 5B).

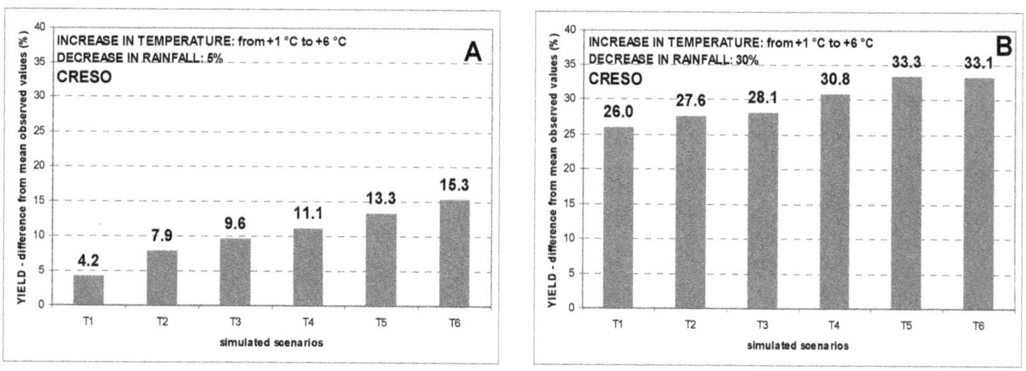

Figure 5. *Cont.*

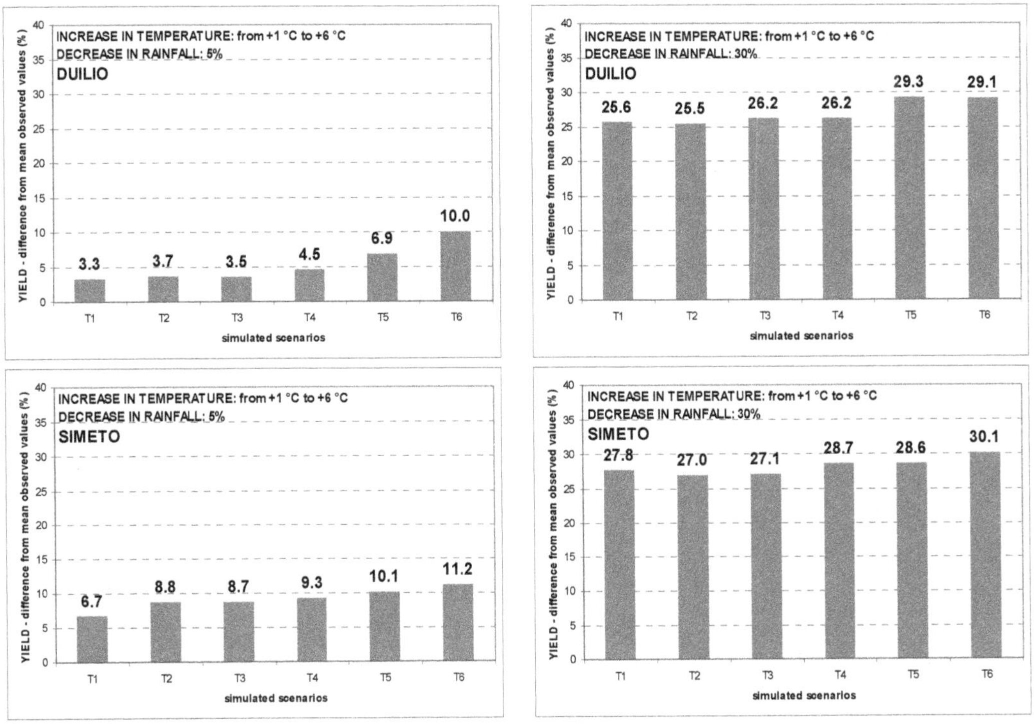

Figure 5. Percentage decline of grain yield over the period 1989–2004 at the experimental site of Ussana for climate change scenarios characterized by increasing temperature (from +1 °C to +6 °C) and rainfall reduction by 5% (**A**) and 30% (**B**).

In summary, the overall simulated effect of climate change scenarios characterized by increasing temperature and decreasing rainfall is a gradual reduction in grain yield for all cultivars and sites. This effect increases from the mildest to the worst-case scenarios. Interestingly, the CERES-Wheat model predicted greater grain yield reductions at the low-yielding site of Ussana than in the fertile soil of Benatzu. In particular, the overall average grain yield reduction for the three cultivars in all scenarios was equal to 16.2% and 19.0% at Benatzu and Ussana, respectively.

The overall simulated impact of increasing temperatures and decreasing rainfall on kernel weight showed an opposite trend. CERES-Wheat simulations showed that kernel weight tends to increase slightly and this response is greater when annual rainfall amount decreases by 5% (Figures 6A and 7A). In addition, the slight increase in kernel weight is greater at Ussana and this confirms the trend in observed data (Figures 6 and 7 for Benatzu and Ussana experimental sites, respectively). No remarkable trend from the analysis of the different responses of each variety emerges.

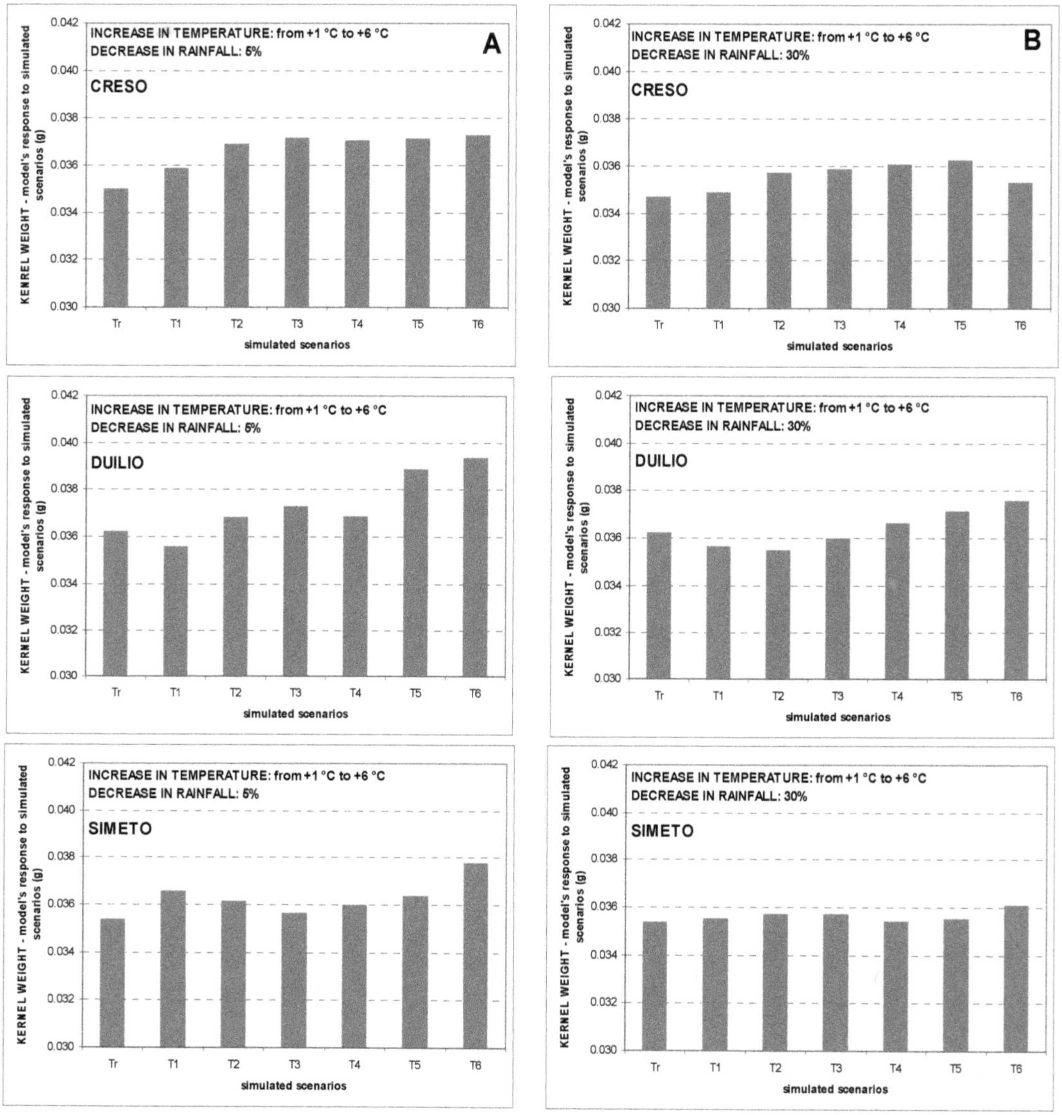

Figure 6. Average seed weight trends over the period 1990–2004 at the experimental site of Benatzu for climate change scenarios characterized by increasing temperature (from +1 °C to +6 °C) and rainfall reduction by 5% (**A**) and 30% (**B**). Tr = observed data.

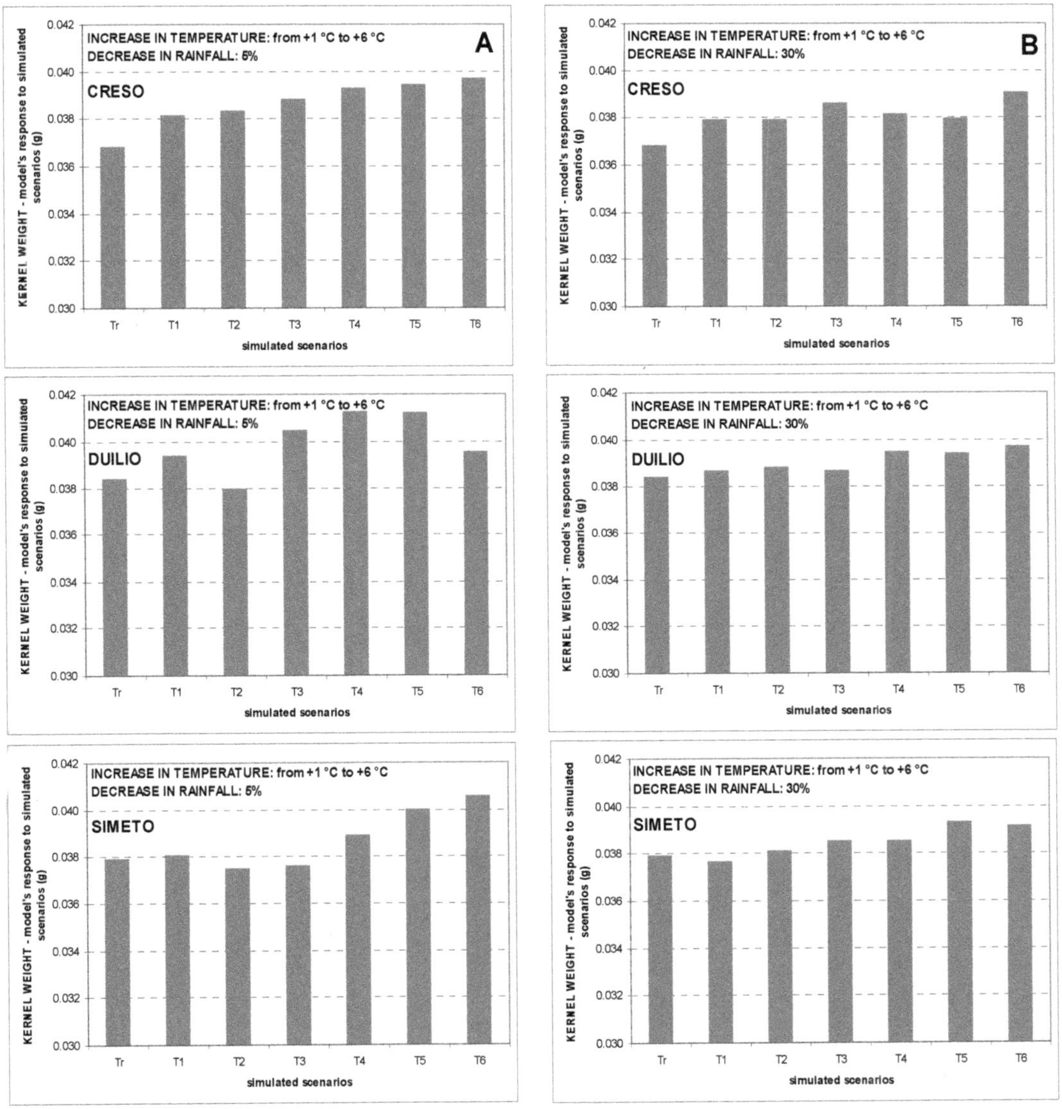

Figure 7. Average seed weight trends over the period 1989–2004 at the experimental site of Ussana for climate change scenarios characterized by increasing temperature (from +1 °C to +6 °C) and rainfall reduction by 5% (**A**) and 30% (**B**). Tr = observed data.

The effects of the 48 climate change scenarios on phenology of durum wheat were determined by comparing predicted and observed anthesis dates of each cultivar. In general, a shortening effect on cycle length of durum wheat was observed. This response probably depends on the modelling approach on phenology used by the CERES-Wheat crop model, which simulates crop development rate as a function of temperature only. Figure 8 illustrates the general shortening effect of climate change scenarios on the crop growing cycle at Benatzu (Figure 8A) and Ussana (Figure 8B) experimental sites.

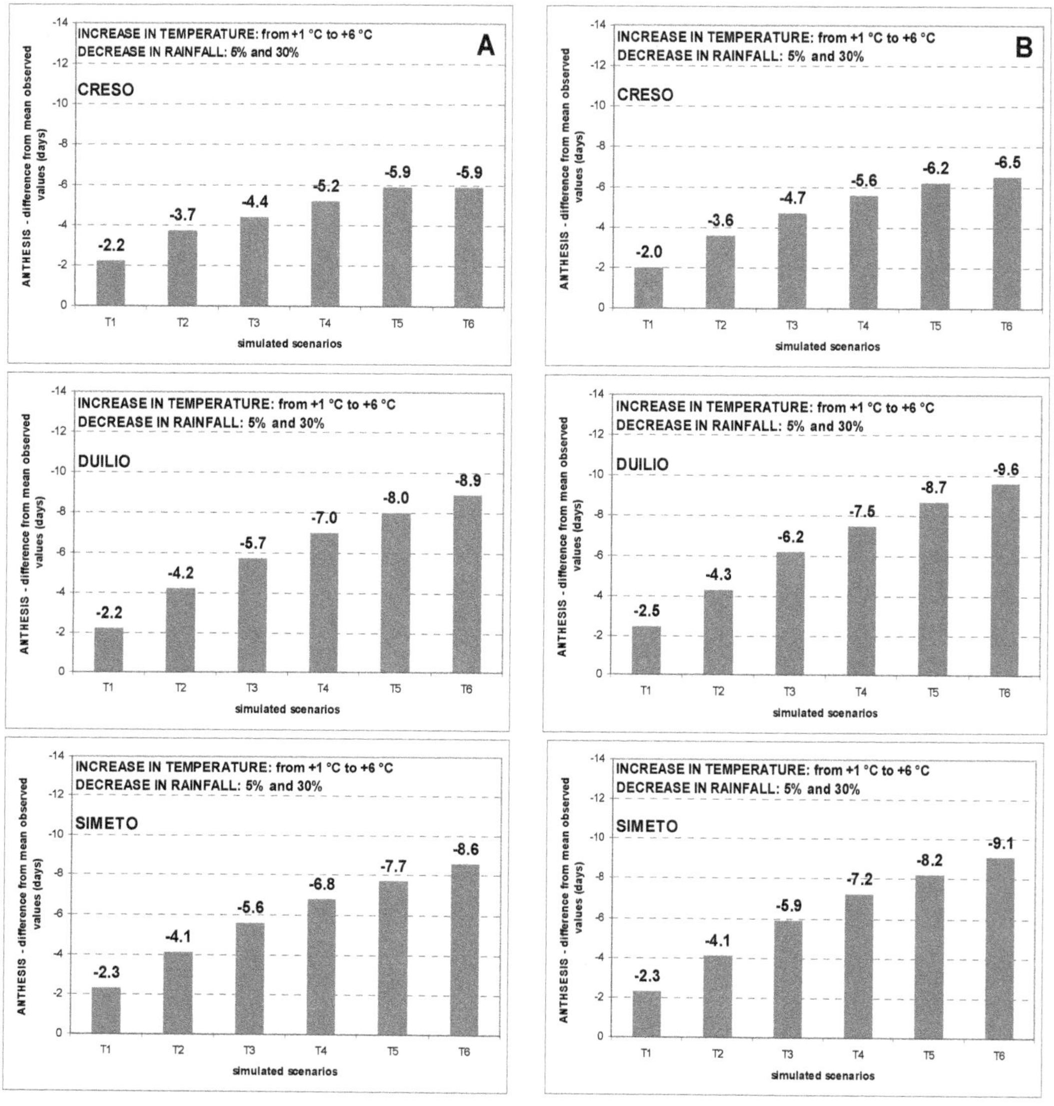

Figure 8. Reduction (in days) of crop growing cycle (from sowing to anthesis) over the period 1990–2004 at Benatzu (**A**) and the period 1989–2004 at Ussana (**B**) for climate change scenarios characterized by increasing temperature (from +1 °C to +6 °C) and rainfall reduction by 5% and 30%.

Based on the greater overall shortening effect at Ussana (medium-low fertility soil) than at Benatzu (high-fertility soil), soil fertility seems to play a remarkable role in reducing the growth duration of durum wheat. In addition, the difference between simulated and observed values increases moving from scenario T1 (temperature increase: +1 °C) to scenario T6 (temperature increase: +6 °C). Examining this shortening effect on each variety, Creso showed a more limited reduction at both sites when compared to the early genotypes Duilio and Simeto.

Table 4 summarizes the simulated impacts of two climate change scenarios (T1_R5 and T6_R30) on grain production and phenology of durum wheat using the CERES-Wheat crop model on three cultivars and two experimental sites.

Table 4. Simulated responses of three durum wheat cultivars (Creso—Cr, Duilio—Du, Simeto—Si) to two climate change scenarios (T1_R5 and T6_R30) at Benatzu (B) and Ussana (U) experimental sites in Sardinia, Italy. T1_R5 and T6_R30 scenarios project an average temperature increase of 1 °C and 6 °C, respectively, and an annual rainfall reduction of 5% and 30%, respectively. Simulation results (SIM) and means (M) of grain yield (kg ha^{-1}), date of 2004 at Benatzu and 1989–2004 at Ussana.

CV	Site	Grain Yield OBS	Grain Yield SIM Scenario T1_R5		Grain Yield SIM Scenario T6_R30		Anthesis OBS	Anthesis SIM Scenario T1_R5		Anthesis SIM Scenario T6_R30		Kernel Weight OBS	Kernel Weight SIM Scenario T1_R5		Kernel Weight SIM Scenario T6_R30	
		(kg ha^{-1})	(kg ha^{-1})	% Change	(kg ha^{-1})	% Change	(Dap)	(Dap)	Dap Change	(Dap)	Dap Change	(g)	(g)	% Change	(g)	% Change
Cr	B	4054	3955	−2.4	2869	−29.2	135	132	−3	127	−8	0.035	0.036	+2.5	0.035	0.0
	U	3700	3543	−4.2	2474	−33.1	141	138	−3	131	−10	0.037	0.038	+3.6	0.039	+6.0
	M	**3877**	**3749**	**−3.3**	**2672**	**−31.2**	**138**	**135**	**−3**	**129**	**−9**	**0.036**	**0.037**	**+3.0**	**0.037**	**+3.0**
Du	B	4573	4449	−2.7	3406	−25.5	129	127	−2	118	−11	0.036	0.036	0.0	0.038	+3.7
	U	3756	3633	−3.3	2662	−29.1	135	132	−3	122	−13	0.038	0.039	+2.6	0.040	+3.4
	M	**4165**	**4041**	**−3.0**	**3484**	**−27.3**	**132**	**130**	**−3**	**120**	**−12**	**0.037**	**0.038**	**+1.3**	**0.039**	**+3.6**
Si	B	4354	4280	−1.7	3206	−26.4	132	129	−3	121	−11	0.035	0.037	+3.3	0.036	+2.0
	U	3831	3575	−6.7	2676	−30.1	138	135	−3	125	−13	0.038	0.038	0.0	0.039	+3.2
	M	**4093**	**3928**	**−4.2**	**2941**	**−28.3**	**135**	**132**	**−3**	**123**	**−12**	**0.037**	**0.038**	**+1.7**	**0.038**	**+2.6**
Mean	B	4327	4228	−2.3	3460	−27.0	132	129	−2.7	122	−10	0.035	0.036	+1.9	0.036	+1.9
	U	3762	3584	−4.7	2604	−30.8	138	135	−3.0	126	−12	0.038	0.038	+2.1	0.039	+4.2
	M	**4045**	**3906**	**−3.5**	**3032**	**−28.9**	**135**	**132**	**−2.9**	**124**	**−11**	**0.037**	**0.037**	**+2.0**	**0.038**	**+3.1**

Legend: OBS—Observed data; SIM—Simulated data, CV—Cultivar, Cr—Creso, Du—Duilio, Si—Simeto.

Creso shows the lowest observed yield potential (observed mean yield: 3877 kg ha^{-1}) and the largest percentage reductions in grain yield (mean percentage reduction: 31.2%) under the worst-case (T6_R30) climate change scenario when compared to Simeto (observed mean yield: 4093 kg ha^{-1}; percentage reduction under T6_R30 scenario: 28.3%) and Duilio (observed mean yield: 4165 kg ha^{-1}; percentage reduction under T6_R30 scenario: 27.3%). In general, Duilio exhibits the smallest simulated grain yield reduction under climate change with a decrease ranging from 3.0% (scenario T1_R5) to 27.3% (scenario T6_R30) and proves to be the most resilient genotype to increasing unfavorable conditions. Furthermore, Ussana was the most vulnerable environment to climate change conditions with a general grain yield reduction of 4.7% and 30.8%, respectively for scenarios T1_R5 and T6_R30, when compared to Benatzu (2.3% and 27.0% for scenarios T1_R5 and T6_R30, respectively).

As for grain size, the largest effects of the two climate change scenarios T1_R5 and T6_R30 on kernel weight were registered for Creso, with a grain weight increase ranging from 2.5% at Benatzu to 6.0% at Ussana. Moreover, at Ussana the simulated percentage kernel weight increase ranged from 2.1% to 4.2% for scenarios T1_R5 and T6_R30, respectively, and was higher than Benatzu (1.9 for the two scenarios, respectively).

The analysis of the differences between observed and simulated anthesis dates under climate change scenarios indicates a general anticipation of anthesis, with some differences among genotypes. In particular, the late genotype Creso shows a general reduction to increasing temperature scenarios, ranging from 3 days at both experimental sites (scenario T1_R5) to 8 and 10 days at Benatzu and Ussana, respectively, for scenario T6_R30. The responses of the early genotypes Simeto and Duilio indicate a slightly larger shortening effect, ranging from 2 and 3 days for scenario T1_R5 to 11 and 13 days at Benatzu and Ussana sites respectively, for scenario T6_T30.

3.5. Molecular Responses to Drought Stress

Previous RT-PCR experiments carried out using RNA extracts from different durum wheat cultivars, including Creso, Duilio and Simeto, showed an intense band at 500 bp, instead of the expected 450 bp, and two faint bands at 450 bp and 580 bp, respectively (Figure 9) [73].

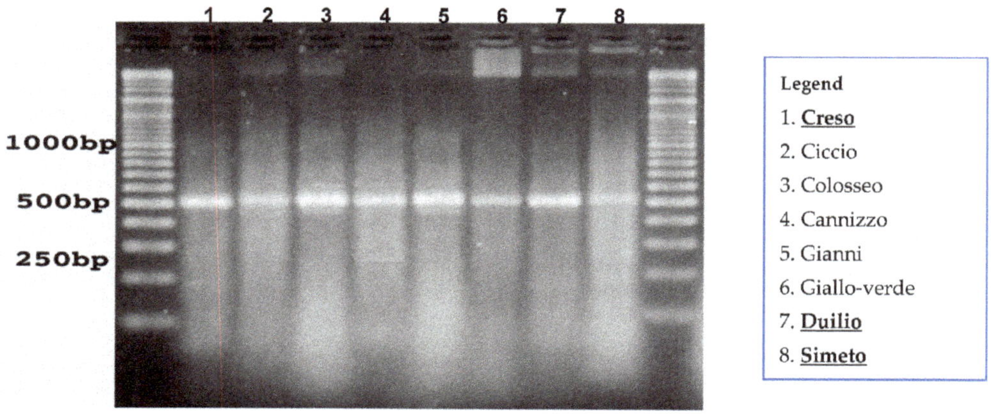

Figure 9. The expression pattern of RT-PCR with primer *drebfor1* e *drebrev1* of some durum wheat cultivars, including Creso (1), Duilio (7) and Simeto (8) analyzed at the 4th day without water.

Moreover, sequencing and aligning with TC85717 500 bp bands from all these cultivars, the presence of a short insert 53-bp was detected, revealing a complete homology with transcripts found in barley, homologous to *DREB2* genes and related to drought. The transcript isolated in barley derives from an alternative splicing of a gene, named *HvDRF1*, where Hv stands for *Hordeum vulgare*, generating three transcripts. The primer pair used for these experiments was compatible with these transcripts and produced fragments at about 580 bp, 500 bp and 450 bp, which is the same pattern observed in durum wheat. This result led to the conclusion that in durum wheat a homologous gene to *HvDRF1* is present and it was named Triticum durum Dehydration-Responsive Factor 1 (*TdDRF1*). Further studies revealed that this gene produces three transcripts by alternative splicing: *TdDRF1.1*, consisting of four exons, from E1 to E4; *TdDRF1.2*, consisting of three exons E1, E2 and E4; and *TdDRF1.3*, consisting of two exons E1 and E4 [74]. This gene and its three isoforms play a crucial role in conditioning and modulating the responses of cultivars to drought. In all genotypes, the *TdDRF1.2* transcript was always expressed at higher levels, the *TdDRF1.1* transcript was the least expressed and the TdDRF1.3 transcript was intermediate between TdDRF1.2 and *TdDRF1.1* transcripts. These results suggest a correlation between water stress and the expression profile of the *TdDRF1* gene and its transcripts.

4. Discussion

4.1. Meteorological Trends

The analysis of the historical weather data set covering the study area over the period 1974–2004 confirmed an overall trend with increasing temperatures and decreasing and/or more erratic precipitations. Mean temperatures (Figure 1A) showed an increasing rate in agreement with the observed trend in Europe during the last three decades [75,76]. In addition, our results confirm a negative yearly rainfall trend in the Mediterranean area [77]. The different trends shown in Figures 1A and 2B have some relevant agricultural implications: (i) increasing autumn rainfall in rain-fed durum growing areas leads to a greater water storage in the soil but may also delay sowing, especially when associated with increasing mean precipitation [78]; (ii) decreasing rainfall in winter reduces soil moisture with detrimental effects on water uptake especially if combined with limited root growth due to increased temperatures and delayed sowing; (iii) increased temperatures or heat shocks in late spring may abruptly interrupt translocation of photosynthates during grain filling thereby exposing caryopses to the risk of 'haying-off' [14,79,80]. All these points may dramatically result in increased vulnerability particularly in the agricultural systems of the Mediterranean Region [81,82].

4.2. Calibration, Validation and Evaluation of CERES-Wheat Model Performances

Calibration and validation of the CERES-Wheat crop model in the study area was already discussed in [45], where full details are available. Importantly, this study emphasizes the crucial importance of using data from long-term experiments [58] to overcome poor performance of the model due to deficiencies in model inputs and experimental observations as well as to allow a proper calibration. In general, the values of the genetic coefficients determined in this current study are similar to those obtained by the few other ones conducted on durum wheat [60,61], with the exception of parameters G1, G2 and G3. Moreover, the good and excellent results of the model in predicting grain yield and phenology, respectively, confirm the observations of [58] in their review of the performance of CERES-Rice and CERES-Wheat models in rice-wheat systems of South Asia, China and Southeast Australia.

The model proved to be less satisfactory in the case of kernel weight. This was probably due to modelling inaccuracies in simulating underlying physiological processes under stressed and non-stressed conditions [83]. Interestingly, the combined overestimate of grain yield and the underestimate of kernel weight resulting in an overestimate of the number of kernel per unit area had already been remarked in previous studies [45]. Hence, these systematic errors might be due to either inconsistent estimation of the number of grains or differences between durum wheat and bread wheat. Of note, the analysis by site revealed a better performance of the model at Benatzu when compared to Ussana for both grain yield and kernel weight. This is likely due to the low fertility of Ussana soil making this site drought-prone and with a greater frequency of very low yields. In this context, the poorer performance of the CERES-Wheat model under low-yielding conditions was already known [58]. In addition, the tendency of CERES-Wheat to overestimate grain yield under water shortage conditions has already been underlined [84]. Finally, the model accuracy in predicting anthesis dates did not show any remarkable differences between sites.

4.3. Climate Change Scenarios: Cultivar and Molecular Responses

The negative effect of increasing temperatures and decreasing rainfall on simulated grain yields at the two experimental sites is clear (Figure 3). Interestingly, Ussana site, less fertile and negatively affected by rainfall decrease and water scarcity, showed the greater yield reductions when compared to Benatzu (Figures 4 and 5). In summary, the yield reductions observed in our simulations are mostly consistent with other global [85–89] and regional studies [22,35,36], excluding the CO_2 effect that was not taken into consideration in this study, limiting evaluation of the cultivar responses to climate stimuli.

Concerning grain quality, the slight positive effect of increasing temperature and decreasing rainfall on kernel weight (Figures 6 and 7) is in contrast with findings on bread wheat showing a weight reduction due to high temperatures [39], heat shocks [90] and water stress [91] during grain filling. In all likelihood, these contrasting results may be associated with a lower correspondence between observed and predicted data for this trait during calibration and validation of the CERES-Wheat model [45].

As for phenology, the shortening of the growing period highlighted in this study is in agreement with other studies [19,37,92]. Interestingly, the reduction in growth duration from sowing to anthesis was larger for the early cultivars Duilio and Simeto when compared to the late cultivar Creso at both sites (Figure 8), confirming the adaptive role of earliness for durum wheat in drought prone environments [93]. Remarkably, the early genotypes Duilio and Simeto had a better yield performance than the late genotype Creso in both observed and predicted data (Table 4). Therefore, the CERES-Wheat crop model seems to capture fairly well the greater resilience shown by early genotypes in rain-fed Mediterranean conditions.

Regarding cultivar choice, this study confirms its potential key role as a farm-level adaptation measure to reduce the negative impacts of climate change on crop production [19]. In particular, an estimated avoidance of 10–15% yield reduction due to cropping adaptations such as changing cultivars and sowing times has been reported in the literature [15]. Thus far, a little effort has been made to understand the effect of cultivar choice and its role in tackling the detrimental effects of increasing temperatures and decreasing (or more erratic) rainfall on crop production. Our study shows a negative impact of harsh scenarios (i.e., increased air temperatures and decreased rainfall) on grain yield for all cultivars and sites. However, this detrimental effect can be mitigated by: (i) early sowing and (ii) replacing late genotypes with early ones. Concerning the latter point, our simulations show a percentage grain yield reduction from −31% (−29.2% and −33.1% at Ussana and Benatzu, respectively) for Creso (late cultivar) to −28.3% (−26.4% and 30.1% at Ussana and Benatzu, respectively) for Simeto (early cultivar) and −27.3% (−25.5% and −29.1% at Ussana and Benatzu, respectively) for Duilio (early cultivar) (Figures 4 and 5). Comparing the average simulated grain yield results of the most drought prone cultivar (i.e., Creso) with the most resilient one (i.e., Duilio), a percentage gain of 3.9% in grain yield has been registered in favor of the latter. From this perspective, targeting cultivars onto different environments and climate conditions is one of the main adaptation strategies to climate change [22,94]. Furthermore, this study has another important implication: the cultivars considered for this study were released in Italy between thirty and forty years ago in different environmental conditions when compared to now. This means that: (i) a plethora of higher-yielding and better adapted cultivars is now available for current growing conditions; (ii) the importance of plant breeding in selecting superior genotypes ensuring good yield performances and yield stability in climate change conditions is paramount. Therefore, the role of cultivar choice in the short term and of plant breeding in the long term to tackle the detrimental effects of climate change on yield production and stability must be fully emphasized.

A last crucial issue is related to molecular responses in climate change conditions. Since global warming, heatwaves, droughts and a general decreasing rainfall trend are projected in the Mediterranean areas [1], identifying, sequencing and characterizing stress-inducible genes becomes essential to develop molecular markers for marker assisted breeding. Therefore, conventional breeding techniques and biotechnologies may increase the effectiveness of selection, allowing high-yielding and drought resistant genotypes. Focusing on a molecular approach, the results presented in this study confirm the important role of *DREB* genes in abiotic stress conditions as shown by Liu et al., in 1998 [65]. Furthermore, the importance of the dehydration-responsive factor gene (*TdDRF1*) has been confirmed both in greenhouse and in field conditions in other durum wheat and triticale cultivars [73]. Moreover, the link between grain yield, drought tolerance and specific polimorphism of the *TdDRF1* gene has been demonstrated in recent studies [95] and a correlation between grain yield and an increased expression of *TdDRF1.3* transcript in some drought tolerant and rustic durum wheat and triticale cultivars was found [96]. However, other studies must be addressed to explore the molecular mechanisms of regulation of the *TdDRF1* gene expression as well as the contribution of other genes.

5. Conclusions

This study highlights the importance of a multidisciplinary approach involving the use of crop modelling and biotechnology in order to predict and evaluate the performances of durum wheat genotypes under climate change conditions. Concerning crop modelling, CERES-Wheat proved to be an effective tool when used to predict grain yield, anthesis date and, to a lesser extent, kernel weight. The impact of climate change scenarios on grain yield is more negative, moving from mild to severe scenarios for all genotypes, but reductions are to some extent mitigated for Simeto and namely Duilio (early genotypes) in comparison with Creso (late genotype). On the other hand, kernel weight tends to increase slightly in response to increasing temperatures and decreasing rainfall in particular under mild climate change scenarios. All genotypes showed a reduction of their crop growing cycle as a consequence of increasing temperatures, with Duilio and Simeto revealing to be more resilient than Creso. The detrimental joint effect of simulated increasing temperatures and decreasing rainfall is also affected by soil fertility, with a stronger impact in low-yielding potential soils.

The predictive responses of the CERES-Wheat model can be also interpreted in the light of molecular responses of durum wheat cultivar to drought stress. In this context, the role of *DREB* genes, with a special focus on *TdDRF1*, in conditioning the resistance of genotypes to drought conditions must be underlined.

Furthermore, our analysis indicates that CERES-Wheat crop model responses are highly consistent with observations from most rain-fed durum wheat growing areas of the Mediterranean Region. By showing that early genotypes can be better adapted to increasing temperature and decreasing rainfall, the CERES-Wheat model proves to be a reliable tool to determine the impact of climate change on crops and can help to underline and quantify the simulated effects of cultivar choice to tackle downward trends in grain yield, particularly in the Mediterranean rain-fed areas. Hence, CERES-Wheat can be successfully used to support adaptation strategies such as targeting cultivars onto specific environments or to guide selection decisions in crop breeding programs, also implying the contribution of a molecular approach aiming at developing molecular markers for the resistance to abiotic and drought stresses.

Author Contributions: Conceptualization, M.D. and P.D.; methodology, M.D. and C.C.; validation, M.D. and C.C.; data curation, M.D.; writing—original draft preparation, M.D. and V.M.; writing—review and editing, M.D. and V.M.; supervision, P.D. All authors have read and agreed to the published version of the manuscript.

Funding: This research received no external funding.

Institutional Review Board Statement: Not applicable.

Informed Consent Statement: Not applicable.

Data Availability Statement: The data presented in this study are available on request from the corresponding author.

Conflicts of Interest: The authors declare no conflict of interest. The funders had no role in the design of the study; in the collection, analyses, or interpretation of data; in the writing of the manuscript, or in the decision to publish the results.

References

1. IPCC. Summary for policymakers. In *Climate Change 2021: The Physical Science Basis. Contribution of Working Group I to the Sixth Assessment Report of the Intergovernmental Panel on Climate Change*; Masson-Delmotte, V., Zhai, P., Pirani, A., Connors, S.L., Péan, C., Berger, S., Caud, N., Chen, Y., Goldfarb, L., Gomis, M.I., et al., Eds.; 2021; *in press*.
2. IPCC. Summary for policymakers. In *Climate Change and Land: An IPCC Special Report on Climate Change, Desertification, Land Degradation, Sustainable Land Management, Food Security, and Greenhouse Gas Fluxes in Terrestrial Ecosystems*; Shukla, P.R., Skea, J., Buendia, E.C., Masson-Delmotte, V., Pörtner, H.-O., Roberts, D.C., Zhai, P., Slade, R., Connors, S., van Diemen, R., Eds.; 2019; *in press*.
3. Jägermeyr, J.; Müller, C.; Ruane, A.C.; Elliott, J.; Balkovic, J.; Castillo, O.; Faye, B.; Foster, I.; Folberth, C.; Franke, J.A.; et al. Climate impacts on global agriculture emerge earlier in new generation of climate and crop models. *Nat. Food* **2021**, *2*, 873–885. [CrossRef]
4. Ainsworth, E.A.; Long, S.P. What have we learned from 15 years of free-air CO_2 enrichment (FACE)? A meta-analysis of the responses of photosynthesis, canopy properties and plant production to rising CO_2. *New Phytol.* **2005**, *165*, 351–372. [CrossRef] [PubMed]
5. Tubiello, F.N.; Amthor, J.S.; Boote, K.J.; Donatelli, M.; Easterling, W.; Fischer, G.; Gifford, R.M.; Howden, M.; Reilly, J.; Rosenzweig, C. Crop response to elevated CO_2 and world food supply: A comment on "Food for Thought..." by Long et al.; Science 312:1918–1921, 2006. *Eur. J. Agron.* **2007**, *26*, 215–223. [CrossRef]
6. Toreti, A.; Deryng, D.; Tubiello, F.N.; Müller, C.; Kimball, B.A.; Moser, G.; Boote, K.; Asseng, S.; Pugh, T.A.M.; Vanuytrecht, E.; et al. Narrowing uncertainties in the effects of elevated CO_2 on crops. *Nat. Food* **2020**, *1*, 775–782. [CrossRef]
7. Baker, J.T. Yield responses of southern U.S. rice cultivars to CO_2 and temperature. *Agric. For. Meteorol.* **2004**, *122*, 129–137. [CrossRef]
8. Caldwell, C.R.; Britz, S.J.; Mirecki, R.M. Effect of temperature, elevated carbon dioxide, and drought during seed development on the isoflavone content of dwarf soybean [Glycine max (L.) Merrill] grown in controlled environments. *J. Agric. Food Chem.* **2005**, *53*, 1125–1129. [CrossRef]
9. Thomas, J.M.G.; Boote, K.J.; Allen, L.H., Jr.; Gallo-Meagher, M.; Davis, J.M. Elevated temperature and carbon dioxide effects on soybean seed composition and transcript abundance. *Crop Sci.* **2003**, *43*, 1548–1557. [CrossRef]

10. Beach, R.H.; Sulser, T.B.; Crimmins, A.; Cenacchi, N.; Cole, J.; Fukagawa, N.K.; Mason-D'Croz, D.; Myers, S.; Sarofim, M.C.; Smith, M.; et al. Combining the effects of increased atmospheric carbon dioxide on protein, iron, and zinc availability and projected climate change on global diets: A modelling study. *Lancet Planet. Health* **2019**, *3*, e307–e317. [CrossRef]
11. Fernando, N.; Panozzo, J.; Tausz, M.; Norton, R.; Fitzgerald, G.; Khan, A.; Seneweera, S. Rising CO_2 concentration altered wheat grain proteome and flour rheological characteristics. *Food Chem.* **2015**, *170*, 448–454. [CrossRef]
12. Zhou, B.; Zhai, P.; Chen, Y.; Yu, R. Projected changes of thermal growing season over Northern Eurasia in a 1.5 °C and 2 °C warming world. *Environ. Res. Lett.* **2018**, *13*, 035004. [CrossRef]
13. Högy, P.; Wieser, H.; Köhler, P.; Schwadorf, K.; Breuer, J.; Franzaring, J.; Muntifering, R.; Fangmeier, A. Effects of elevated CO_2 on grain yield and quality of wheat: Results from a 3-year free-air CO_2 enrichment experiment. *Plant Biol.* **2009**, *11*, 60–69. [CrossRef] [PubMed]
14. Nuttall, J.G.; O'Leary, G.J.; Khimashia, N.; Asseng, S.; Fitzgerald, G.; Norton, R. 'Haying-off' in wheat is predicted to increase under a future climate in south-eastern Australia. *Crop Pasture Sci.* **2012**, *63*, 593–605. [CrossRef]
15. Easterling, W.E.; Aggarwal, P.K.; Batima, P.; Brander, K.M.; Erda, L.; Howden, S.M.; Kirilenko, A.; Morton, J.; Soussana, J.-F.; Schmidhuber, J.; et al. Food, fibre and forest products. In *Climate Change 2007: Impacts, Adaptation and Vulnerability. Contribution of Working Group II to the Fourth Assessment Report of the Intergovernmental Panel on Climate Chang*; Parry, M.L., Canziani, O.F., Palutikof, J.P., van der Linden, P.J., Hanson, C.E., Eds.; Cambridge University Press: Cambridge, UK, 2007; pp. 273–313.
16. Giannakopoulos, C.; Le Sager, P.; Bindi, M.; Moriondo, M.; Kostopoulou, E.; Goodess, C.M. Climatic changes and associated impacts in the Mediterranean resulting from global warming. *Glob. Planet. Chang.* **2009**, *68*, 209–224. [CrossRef]
17. van Ittersum, M.K.; Howden, S.M.; Asseng, S. Sensitivity of productivity and deep drainage of wheat cropping system in a Mediterranean environment to changes in CO_2, temperature and precipitation. *Agric. Ecosyst. Environ.* **2003**, *97*, 255–273. [CrossRef]
18. Moriondo, M.; Giannakopoulos, C.; Bindi, M. Climate change impact assessment: The role of climate extremes in crop yield simulation. *Clim. Chang.* **2011**, *104*, 679–701. [CrossRef]
19. Parry, M.; Rosenzweig, C.; Livermore, M. Climate change, global food supply and risk of hunger. *Philos. Trans. R. Soc. Lond. Ser. B* **2005**, *360*, 2125–2138. [CrossRef]
20. Porter, J.R.; Gawith, M. Temperatures and the growth and development of wheat: A review. *Eur. J. Agron.* **1999**, *10*, 23–36. [CrossRef]
21. Rötter, R.; van de Geijn, S.C. Climate change effects on plant growth, crop yield and livestock. *Clim. Chang.* **1999**, *43*, 651–681. [CrossRef]
22. Tubiello, F.N.; Donatelli, M.; Rosenzweig, C.; Stockle, C.O. Effects of climate change and elevated CO_2 on cropping systems: Model predictions at two Italian locations. *Eur. J. Agron.* **2000**, *13*, 179–189. [CrossRef]
23. Hristov, J.; Toreti, A.; Pérez Domínguez, I.; Dentener, F.; Fellmann, T.; Elleby, C.; Ceglar, A.; Fumagalli, D.; Niemeyer, S.; Cerrani, I.; et al. *Analysis of Climate Change Impacts on EU Agriculture by 2050*; EUR 30078EN; Publications Office of the European Union: Luxembourg, 2020; ISBN 978-92-76-10617-3. [CrossRef]
24. Ewert, F.; Rounsevell, M.D.A.; Reginster, I.; Metzger, M.J.; Leemans, R. Future scenarios of European agricultural land use I. Estimating changes in crop productivity. *Agric. Ecosyst. Environ.* **2005**, *107*, 101–116. [CrossRef]
25. Olesen, J.E.; Jensen, T.; Petersen, J. Sensitivity of field-scale winter wheat production in Denmark to climate variability and climate change. *Clim. Res.* **2000**, *15*, 221–238. [CrossRef]
26. Bray, E.A. Genes commonly regulated by water-deficit stress in Arabidopsis thaliana. *J. Exp. Bot.* **2004**, *55*, 2331–2341. [CrossRef] [PubMed]
27. Smith, P.; Martino, D.; Cai, Z.; Gwary, D.; Janzen, H.; Kumar, P.; McCarl, B.; Ogle, S.; O'Mara, F.; Rice, C.; et al. Agriculture. In *Climate Change 2007: Mitigation. Contribution of Working Group III to the Fourth Assessment Report of the Intergovernmental Panel on Climate Change*; Metz, B., Davidson, O.R., Bosch, P.R., Dave, R., Meyer, L.A., Eds.; Cambridge University Press: Cambridge, UK; New York, NY, USA, 2007; pp. 497–540.
28. Maracchi, G.; Sirotenko, O.; Bindi, M. Impacts of present and future climate variability on agriculture and forestry in the temperate regions: Eur. *Clim. Chang.* **2005**, *70*, 117–135. [CrossRef]
29. Cramer, W.; Guiot, J.; Fader, M.; Garrabou, J.; Gattuso, J.P.; Iglesias, A.; Lange, M.A.; Lionello, P.; Llasat, M.C.; Paz, S.; et al. Climate change and interconnected risks to sustainable development in the Mediterranean. *Nat. Clim. Chang.* **2018**, *8*, 972–980. [CrossRef]
30. Lionello, P.; Scarascia, L. The relation between climate change in the Mediterranean region and global warming. *Reg. Environ. Chang.* **2018**, *18*, 1481–1493. [CrossRef]
31. IPCC. *Climate Change 2014: Impacts, Adaptation, and Vulnerability. Part B: Regional Aspects. Contribution of Working Group II to the Fifth Assessment Report of the Intergovernmental Panel on Climate Change*; Barros, V.R., Field, C.B., Dokken, D.J., Mastrandrea, M.D., Mach, K.J., Bilir, T.E., Chatterjee, M., Ebi, K.L., Estrada, Y.O., Genova, R.C., Eds.; Cambridge University Press: Cambridge, UK; New York, NY, USA, 2014; p. 688.
32. Miflin, B. Technologies for crop improvement in the 21st century. In *Durum Wheat Improvement in the Mediterranean Region: New Challenges*; Royo, C., Nachit, M.M., Di Fonzo, N., Araus, J.L., Eds.; CIHEAM: Zaragoza, Spain, 2000; pp. 19–25.

33. Gueye, L.; Bzioul, M.; Johnson, O. Water and sustainable development in the countries of Northern Africa: Coping with challenges and scarcity. In *Assessing Sustainable Development in Africa. Africa's Sustainable Development Bulletin*; Economic Commission for Africa: Addis Ababa, Ethiopia, 2005; pp. 24–28.
34. IGC. International Grains Council. 2014. Available online: http://www.igc.int/en/default.aspx (accessed on 1 December 2021).
35. Ferrise, R.; Moriondo, M.; Bindi, M. Probabilistic assessment of climate change impacts on durum wheat in the Mediterranean region. *Nat. Hazards Earth Syst. Sci.* **2011**, *11*, 1293–1302. [CrossRef]
36. Ventrella, D.; Charfeddine, M.; Moriondo, M.; Rinaldi, M.; Bindi, M. Agronomic adaptation strategies under climate change for winter durum wheat and tomato in southern Italy: Irrigation and nitrogen fertilization. *Reg. Environ. Chang.* **2012**, *12*, 407–419. [CrossRef]
37. Mereu, V.; Gallo, A.; Trabucco, A.; Carboni, G.; Spano, D. Modeling high-resolution climate change impacts on wheat and maize in Italy. *Clim. Risk Manag.* **2021**, *33*, 100339. [CrossRef]
38. Chowdhury, S.I.; Wardlaw, I.F. The effect of temperature on kernel development in cereals. *Aust. J. Agric. Res.* **1978**, *29*, 205–223. [CrossRef]
39. Gibson, L.R.; Paulsen, G.M. Yield components of wheat grown under high temperature stress during reproductive growth. *Crop Sci.* **1999**, *39*, 1841–1846. [CrossRef]
40. Passarella, V.S.; Savin, R.; Slafer, G.A. Breeding effects of barley grain weight and quality to events of high temperature during grain filling. *Euphytica* **2005**, *141*, 41–48. [CrossRef]
41. Wardlaw, I.F.; Wrigley, C.W. Heat tolerance in temperate cereals: An overview. *Aust. J. Plant Physiol.* **1994**, *21*, 695–703. [CrossRef]
42. Bindi, M.; Olesen, J.E. The responses of agriculture in Europe to climate change. *Reg. Environ. Chang.* **2011**, *11*, 151–158. [CrossRef]
43. Donatelli, M.; van Ittersum, M.K.; Bindi, M.; Porter, R.J. Modelling cropping systems-highlights of the symposium and preface to the special issues. *Eur. J. Agron.* **2002**, *18*, 187–197. [CrossRef]
44. Bindi, M.; Fibbi, L.; Gozzini, B.; Orlandini, S.; Miglietta, F. Modelling the impact of climate scenarios on yield and yield variability of grapevine. *Clim. Res.* **1996**, *7*, 213–224. [CrossRef]
45. Dettori, M.; Cesaraccio, C.; Motroni, A.; Spano, D.; Duce, P. Using CERES-Wheat to simulate durum wheat production and phenology in Southern Sardinia, Italy. *Field Crop. Res.* **2011**, *120*, 179–188. [CrossRef]
46. Guereña, A.; Ruiz-Ramos, M.; Díaz-Ambrona, C.H.; Conde, J.; Mínguez, M.I. Assessment of climate change and agriculture in Spain using climate models. *Agron. J.* **2001**, *93*, 237–249. [CrossRef]
47. Moriondo, M.; Bindi, M.; Kundzewicz, Z.W.; Szwed, M.; Chorynski, A.; Matczak, P.; Radziejewski, M.; McEvoy, D.; Wreford, A. Impact and adaptation opportunities for European agriculture in response to climatic change and variability. *Mitig. Adapt. Strateg. Glob. Chang.* **2010**, *15*, 657–679. [CrossRef]
48. Moriondo, M.; Bindi, M.; Fagarazzi, C.; Ferrise, R.; Trombi, G. Framework for high-resolution climate change impact assessment on grapevines at a regional scale. *Reg. Environ. Chang.* **2011**, *3*, 553–567. [CrossRef]
49. Toscano, P.; Ranieri, M.; Matese, A.; Vaccari, F.P.; Gioli, B.; Zaldei, A.; Silvestri, M.; Ronchi, C.; La Cava, P.; Porter, J.R.; et al. Durum wheat modeling: The Delphi system, 11 years of observations in Italy. *Eur. J. Agron.* **2012**, *43*, 108–118. [CrossRef]
50. Dettori, M.; Cesaraccio, C.; Duce, P. Simulation of climate change impacts on production and phenology of durum wheat in Mediterranean environments using CERES-Wheat model. *Field Crop. Res.* **2017**, *206*, 43–53. [CrossRef]
51. Mereu, V.; Gallo, A.; Spano, D. Optimizing Genetic Parameters of CSM-CERES Wheat and CSM-CERES Maize for Durum Wheat, Common Wheat, and Maize in Italy. *Agronomy* **2019**, *9*, 665. [CrossRef]
52. Yin, X.; van Laar, H.H. *Crop Systems Dynamics: An Ecophysiological Simulation Model for Genotype-by-Environment Interactions*; Wageningen Academic Publishers: Wageningen, The Netherlands, 2005; p. 168.
53. Jones, J.W.; Hoogenboom, G.; Porter, C.; Boote, K.; Batchelor, W.; Hunt, L.A.; Singh, U.; Gijsman, A.; Ritchie, J. The DSSAT cropping system model. *Eur. J. Agron.* **2003**, *18*, 235–265. [CrossRef]
54. Ritchie, J.T.; Otter, S. Description and performance of CERES-wheat: Use-oriented wheat yield model. In *ARS Wheat Yield Project. Agricultural Research Service*; Willis, O.W., Ed.; Department of Agriculture: Washington, DC, USA, 1985; pp. 159–175.
55. Ritchie, J.T.; Godwin, D.C.; Otter-Nacke, S. *CERES-Wheat*; University of Texas Press: Austin, TX, USA, 1988.
56. Ritchie, J.T.; Singh, U.; Godwin, D.C.; Bowen, W.T. Cereal growth, development and yield. In *Understanding Options for Agricultural Production*; Tsuji, Y.G., Hoogenboom, G., Thornton, P.K., Eds.; Kluwer Academic Publishers: Dordrecht, The Netherlands, 1998; pp. 79–98.
57. Mathews, R.; Blackmore, B.S. Using crop simulation models to determine optimum management practices in precision agriculture. In Proceedings of the 1st European Conference on Precision Agriculture, Warwick, UK, 8–10 September 1997; pp. 413–420.
58. Timsina, J.; Humphreys, E. Performance of CERES-Rice and CERES-Wheat models in rice-wheat systems: A review. *Agric. Syst.* **2006**, *90*, 5–31. [CrossRef]
59. Pecetti, L.; Hollington, P.A. Application of the CERES-Wheat simulation model to durum wheat in two diverse Mediterranean environments. *Eur. J. Agron.* **1997**, *6*, 125–139. [CrossRef]
60. Rezzoug, W.; Gabrielle, B.; Suleiman, A.; Benabdeli, K. Application and evaluation of the DSSAT-wheat in the Tiaret region of Algeria. *Afr. J. Agric. Res.* **2008**, *3*, 284–296.
61. Rinaldi, M. Water availability at sowing and nitrogen management of durum wheat: A seasonal analysis with the CERES-Wheat model. *Field Crop. Res.* **2004**, *89*, 27–37. [CrossRef]

62. Dalla Marta, A.; Orlando, F.; Mancini, M.; Guasconi, F.; Motha, R.; Qu, J.; Orlandini, S. A simplified index for an early estimation of durum wheat yield in Tuscany (Central Italy). *Field Crop. Res.* **2015**, *170*, 1–6. [CrossRef]
63. USDA. *Soil Survey Staff. Natural Resources Conservation Service*; Keys to Soil Taxonomy: Washington, DC, USA, 2002.
64. Donatelli, M.; Bellocchi, G.; Fontana, F. RadEst3.00: Software to estimate daily radiation data from commonly available meteorological variables. *Eur. J. Agron.* **2003**, *18*, 369–372. [CrossRef]
65. Liu, Q.; Kasuga, Y.; Sakuma, H.; Abe, S.; Miura, K.; Yamaguchi-Shinozaki, K. Two Transcription factors, DREB1 and DREB2 with EREBP/AP2 DNA binding domain separate two cellular transduction pathways in drought—And low-temperature responsive gene expression, respectively, in Arabidopsis. *Plant Cell* **1998**, *10*, 1391–1406. [CrossRef]
66. Galeffi, P.; Latini, A.; Kumar, S.; Sperandei, M.; Cavicchioni, G.; Palmieri, E.; Cantale, C.; Iannetta, M. Functional genomic analyses of DROUGHT stress in various Italian durum wheat varieties. In Proceedings of the ICABR 8th International Conference on Agricultural Biotechnology: International Trade and Domestic Production, Ravello, Italy, 8–11 July 2004.
67. Godwin, D.; Ritchie, J.; Singh, U.; Hunt, L. *A User's Guide to CERES-Wheat v2.1*, 2nd ed.; International Fertilizer Development Center: Muscle Shoals, AL, USA, 1990; pp. 1–131.
68. Boote, K.J.; Jones, J.W.; Hoogenboom, G.; Pickering, N.B. The CROPGRO model for grain legumes. In *Understanding Options for Agricultural Production*; Tsuji, G.Y., Hoogenboom, G., Thornton, P.K., Eds.; Kluwer Academic Publisher: Dordrecht, The Netherlands, 1998; pp. 99–128.
69. Loague, K.; Green, R.W. Statistical and graphical methods for evaluating solute transport models: Overiview and application. *J. Contam. Hydrol.* **1991**, *7*, 51–73. [CrossRef]
70. Xevi, E.; Gilley, J.; Feyen, J. Comparative study of two crop yield simulation models. *Agric. Water Manag.* **1996**, *30*, 155–173. [CrossRef]
71. Giorgi, F.; Lionello, P. Climate change projections for the Mediterranean region. *Glob. Planet. Chang.* **2008**, *63*, 90–104. [CrossRef]
72. Gualdi, S.; Somot, S.; Li, L.; Artale, V.; Adani, M.; Bellucci, A.; Braun, A.; Calmanti, S.; Carillo, A.; Dell'Aquila, A.; et al. The CIRCE Simulations Regional Climate Change Projections with Realistic Representation of the Mediterranean Sea. *Bull. Am. Meteorol. Soc.* **2013**, *94*, 65–81. [CrossRef]
73. Galeffi, P.; Latini, A.; Rasi, C.; Sperandei, M.; Cavicchioni, G.; Palmieri, E.; Sharma, S.; Cantale, C.; Iannetta, M.; Dettori, M.; et al. *Molecular Responses to Drought Stress: Expression of TdDRF1 Gene in Several Durum Wheat Varieties in Controlled Greenhouse and in Field Conditions*; Interdrought II: Rome, Italy, 2005.
74. Latini, A.; Rasi, C.; Sperandei, M.; Cantale, C.; Iannetta, M.; Dettori, M.; Ammar, K.; Galeffi, P. Identification of a DREB-related gene in Triticum durum and its expression under water stress conditions. *Ann. Appl. Biol.* **2007**, *150*, 187–195. [CrossRef]
75. Alcamo, J.; Moreno, J.M.; Nováky, B.; Bindi, M.; Corobov, R.; Devoy, R.J.N.; Giannakopoulos, C.; Martin, E.; Olesen, J.E.; Shvidenko, A. Europe. In *Climate Change 2007: Impacts, Adaptation and Vulnerability. Contribution of Working Group II to the Fourth Assessment Report of the Intergovernmental Panel on Climate Change*; Parry, M.L., Canziani, O.F., Palutikof, J.P., van der Linden, P.J., Hanson, C.E., Eds.; Cambridge University Press: Cambridge, UK, 2007; pp. 541–580.
76. Jones, P.D.; Moberg, A. Hemispheric and large-scale surface air temperature variations: An extensive revision and update to 2001. *J. Clim.* **2003**, *16*, 206–223. [CrossRef]
77. Norrant, C.; Douguédroit, A. Monthly and daily precipitation trends in the Mediterranean. *Theor. Appl. Climatol.* **2006**, *83*, 89–106. [CrossRef]
78. Klein Tank, A.M.G.; Wijngaard, J.B.; Konnen, G.P.; Bohm, R.; Demaree, G.; Gocheva, A.; Mileta, M.; Pashiardis, S.; Hejkrlik, L.; Kern-Hansen, C.; et al. Daily dataset of 20th-century surface air temperature and precipitation series for the European Climate Assessment. *Int. J. Climatol.* **2002**, *22*, 1441–1453. [CrossRef]
79. van Herwaarden, A.F.; Farquar, G.D.; Angus, J.F.; Richards, R.A.; Howe, G.N. 'Haying-off', the negative grain yield response of dryland wheat to N fertilizer. I. Biomass, grain yield, and water use. *Aust. J. Agric. Res.* **1998**, *49*, 1067–1081. [CrossRef]
80. van Herwaarden, A.F.; Richards, R.A.; Farquar, G.D.; Angus, J.F. 'Haying-off', the negative grain yield response of dryland wheat to N fertilizer. III. The influence of water deficit and heat shock. *Aust. J. Agric. Res.* **1998**, *49*, 1095–1110. [CrossRef]
81. Metzger, M.J.; Leemans, R.; Schröter, D.; Cramer, W. ATEAM consortium. In *The ATEAM Vulnerability Mapping Tool. Quantitative Approaches in System Analysis No. 27*; Wageningen, C.T., Ed.; de Witt Graduate School for Production Ecology and Resource Conservation: Wageningen, The Netherlands, 2004.
82. Olesen, J.E.; Bindi, M. Consequences of climate change for European agricultural productivity, land use and policy. *Eur. J. Agron.* **2002**, *16*, 239–262. [CrossRef]
83. Langensiepen, M.; Hanus, H.; Schoop, P.; Gräsle, W. Validating CERES-wheat under North-German environmental conditions. *Agric. Syst.* **2008**, *97*, 34–37. [CrossRef]
84. Staggenborg, S.A.; Vanderlip, R.L. Crop simulation models can be used as dryland cropping systems research tools. *Agron. J.* **2005**, *97*, 378–384. [CrossRef]
85. Fischer, G.; Shah, M.; Tubiello, F.N.; van Velthuizen, H. Socio-economic and climate change impacts on agriculture: An integrated assessment, 1990–2080. *Philos. Trans. R. Soc. Lond. B Biol. Sci.* **2005**, *360*, 2067–2083. [CrossRef]
86. Parry, M.L.; Rosenzweig, C.; Iglesias, A.; Livermore, M.; Fischer, G. Effects of climate change on global food production under SRES emissions and socio-economic scenarios. *Glob. Environ. Chang.* **2004**, *14*, 53–67. [CrossRef]
87. Rosenzweig, C.; Parry, M.L. Potential impact of climate change on world food supply. *Nature* **1994**, *367*, 133–138. [CrossRef]

88. Rosenzweig, C.; Parry, M.L.; Fischer, G.; Frohberg, K. *Climate Change and World Food Supply*; Research Report No. 3; Environmental Change Unit, University of Oxford: Oxford, UK, 1993; p. 28.
89. Tubiello, F.N.; Fischer, G. Reducing climate change impacts on agriculture: Global and regional effects of mitigation, 2000–2080. *Technol. Forecast. Soc. Chang.* **2007**, *74*, 1030–1056. [CrossRef]
90. Wardlaw, I.F.; Blumenthal, C.; Larroque, O.; Wrigley, C.W. Contrasting effect of chronic heat stress and heat shock on kernel weight and flour quality in wheat. *Funct. Plant Biol.* **2002**, *29*, 25–34. [CrossRef]
91. Khanna-Chopra, R.; Rao, P.S.S.; Maheswari, M.; Xiaobing, L.; Shivshankar, K.S. Effect of water deficit on accumulation of dry matter, carbon and nitrogen in the kernel of wheat genotypes differing in yield stability. *Ann. Bot.* **1994**, *74*, 503–511. [CrossRef]
92. Moriondo, M.; Bindi, M. The impact of climate change on the phenology of typical Mediterranean crops. *Ital. J. Agrometeorol.* **2009**, *3*, 5–12.
93. Annicchiarico, P.; Pecetti, L. Contribution to some agronomic traits to durum wheat performance in a dry Mediterranean region of Northern Syria. *Agronomie* **1993**, *13*, 25–34. [CrossRef]
94. Annicchiarico, P. Coping with and exploiting genotype-by-environment interactions. In *Plant Breeding and Farmer Participation*; Ceccarelli, S., Guimarães, E.P., Weltzien, E., Eds.; Food and Agriculture Organization of the United Nations: Rome, Italy, 2009; pp. 519–564.
95. Cantale, C.; Di Bianco, D.; Thiyagarajan, K.; Ammar, K.; Galeffi, P. B-genome specific polymorphism in the TdDRF1 gene is in relationship with grain yield. *Planta* **2018**, *247*, 459–469. [CrossRef]
96. Latini, A.; Sperandei, M.; Cantale, C.; Arcangeli, C.; Ammar, K.; Galeffi, P. Variability and expression profile of the DRF1 gene in four cultivars of durum wheat and one triticale under moderate water stress conditions. *Planta* **2013**, *237*, 967–978. [CrossRef]

genes MDPI

Article

Juvenile Heat Tolerance in Wheat for Attaining Higher Grain Yield by Shifting to Early Sowing in October in South Asia

Uttam Kumar [1,2], Ravi Prakash Singh [3], Susanne Dreisigacker [3], Marion S. Röder [4], Jose Crossa [3], Julio Huerta-Espino [5], Suchismita Mondal [3], Leonardo Crespo-Herrera [3], Gyanendra Pratap Singh [6], Chandra Nath Mishra [6], Gurvinder Singh Mavi [7], Virinder Singh Sohu [7], Sakuru Venkata Sai Prasad [8], Rudra Naik [9], Satish Chandra Misra [10] and Arun Kumar Joshi [1,2,*]

1 Borlaug Institute for South Asia (BISA), NASC Complex, DPS Marg, New Delhi 110012, India; u.kumar@cgiar.org
2 International Maize and Wheat Improvement Center (CIMMYT), NASC Complex, DPS Marg, New Delhi 110012, India
3 International Maize and Wheat Improvement Center (CIMMYT), El Batan 56237, Mexico; R.Singh@cgiar.org (R.P.S.); s.dreisigacker@cgiar.org (S.D.); J.CROSSA@cgiar.org (J.C.); s.mondal@cgiar.org (S.M.); l.crespo@cgiar.org (L.C.-H.)
4 Leibniz Institute of Plant Genetics and Crop Plant Research (IPK), 06466 Gatersleben, Germany; roder@ipk-gatersleben.de
5 Campo Experimental Valle de Mexico-INIFAP, Carretera los Reyes-Texcoco, Coatlinchan 56250, Mexico; j.huerta@cgiar.org
6 ICAR-Indian Institute of Wheat and Barley Research (IIWBR), ICAR, Karnal 132001, India; gyanendrapsingh@hotmail.com (G.P.S.); Chandra.Mishra@icar.gov.in (C.N.M.)
7 Plant Breeding and Genetics Department, Punjab Agricultural University, Ludhiana 141004, India; mavig666@pau.edu (G.S.M.); sohuvs@pau.edu (V.S.S.)
8 Regional Research Station, Indian Agricultural Research Institute, Indore 542001, India; sprasad98@gmail.com
9 Department of Genetics and Plant Breeding, University of Agricultural Sciences, Krishi Nagar, Dharwad 580005, India; rvnaikgpb@gmail.com
10 Genetics and Plant Breeding Group, Agharkar Research Institute, Pune 411004, India; wheat.bisa@gmail.com
* Correspondence: a.k.joshi@cgiar.org

Citation: Kumar, U.; Singh, R.P.; Dreisigacker, S.; Röder, M.S.; Crossa, J.; Huerta-Espino, J.; Mondal, S.; Crespo-Herrera, L.; Singh, G.P.; Mishra, C.N.; et al. Juvenile Heat Tolerance in Wheat for Attaining Higher Grain Yield by Shifting to Early Sowing in October in South Asia. *Genes* **2021**, *12*, 1808. https://doi.org/10.3390/genes12111808

Academic Editor: Patrizia Galeffi

Received: 30 September 2021
Accepted: 6 November 2021
Published: 18 November 2021

Abstract: Farmers in northwestern and central India have been exploring to sow their wheat much earlier (October) than normal (November) to sustain productivity by escaping terminal heat stress and to utilize the available soil moisture after the harvesting of rice crop. However, current popular varieties are poorly adapted to early sowing due to the exposure of juvenile plants to the warmer temperatures in the month of October and early November. Therefore, a study was undertaken to identify wheat genotypes suited to October sowing under warmer temperatures in India. A diverse collection of 3322 bread wheat varieties and elite lines was prepared in CIMMYT, Mexico, and planted in the 3rd week of October during the crop season 2012–2013 in six locations (Ludhiana, Karnal, New Delhi, Indore, Pune and Dharwad) spread over northwestern plains zone (NWPZ) and central and Peninsular zone (CZ and PZ; designated as CPZ) of India. Agronomic traits data from the seedling stage to maturity were recorded. Results indicated substantial diversity for yield and yield-associated traits, with some lines showing indications of higher yields under October sowing. Based on agronomic performance and disease resistance, the top 48 lines (and two local checks) were identified and planted in the next crop season (2013–2014) in a replicated trial in all six locations under October sowing (third week). High yielding lines that could tolerate higher temperature in October sowing were identified for both zones; however, performance for grain yield was more promising in the NWPZ. Hence, a new trial of 30 lines was planted only in NWPZ under October sowing. Lines showing significantly superior yield over the best check and the most popular cultivars in the zone were identified. The study suggested that agronomically superior wheat varieties with early heat tolerance can be obtained that can provide yield up to 8 t/ha by planting in the third to fourth week of October.

Keywords: early heat stress; *Triticum aestivum*; heat tolerance; VRN; PPD; photoperiod

1. Introduction

Wheat is a strategic, staple crop in India [1] and South Asia [2]. It holds particular significance for women from marginal and small farming households, who contribute significantly to wheat production systems and livestock management. The Indo-Gangetic Plains of South Asia are considered crucial for meeting foot security needs of a huge population of >900 million people. The total population of the four South Asian countries (India, Nepal, Bangladesh and Pakistan) is 1971 million as of November 2021 (http://www.worldometers.info/world-population/southern-asia-population/, accessed on 11 November 2021). At present, the Gangetic Plains of South Asia are considered optimal for wheat farming, but may become sub-optimal by 2050 due to climate change [3]. On top of that, if the speed of decline in ground water table remains as it is today, this may be a major cause of food insecurity and affect the livelihood of farmers [4]. Heat tolerance of crops varies greatly, and wheat is among the most sensitive of the major staples. A study reported that yield losses in wheat for each 1 °C temperature may fall between 3 and 17% for north western India and Pakistan [5].

To sustain wheat productivity, farmers in northwestern and central India are shifting to earlier sowings (in the second fortnight of October, immediately after rice) to take benefits of residual moisture of the previous monsoon and to allow crop to mature much earlier than the end of March to beginning of April, a period when terminal heat stress becomes a major issue. In the 10 million ha rice-wheat system of South Asia, there is often adequate residual soil moisture at the end of the irrigated rice season, but these soils dry out by mid-November, the current optimum sowing date, requiring additional irrigation to achieve uniform wheat germination and establishment [6]. Furthermore, water tables cyclically fall so low in central India by late January that pumping is prohibitively expensive or not possible [7]. Early sowing using residual moisture allows farmers to save one irrigation and thus increases water productivity, especially when combined with other agronomic practices. However, most of the present wheat cultivars in farmers fields are not well adapted to early season sowing (3rd week of October), since temperatures are quite warm in this period and affect wheat crop by forcing it to grow much faster, accumulate lesser biomass and produce lower yield. Adapted varieties will need to be tolerant to early as well as late-season warmer temperatures [8].

The present knowledge on genetic control of adaptation response and heat tolerance is lacking. Major genes underlying adaptation have been identified in recent years, namely those controlling vernalization response, photoperiod sensitivity and development rate "earliness per se" [9–12]. However, the major gaps in the knowledge of the effects of possible allele combination on adaptation in specific environments. The *Vrn* genes control difference of spring and winter wheat by defining the chilling hours required by the wheat plant to be able to flower, while the *Ppd* genes play an important part in delaying flowering time in the spring after vernalization requirement has been satisfied. The *Eps* loci may influence more subtle effects in the life cycle for regional adaptation. QTL studies have described various regions of potential *Eps* genes [13,14] with some chromosomes (4B, 6A, 7D) showing this more frequently in a diverse germplasm [15]. *Vrn*, *Ppd* and *Eps* genes additionally present epistatic interactions [16,17]. Large numbers of allele combinations will therefore be involved in determining the regulation of growth habit and optimal adaptation to a certain environment.

CIMMYT's contribution to the Green Revolution is well known. Still today, advanced germplasm lines developed and distributed through international trials and nurseries by CIMMYT are grown annually on more than 100 locations in the world [18]. Analyses of some of these International Yield Trials indicate that grain yields of the best new genotypes are significantly higher than the local checks [19,20]. However, no efforts were made so far to breed for early heat tolerance under October sowing. Therefore, the present investigation was initiated to identify wheat varieties and adapted germplasm for early (October) sowing under warmer temperatures that exploit the residual moisture and escape terminal heat stress.

2. Materials and Methods

The term 'early (October) sowing' used in this manuscript refers to sowing under third week of October. Weather data were recorded for the three years of experimentations in Northwest India is given in Figure 1 as an example.

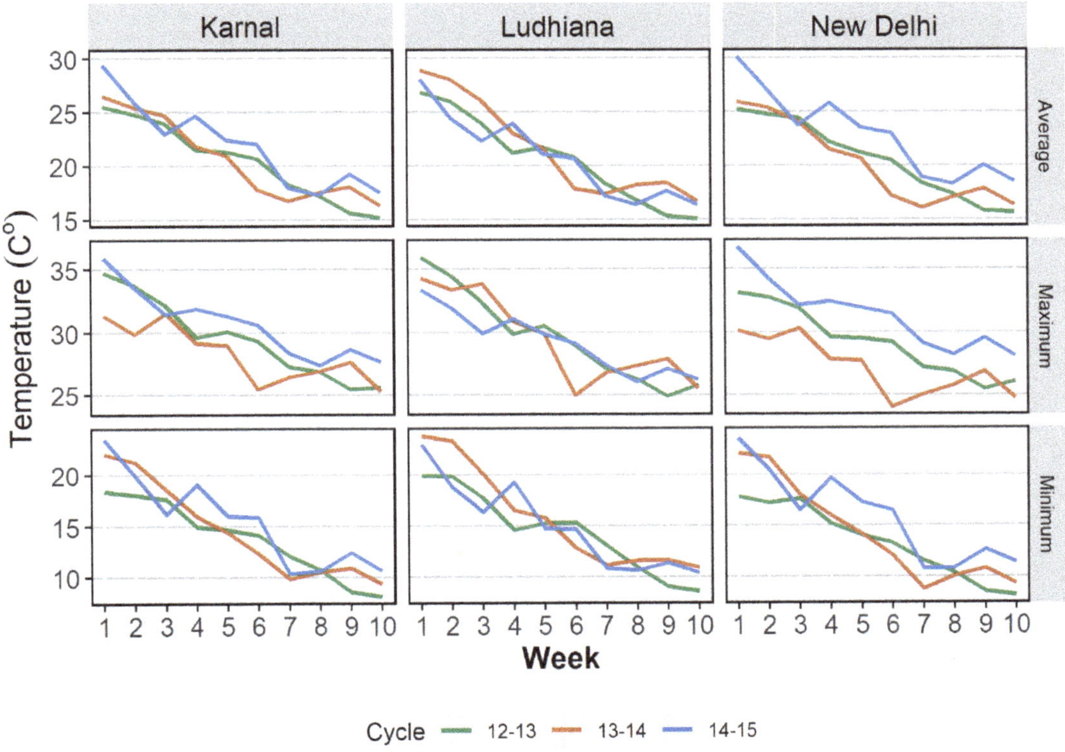

Figure 1. Weather parameters in the three (2012–2015) years of testing in northwest India.

2.1. Experiment 1: Field Trials of 3222 Diverse Wheat Genotypes

A set of 3322 diverse high-yielding bread wheat lines showing a range of maturity and improved wheat varieties were prepared at CIMMYT headquarter in Mexico and provided to national partners. The details of lines with pedigree and other information is provided as supplementary file (Table S1). The lines were planted in an augmented design in six locations (Ludhiana, Karnal, New Delhi, Indore, Pune and Dharwad) spread over NWPZ and CPZ of India under early (October) sowing in year 2012. The date of sowing in these locations varied from October 19–22. Five checks (Super 152, PBW 343, Baj, Munal#1 and Danphe#1) were planted every 20th plot throughout the trial. Since the objective was to identify wheat lines that can demonstrate superior agronomic performance under early (October) sowing, no trial was conducted under normal (November) sowing.

Planting of trial was done by hand in paired rows/line by keeping plot size of 2 m length. Row to row spacing was 25 cm while plant to plant by 3–5 cm at all the locations. The agronomic practices adopted were those recommended for normal fertility (120 kg N: 60 kg P_2O_5: 40 kg K_2O ha^{-1}). As per recommended practice, K_2O and P_2O_5 were applied only at the time of sowing, while nitrogen was split in to three stages: 60 kg N ha^{-1} at sowing, 30 kg N ha^{-1} at first irrigation (21 days after sowing) and 30 kg N ha^{-1} at the second irrigation (45 days after sowing). A total of 5 irrigations were given in NWPZ trials while 4 were given in CPZ, the first being on the 21st day after sowing. Data were recorded for five traits: days to heading (DH), plant height (PH), thousand-grain weight (TGW), grain yield (GY) and canopy temperature (CT).

2.2. Experiment 2: Genotyping of Germplasm for Ppd and Vrn Genes

To understand the distribution of *Ppd* and *Vrn* genes in the germplasm evaluated for early (October) sowing, these genotypes were screened for *Ppd* and *Vrn* genes described below. This was achieved for 3209 lines tested in the first year. This included 2748 lines from CIMMYT and 461 lines from national programs.

The DNA samples were provided by CIMMYT, while molecular mapping was done at Leibniz Institute of Plant Genetics and Crop Plant Research (IPK), Germany.

2.2.1. Genotyping of Ppd-D1

The photoperiodism gene *Ppd-D1* on chromosome 2DS is the major photoperiod response locus in wheat and codes for a gene of the pseudo-response regulator (PRR) family [9]. A semi-dominant mutation widely used in the 'green revolution' converts wheat from a long day (LD) to a photoperiod insensitive (day neutral) plant. Varieties with the photoperiod insensitive *Ppd-D1a* allele, which causes early flowering in short day (SD) or LDs had a 2 kb deletion upstream of the coding region [9]. Specific primers monitoring the presence or absence of this deletion can be used to distinguish wildtype plants (photoperiod sensitive) from plants carrying the mutant allele *Ppd-D1a* (photoperiod insensitive). Therefore, genotyping of all the lines was done with the *Ppd-D1* specific primers (Figure S1).

2.2.2. Genotyping of Vrn-1 Genes

The series of Vrn-1 genes, Vrn-A1, Vrn-B1 and Vrn-D1 is located on chromosomes 5AL, 5BL and 5DL, respectively. The gene product is most likely the MADS box gene AP1 [21]. The Vrn genes determine the vernalization requirement for wheat. In winter wheat, all three loci are usually found in recessive state, while in spring wheat one or several loci contain dominant alleles [22]. The INDEL markers specific to all three genes were available, which were used in genotyping of CIMMYT and country lines (Figures S2–S4).

2.2.3. Genotyping of Vrn-B3

The vernalization gene *VRN3* encodes a RAF kinase inhibitor-like protein with high homology to *Arabidopsis* protein *FLOWstERING LOCUS T* (FT) [23,24]. Gene *Vrn-B3* is located on the short arm of wheat chromosome 7B. It determines, besides the loci *Vrn-A1*, *Vrn-B1* and *Vrn-D1*, the spring or winter-type of wheat varieties. In winter wheat, all four genes are usually present in a recessive state [22]. When one or several of the four loci are present in a dominant state, the variety can be considered a spring variety. The genotyping was conducted according to the protocol of [22] (Figure S5).

2.2.4. Genotyping of Photoperiod Insensitive Ppd-A1a Mutations

In hexaploid wheat, mutations conferring photoperiod insensitivity have been mapped on the 2B (*Ppd-B1*) and 2D (*Ppd-D1*) chromosomes. The mutation in A-genome of hexaploid wheat is lacking so far. However, the mutations by deletions in PRR (pseudo response regulator) gene of A-genome in tetraploid wheat were associated with photoperiod insensitivity. We applied a marker set developed for tetraploid wheat [25] to the CIMMYT and country lines (Figure S6).

2.2.5. Genotyping of SSR-Marker GWM4167 Associated to Ppd-B1

Since, for the gene *Ppd-B1*, no gene specific markers were available during genotyping, we decided to test a linked SSR-Marker. In the study by Zanke et al. (2014), it was shown that the SSR-marker allele GWM4167-217bp was associated with a delay in flowering in European winter wheat. GWM4167 maps to a similar position like *Ppd-B1* and it was assumed that the effect may be due to *Ppd-B1* [26].

2.2.6. Genotyping of SSR-Marker GWM291 Associated to Vrn-A2

VRN2 is a dominant repressor of flowering and down-regulated by vernalization. The *VRN2* region includes two similar ZCCT genes encoding proteins with a putative zinc finger and a CCT domain that have no clear homologs in Arabidopsis [23,24]. *Vrn-A2* is located on the distal end of chromosome 5AL, and no mutants derived from this gene have been described as markers. However, a strong effect of the microsatellite allele GWM291-176bp for decreasing the time to heading was described in European winter wheat [26]. Since GWM291 is in the same chromosomal region as *Vrn-A2*, the effects were attributed to *Vrn-A2* [26].

2.2.7. Marker Ppd-B1_R36-F31 Detecting Copy Number Variation at Locus Ppd-B1

For the insensitivity locus *Ppd-B1* on chromosome 2BS so far, no candidate mutations in the gene sequence have been described. Recent research showed that alleles with an increased copy number of *Ppd-B1* confer an early flowering day neutral phenotype [27]. Specific PCR primers detecting the junction between intact *Ppd-B1* copies as described in the varieties 'Sonora' and 'Timstein' were used to identify varieties with several copies of *Ppd-B1*.

2.3. Experiment 3: Field Trials of Adapted Breeding Materials under October Sowing in NWPZ and CPZ of India in Crop Season 2013–2014

Wheat lines, including local check, found adapted and high-yielding for early sowing in Northwestern Mexico, and NWPZ and CPZ of India were tested in year 2013–2014 in replicated yield trials at the same six locations in India to identify best adapted germplasm. Each trial comprised of 48 genotypes and two local checks that were the best locally adapted varieties in the two zones. Each trial had 3 replications and was arranged in an α lattice design. Planting was done on the second fortnight of October with date of sowing falling between October 19–22. Standard plot size of 6 rows of 6 m with row to row spacing of 20 cm was used. Agronomic management was used as described in experiment 1. Data was recorded for grain yield (GY) and for other traits; days to heading (DH), days to maturity (DM), plant height (PH) and thousand grains weight (TGW). Like experiment 1, the objective was to identify wheat lines that can demonstrate superior agronomic performance under early (October) sowing, hence no trial was conducted under normal (November) sowing.

2.4. Experiment 4: Field Trials of Adapted Breeding Materials under October Sowing in NWPZ of India in Crop Season 2014–2015

Since performance of wheat lines was much better in NWPZ under October sowing, a set of 30 lines was tested in crop season 2014–2015 in replicated yield trials at the same three locations of NWPZ India (Karnal, Ludhiana and Hisar) to further identify adapted genotypes for October sowing. Each trial was comprised of 28 genotypes and two local checks (HD 2967 and DPW 621-50) that were the best locally adapted varieties in NWPZ. Each trial had 3 replications and was arranged in an α lattice design. Planting was done on the second fortnight of October, i.e., between October 19–21. Standard plot size of 6 rows of 6 m with row to row spacing of 20 cm was used. Agronomic management was used as described in the previous experiment. Observations were recorded for yield and yield traits as mentioned in experiment 3.

2.5. Statistical Analysis

For each trial, data analysis was done using R software following a mixed model approach and using the adjusted means for each genotype at individual as well multiple locations in each time of the year. In the analysis of variance and the variance estimates, the model was as shown below.

$$Y = \text{checks} + \text{location} + \text{checks} \times \text{location} + \text{genotypes} + \text{genotypes} \times \text{location} + \text{error}$$

Least square means were estimated for the lines in the trial. The mean grain yield of a genotype was also expressed as percent of the local check using the following formula:

$$\%GY = \left(\frac{GYg}{GYc}\right) \times 100$$

where GYg is the mean grain yield of a line and GYc is the mean grain yield of the local check.

The analysis of variance for all traits was done together in the first year using all six locations. However, in the next two years, analysis was done separately for each zone for each of the two years using data of three locations for each of the two zones. To identify superior lines across locations, we performed stability analysis using the SREG model [28] for the response of the lines on the combination of the 5 sites for yield, and yield related traits (DH, DM, PM and TGW).

The sites regression model (SREG) was:

$$\overline{y}_{ij.} = \mu_j + \sum_{k=1}^{t} \lambda_k \alpha_{ik} \gamma_{jk} + \overline{\varepsilon}_{ij.} \quad (1)$$

where, $\overline{y}_{ij.}$ is the mean of the ith cultivar in the jth environment for g cultivars and e sites (i = 1, 2, ... , g and j = 1, 2, ... , e); μ_j is the site mean; λ_k ($\lambda_1 \geq \lambda_2 \geq \ldots \geq \lambda_t$) are scaling constants (singular values) that allow the imposition of orthonormality constraints on the singular vectors for cultivars, $\alpha_k = (\alpha_{1k}, \ldots, \alpha_{gk})'$ and sites, $\gamma_k = (\gamma_{1k}, \ldots, \gamma_{ek})'$, such that $\sum_i \alpha_{ik}^2 = \sum_j \gamma_{jk}^2 = 1$ and $\sum_i \alpha_{ik}\alpha_{ik'} = \sum_j \gamma_{jk}\gamma_{jk'} = 0$ for $k \neq k'$; α_{ik} and γ_{jk}, for k = 1, 2, 3, ... , are called "primary," "secondary," "tertiary," ... , effects of ith cultivar and the jth site, respectively; $\overline{\varepsilon}_{ij.}$ is the residual error assumed to be normally and independent distributed with 0 means and variance σ^2/r (where σ^2 is the pooled error variance and r is the number of replicates). The number of bilinear terms is $t \leq \min(g, e)$. Estimates of the multiplicative parameters in the kth bilinear term are obtained as the kth component of the deviations from the additive part of the model. In the SREG model, only the main effects of cultivars plus the G×E are absorbed into the bilinear terms.

3. Results

3.1. Performance of 3322 Wheat Genotypes Set for Agronomic Traits in the Early (October) SOWN Conditions

The 3322 genotypes that included CIMMYT breeding lines and spring wheat cultivars collected from different countries, displayed significant variation for grain yield and yield traits: DH, DM, PH and TGW (Table 1). The average range of different traits over locations was very high, for instance 1.1–8.2 t/ha for grain yield, 53–112 for DH, 70–143 for DM, 36–121 cm for PH and 18–53 g for TGW (Table 2). The DH, an important trait for early sowing (Figure 2) indicated that the vegetative phase at locations in the NWPZ was longer than CPZ locations for all wheat genotypes tested. The longest vegetative phase was observed at Karnal, while the shortest at Dharwad (Figure 2). The best performing lines displayed almost similar ranges for the heading date in both zones (NWPZ and CPZ) of India. However, the best lines from CZ showed a slightly earlier heading day compared to lines selected from NWPZ.

Table 1. Analysis of variance of five traits in 3226 genotypes of wheat when evaluated under six locations of India under early (October) sowing conditions.

Source	df	Mean Sum of Squares				
		GY	Heading	Maturity	Pl Height	TGW
Genotype	3326	173,725 **	2661.8 **	2574.6 **	1125.1 **	1339.0 **
Error	15,713	31,919	82.8	415.3	77.9	41.4
F value		5.44	32.14	6.20	14.44	32.33

** Significant at $p < 0.0001$.

Table 2. Mean and variance for the five traits for 3226 genotypes of wheat when evaluated under six locations of India under early (October) sowing conditions.

	GY per Plot (t/ha)	Heading (Days)	Maturity (Days)	Pl Height (cm)	TGW (g)
Range	1.1–8.2	53–112	70–143	36–121	18–53
Var_Genotype	1.10	12.19	2.54	22.17	6.69
Var_Resid	324.9586	82.82	415.33	77.94	41.42
Mean	4.4651	83.72	124.37	90.00	43.83
LSD	1.4	7	16	7	5
CV	13.9	10.9	6.4	9.8	14.7
Heritability	0.200	0.4689	0.354	0.6306	0.4922

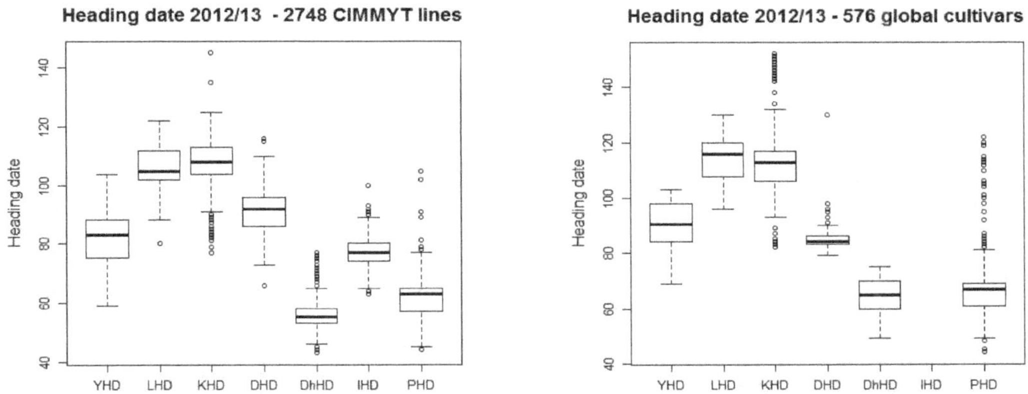

Figure 2. Box-plot for heading date of 3222 CIMMYT lines and global cultivars in Mexico and six different environments in India in 2012–2013. Y: Ciudad Obregon, L: Ludhiana, India, K: Karnal. D: Delhi, Dh: Dhawar. I: Indore, P: Pune. The standard error bars are shown as dotted lines and circles as outliers. The horizontal bar in the box is the median or 50th percentile.

3.2. Genotyping of Germplasm for Ppd and Vrn Genes

The result of genotyping were obtained as follows:

For the *Ppd-D1* specific primers, there were 93.6% of the insensitive allele in the CIMMYT lines, and 87% in the country lines (Figure S1). In the cases of *Vrn-A1, Vrn-B1* and *Vrn-D1*, 2.8%, 11.1% and 1.1% of recessive alleles were discovered in the CIMMYT lines, and 35.4%, 34.9% and 38.4% in the country lines, respectively (Figure S2–S4). The percentage of heterozygotes ranged from 0.58% to 4.77%. We studied the effect of known Ppd and Vrn genes on the heading days (Figure 3, Table S3). The Ppd-D1a gene was very significantly associated with heading in Dharwar (p-value = 3.63×10^{-43}), DWR Karnal (p-value = 2.20×10^{-18}), Indore (p-value = 3.72×10^{-19}), Ludhiana (p-value = 2.89×10^{-22}), Obregon (p-value = 9.31×10^{-36}) and Pune (p-value = 1.31×10^{-42}). Similarly, the VRN-A1 gene was significantly associated with heading in DWR Karnal (p-value = 2.19×10^{-9}), Ludhiana (p-value = 2.68×10^{-6}), Dharwar (p-value = 8.53×10^{-21}), Obregon (p-value = 2.66×10^{-17}) and Pune (p-value = 5.36×10^{-20}). The VRN-B1 gene was also associated with heading in Dharwar (p-value = 1.42×10^{-3}). We also observed clear differences in the days to heading means of the lines with sensitive and insensitive alleles at the Ppd-D1a gene and dominant and recessive alleles at the VRN-A1 and VRN-B1 genes.

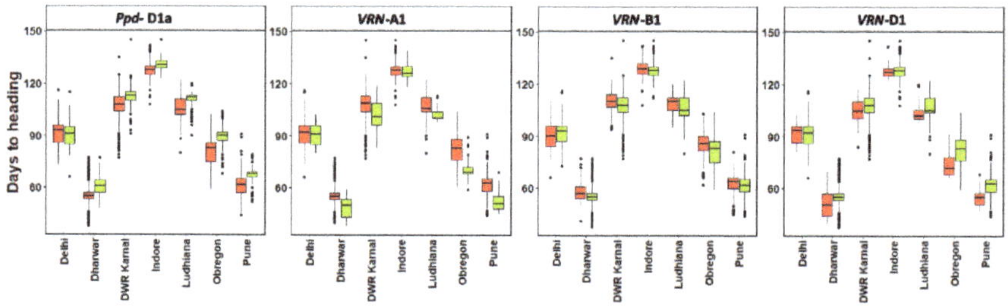

Figure 3. Effect of known Ppd and Vrn genes on the heading days. The horizontal bar in the box is the median or 50th percentile. The recessive allele is denoted by zero while dominant is denoted by 1.

In the case of *Vrn-B3*, of 2751 CIMMYT lines, 2730 lines showed recessive alleles. Likewise, in 461 country lines, 450 were recessive (Figure S5). There were no heterozygotes in CIMMYT lines, while four country lines showed this feature.

For Ppd-A1a, a total of nine insensitive genotypes and six heterozygous genotypes were detected in the CIMMYT lines (Figure S6).

Genotyping of GWM4167 (=WMS4167) associated with Ppd-B1 resulted in eight alleles for the CIMMYT lines (Figure S7) and seven alleles for the country lines (Figure S8). The flowering-delaying allele GWM4167-217bp was present in 589 CIMMYT lines and 51 country lines.

The genotyping of the Vrn-A2 linked GWM291 (=WMS291) resulted in eight alleles for the CIMMYT lines (Figure S9) and in 17 alleles for the country lines (Figure S10). The beneficial allele GWM291-176bp was present in 14 lines of the CIMMYT population and four lines of the country lines.

For the insensitivity locus *Ppd-B1* on chromosome 2BS, the insensitive genotype was discovered in 1206 (43.8%) of the CIMMYT lines and in 179 (38.8%) of the country lines (Figure S11).

3.3. Performance of Elite Lines in the Early (October) Sown Conditions

Based on selection and testing of best lines from the original set of 3322 diverse genotypes, the results obtained are given below.

The details summary of allele types, numbers and frequency based on DNA analysis of markers associated with Vrn and Ppd genes is given in Table S2.

3.3.1. Evaluation during 2013–2014 Crop Season—NWPZ

In the NWPZ zone, significant variation for yield and yield traits was observed among the 50 lines tested in the October sown trial at the three locations (Ludhiana, Karnal and New Delhi) (Table 3). This variation was also supported by the box plot for grain yield and yield traits (Figure 4). The biplot and dendrogram drawn for the three locations of NWPZ showed that the three locations showed different behavior for grain yield and plant height, but TGW and DM expressed high similarity in Ludhiana and Karnal (Figure S12).

Table 3. Analysis of variance of grain yield and yield traits of wheat genotypes when evaluated under three locations of NWPZ of India under early (October) sowing conditions in 2013–2014 and 2014–2015.

Source	df	Mean Sum of Squares (50 Genotypes 2013–2014)			
		GY	Heading	Pl Height	TGW
Loc	2	3527.04 **	7650.75 **	1614.72 **	4.06
Rep (Loc)	3	62.04	2.00	3.96	9.83
Genotype	49	172.66 **	28.18 **	63.42 **	15.33 **
Loc × Genotype	98	111.76 **	8.44 **	31.08 **	15.29 **

Table 3. *Cont.*

Source	df	Mean Sum of Squares (50 Genotypes 2013–2014)			
		GY	**Heading**	**Pl Height**	**TGW**
		Mean Sum of Squares (30 Genotypes 2014–2015)			
Loc	2	6591.66 **	9669.93	7957.21 **	988.24 **
Rep (Loc)	3	77.65	4.12	8.78	4.13
Genotype	30	132.95 **	10.94	3.77	9.20 **
Loc × Genotype		74.14 **	13.29	33.20 **	16.72 **

** Significant at $p < 0.0001$.

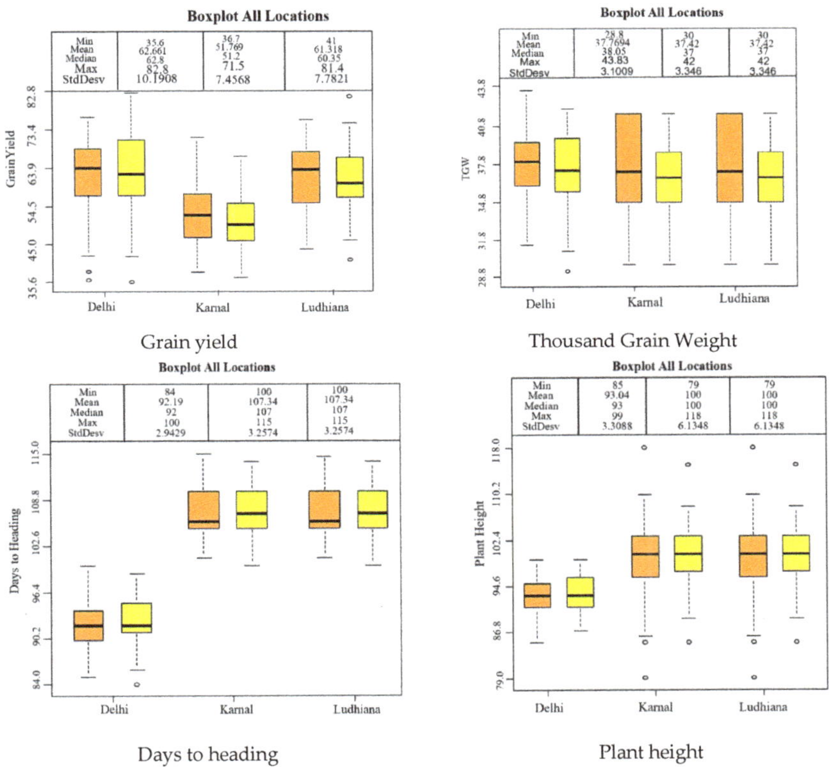

Figure 4. Box plot for grain yield and yield traits in early sown trial of 50 lines tested in three locations (Ludhiana, Karnal and New Delhi) of NWPZ of India in 2013–2014. The standard error bars are shown as dotted lines and circles as outliers. The horizontal bar in the box is the median or 50th percentile.

The biplot for grain yield and yield traits showed that under October sowing of 2013–2014 in three locations of NWPZ, the most stable performance for grain yield was shown by genotypes: BMZ-NW-9, BMZ-NW-4 and the check variety BMZ-NW-7 (DPW 621-50) (Figures S12 and S13). However, the numerical value of mean grain yield showed that the best genotype was BMZ-NW-16, which was significantly superior to the check DPW 621-50, which performed better than the other check HD 2967, the most dominant variety of NWPZ covering more than half of the wheat area under this zone (Table 4). Although no other genotype was significantly superior to DPW 621-50, ten genotypes tested under October sowing in NWPZ were found significantly superior to HD 2967 (Table 4).

Table 4. Performance of genotypes for grain yield under early sowing in 2013–2014 in three locations of the North West Plains Zone of India.

No.	Delhi			Karnal			Ludhiana			Mean of Locations		
	Line	GY t/ha	% Gain over LC	Line	GY t/ha	% Gain over LC	Line	GY t/ha	% Gain over LC	Line	GY t/ha	% Gain over LC
1	BMZ-NW-04	7.78	10.4	BMZ-NW-07	6.91	44.6	BMZ-NW-16	7.85	52.1	BMZ-NW-16	7.35	29.8
2	BMZ-NW-30	7.57	7.4	BMZ-NW-16	6.71	40.4	BMZ-NW-26	7.25	40.5	BMZ-NW-07	7.15	26.3
3	BMZ-NW-07	7.51	6.5	BMZ-NW-40	6.57	37.4	BMZ-NW-35	7.16	38.8	BMZ-NW-09	6.60	16.5
4	BMZ-NW-16	7.49	6.2	BMZ-NW-27	6.22	30.1	BMZ-NW-28	7.15	38.6	BMZ-NW-04	6.50	14.8
5	BMZ-NW-17	7.39	4.8	BMZ-NW-06	6.21	29.9	BMZ-NW-25	7.10	37.6	BMZ-NW-42	6.46	14.1
6	BMZ-NW-43	7.38	4.7	BMZ-NW-41	5.93	24.1	BMZ-NW-17	7.03	36.2	BMZ-NW-29	6.33	11.8
7	BMZ-NW-09	7.36	4.4	BMZ-NW-29	5.82	21.8	BMZ-NW-09	7.02	36.0	BMZ-NW-28	6.32	11.6
8	BMZ-NW-31	7.30	3.5	BMZ-NW-03	5.78	20.9	BMZ-NW-45	6.81	32.0	BMZ-NW-17	6.30	11.2
9	BMZ-NW-42	7.25	2.8	BMZ-NW-20	5.72	19.7	BMZ-NW-22	6.79	31.6	BMZ-NW-30	6.27	10.7
10	BMZ-NW-29	7.11	0.9	BMZ-NW-48	5.65	18.2	BMZ-NW-48	6.78	31.4	BMZ-NW-06	6.26	10.5
11	DPW 621-50 (C)	7.50		DPW 621-50 (C)	6.35		DPW 621-50 (C)	6.97		DPW 621-50 (C)	6.94	
12	HD 2967(C)	7.05		HD 2967(C)	4.78		HD 2967(C)	5.16		HD 2967(C)	5.66	
	LSD 5%	3.1			4.2			4.4			4.3	

Note: LC = Local Check (HD2967) is the most dominant variety of the NWPZ of India.

3.3.2. Evaluation during 2013–2014 Crop Season—CPZ

Like the NWPZ, CPZ zone also showed significant variation for yield and yield traits among the 50 lines tested under early (October) sown conditions at the three locations (Indore, Pune and Dharwad) in 2013–2014 crop season (Table 5). The biplot and dendrogram drawn for the three locations of CPZ showed that the three locations showed different behavior for all four traits, although there was some similarity with Pune and Dharwad except for plant height (Figure S14).

Table 5. Analysis of variance of grain yield and yield traits of wheat genotypes evaluated at three locations of CPZ of India under early (October) sowing conditions in 2013–2014.

Source	df	Mean Sum of Squares (50 Genotypes 2013–2014)			
		GY	Heading	Pl Height	TGW
Loc	2	4890.66 **	17,617.04 **	8631.54 **	7482.93 **
Rep(Loc)	3	136.25	236.11 **	60.85	12.40
Geno	49	5025.41 **	3660.75 **	4553.41 **	3364.82 **
Loc × Geno	98	6924.13 **	1464.96 **	1694.79 **	992.25 **

** Significant at $p < 0.0001$.

The biplot for grain yield and yield traits showed that under October sowing of 2013–2014 in three locations of CPZ, the most stable performance for grain yield was shown by genotypes: BMZ-CPZ-12, BMZ-CPZ-4 and BMZ-CPZ-36 (Figure S15). However, the numerical value of mean grain yield showed that the best genotype was BMZ-CPZ-07 which was significantly superior to the check MACS 6222 which performed much better than the other check GW 322, the most dominant variety of CPZ (Table 6). Although no other genotype was significantly superior to MACS 6222, ten genotypes tested under October sowing in CPZ were significantly superior to GW 322 (Table 6).

Overall, several stable lines with grain yield advantage of more than 1.0 t/ha over the most popular varieties (HD 2967 in NWPZ and GW 322 in CPZ) were obtained in both zones of India (Tables 4 and 6). This was an interesting evidence to demonstrate that there is ample possibility of obtaining wheat genotypes for October sowing and that in future varieties for early sowing, which are supposed to possess early heat tolerance, can be developed and released for farmers.

3.3.3. Performance of Genotypes in 2014–2015 Crop Season in NWPZ

Like the 2013–2014 season, a significant variation was again noted for yield and yield traits in the 2014–2015 season among the 30 lines tested in the October sown trial at the three locations (Ludhiana, Karnal and New Delhi) (Table 3). The biplot and dendrogram drawn for the three locations of NWPZ showed that the three locations showed different behavior for grain yield and plant height, but TGW and DM expressed quite similarly in Ludhiana and Karnal (Figure S16). The PCA between traits within location and suggests strong interaction of environment on heading days (Figure S17)

The impressive performance of some of the new lines in the NWPZ zone under October sowing was again demonstrated in the crop season 2014–2015 (Table 7). The average performance for grain yield and yield traits showed that under October sowing of 2014–2015 in three locations of NWPZ, the most stable performance for grain yield was shown by genotypes: BMZ-NW-16, BMZ-NW-7, BMZ-NW-9, BMZ-NW-4 and BMZ-NW-42 (Table 7). All these lines were significantly superior to the check variety HD 2967, which has been most popular among farmers in the years of these experimentations (Table 7). Their superiority in numerical terms over HD 2967 was in the range of 11–27%. Since performance of genotypes varied across locations as indicated by significant Loc*Geno interaction, genotypes that performed significantly superior to checks differed in each of the three locations.

Table 6. Performance of genotypes for grain yield under early sowing in 2013–2014 in three locations of the Central and Peninsular Plains Zone of India.

No.	Dharwad			Indore			Pune			Mean of Locations		
	Line	GY t/ha	% Gain over GW 322	Line	GY t/ha	% Gain over GW 322	Line	GY t/ha	% Gain over GW 322	Line	GY t/ha	% Gain over GW 322
1	BMZ-CPZ-04	7.78 *	9.1	BMZ-CPZ-07	6.91 *	44.6	BMZ-CPZ-16	7.85 *	52.3	BMZ-CPZ-07	7.00 *	23.1
2	BMZ-CPZ-30	7.57 *	6.1	BMZ-CPZ-16	6.71 *	40.5	BMZ-CPZ-26	7.25	40.5	BMZ-CPZ-16	6.85	20.4
3	BMZ-CPZ-07	7.51	5.4	BMZ-CPZ-40	6.57	37.5	BMZ-CPZ-35	7.16	38.8	BMZ-CPZ-09	6.60	16.0
4	BMZ-CPZ-17	7.39	3.6	BMZ-CPZ-27	6.22	30.2	BMZ-CPZ-28	7.15	38.6	BMZ-CPZ-04	6.50	14.2
5	BMZ-CPZ-43	7.38	3.4	BMZ-CPZ-06	6.21	30.0	BMZ-CPZ-25	7.10	37.8	BMZ-CPZ-42	6.46	13.5
6	BMZ-CPZ-09	7.36	3.2	BMZ-CPZ-41	5.93	24.1	BMZ-CPZ-17	7.03	36.4	BMZ-CPZ-29	6.33	11.2
7	BMZ-CPZ-31	7.30	2.4	BMZ-CPZ-29	5.82	21.9	BMZ-CPZ-09	7.02	36.2	BMZ-CPZ-28	6.32	11.2
8	BMZ-CPZ-42	7.25	1.7	BMZ-CPZ-03	5.78	21.0	BMZ-CPZ-45	6.81	32.0	BMZ-CPZ-17	6.30	10.8
9	BMZ-CPZ-29	7.11	−0.2	BMZ-CPZ-20	5.72	19.7	BMZ-CPZ-22	6.79	31.7	BMZ-CPZ-30	6.27	10.2
10	BMZ-CPZ-11	7.03	−1.4	BMZ-CPZ-48	5.65	18.2	BMZ-CPZ-48	6.78	31.5	BMZ-CPZ-06	6.26	10.1
11	MACS6222 (C)	7.20		MACS6222 (C)	6.35		MACS6222 (C)	6.97		MACS6222 (C)	6.84	
12	GW 322 (C)	7.13		GW 322 (C)	4.78		GW 322 (C)	5.16		GW 322 (C)	5.69	
	LSD 5%	3.7			3.4			4.3			3.3	

Note: The check GW 322 is the most dominant variety of the CPZ of India; * significantly superior over both checks.

Table 7. Performance of genotypes for grain yield under early sowing in 2014–2015 in three locations of the North West Plains Zone of India.

No.	Delhi			Karnal			Ludhiana			Mean of Locations		
	Line	GY t/ha	% Gain over HD 2967	Line	GY t/ha	% Gain over HD 2967	Line	GY t/ha	% Gain over HD 2967	Line	GY t/ha	% Gain over HD 2967
1	BMZ-NW-2015-6	8.00	22.0	BMZ-NW-2015-20	7.74	25.5	BMZ-NW-2015-25	7.92	34.7	BMZ-NW-16	7.89	27.2
2	BMZ-NW-2015-20	7.56	15.3	BMZ-NW-2015-10	7.36	19.4	BMZ-NW-2015-21	7.55	28.3	BMZ-NW-07	7.49	20.8
3	BMZ-NW-2015-17	7.40	12.9	BMZ-NW-2015-23	7.21	16.8	BMZ-NW-2015-26	7.49	27.5	BMZ-NW-09	7.37	18.8
4	BMZ-NW-2015-9	6.99	6.6	BMZ-NW-2015-28	7.03	13.9	BMZ-NW-2015-28	7.23	22.9	BMZ-NW-04	7.08	14.2
5	BMZ-NW-2015-28	6.73	2.7	BMZ-NW-2015-18	6.76	9.6	BMZ-NW-2015-27	7.20	22.5	BMZ-NW-42	6.90	11.2
6	BMZ-NW-2015-24	6.60	0.6	BMZ-NW-2015-25	6.73	9.0	BMZ-NW-2015-12	7.09	20.5	BMZ-NW-29	6.80	9.7
7	BMZ-NW-2015-13	6.42	−2.2	BMZ-NW-2015-24	6.65	7.7	BMZ-NW-2015-5	6.87	16.8	BMZ-NW-28	6.64	7.1
8	BMZ-NW-2015-25	6.39	−2.6	BMZ-NW-2015-22	6.55	6.2	BMZ-NW-2015-18	6.80	15.7	BMZ-NW-17	6.58	6.1
9	BMZ-NW-2015-10	6.29	−4.1	BMZ-NW-2015-16	6.23	0.9	BMZ-NW-2015-3	6.71	14.1	BMZ-NW-30	6.41	3.3
10	BMZ-NW-2015-18	6.23	−5.0	BMZ-NW-2015-6	6.15	−0.3	BMZ-NW-2015-14	6.60	12.2	BMZ-NW-06	6.33	2.0
11	KACHU #1 (C)	6.79		KACHU #1 (C)	5.78		KACHU #1 (C)	6.08		KACHU #1 (C)	6.22	
12	HD 2967 (C)	5.56		HD 2967 (C)	5.17		HD 2967 (C)	4.88		HD 2967 (C)	5.20	
	LSD 5%	4.5			5.3			3.5			4.1	

Note: The check HD2967 is the most dominant variety of the NWPZ of India.

4. Discussion

Since green revolution, the wheat breeding programs in different countries of South Asia have released a significant number of well adapted wheat varieties and thus have been able to sustain wheat production matching with the need of the regions. The collaboration between CIMMYT and South Asian countries have played a major role in this developing varieties adapted to variable conditions in different agro-ecological zones [1,29]. However, these varieties were released for normal (November) or late (December) sowings under different management conditions. In parts of Punjab, farmers used to plant mostly in the first fortnight of November, but a few as early as in the end of October. However, wheat planting much earlier, i.e., in the third week of October due to its various advantages under changing weather patterns continues to be a dream for the farmers. As the monsoon period is the major rainfall period, farmers desired to plant wheat earlier in the season in many parts of northwestern and central India to take advantage of the residual moisture and escape terminal heat stress in the month of March since if planted early, wheat would be past physiological maturity by that time [30]. However, whenever this was attempted, the wheat crop used developed too quickly under the warm/hot early seedling conditions, resulting in reduction in biomass, ear and grain numbers, with resultant decreases in yield. The reason being the prevalent wheat varieties being poorly adapted to warmer temperatures at juvenile stages. In fact, breeding was never done for heat tolerance at the juvenile stage, and hence there was no knowledge on this subject.

The results of this study showed that wheat lines with early heat tolerance can be obtained if diverse breeding lines are exposed to selection under October sowing [30]. The performance of top lines reached to almost 8 t/ha, which is remarkable compared to the performance of Elite Spring Wheat Yield Trial (ESWYT) lines under November sowing, which hardly exceeded 6 t/ha [31]. This performance was observed under similar fertilizer and irrigation management. Hence, the research results are expected to benefit wheat growers and consumers through this new type of wheat germplasm, which can perform well under early sowing. The sustainability of wheat production will be further enhanced by reduced pumping of water. Since there are machines available like Happy Seeder, super seeder and super straw management system in the combines, early planting is possible immediately after harvest [32]. This throws another advantage of reducing straw burning and the consequent emissions of greenhouse gases. Since varieties suitable for early sowing were not available in the past, there was no scope of immediate planting after rice harvest which mostly (about 90%) happens in the month of October. The availability of wheat varieties with early heat tolerance, therefore, is expected to generate all round benefits to the farmers and the sustainability of environment.

The superior performing genotypes under early sowing can be used in crossing program for further strengthening breeding for early heat tolerance in a systematic manner. Hence, the results of this study showing possibility of breeding for early heat tolerance will be applicable to many wheat growing areas worldwide. The ultimate users will be smallholder wheat farmers in both irrigated and rainfed situations in South Asia estimated to exceed 10 million wheat farming households, and through spin-off effects on technology development in other parts of the world [33]. Farmers will have options for more sustainable and productive farming systems, even under conditions of climate change. Intermediate users of the technology will include the All India Co-ordinated Wheat Improvement Project (AICWIP) through national wheat breeding program; and National Agriculture Research System (NARS) breeding programs in development of superior wheat varieties. Most NARS in the developing world have well-established linkages with CIMMYT and receive wheat nurseries and trials annually [20]; CIMMYT spring and winter bread wheat breeding programs that are partners in this project for development of lines for international distribution; and breeding programs in developed countries for use as parents in variety development [34]. This is likely at the organizations involved in this experimentation, but also others, as information and germplasm will be freely available.

The most common vernalization alleles in the project materials were the dominant spring alleles *Vrn-D1a*, followed by *Vrn-B1a*. The Japanese cultivar 'Akakomugi' is thought to be the donor parent of the *Vrn-D1a* allele in CIMMYT wheat [35], which was later transferred into early Green Revolution cultivars like 'Lerma Rojo' and 'Sonora 64.' These two cultivars are also thought to be the potential source of the *Vrn-D1a* allele in South and Southeast Asian wheat [35,36]. Furthermore, it has been concluded that the highest yield was predicted for varieties containing *Vrn-D1a* [37]. While the frequency of both alleles was above 90% in CIMMYT lines the frequency was lower in the set of global cultivars and also in a set of lines released in India [38].

Plant height is one of the crucial traits to understand cultivars superiority under early sown condition. However, we observed a poor correlation (0.109, $p < 0.001$; Figure S17) between plant height and grain yield, indicating that plant height did not play a significant role to get higher yield under early sown condition. However, it has been shown that early planted wheat plants gain slightly more height compared to those normally sown [39]. The dominant spring wheat allele *Vrn-A1a* was almost absent in CIMMYT wheat, but was present in 39% of global cultivar set. When released cultivars in India were evaluated for the *Vrn-1* genes, a high frequency (>60%) for the *Vrn-A1a* allele was observed [40]. No or very little variation in all datasets was observed for *Vrn-A2* and *Vrn-B3*. A slightly higher frequency of the *Vrn-A2* was selected for CZ.

The allele *Ppd-D1a* was the predominant photoperiod insensitive allele in the lines evaluated. The *Ppd-D1a* allele was introduced into CIMMYT wheat since Norman Borlaug shuttled germplasm between two contrasting environments in Mexico and exposed wheat materials to diverse photoperiods and temperatures. The *Ppd-B1* alleles showed the highest variation across datasets of genotypes investigated. A lower frequency of the insensitive allele *Ppd-B1a* was selected for NWPZ and CZ. At the same time, a higher frequency of the *Ppd-B1(217bp)* allele was selected for both NWPZ and CZ. In the set of global wheat cultivars, the frequency of the *Ppd-B1a* allele was similar than in the NWPZ and CZ selections. The frequency of the allele *Ppd-B1(217bp)* was lowest in the global cultivars set. The *Ppd-A1a* alleles were first described in durum wheat and are present CIMMYT bread wheat germplasm due to introgressions from synthetic hexaploid wheat, but with low frequency. Synthetic hexaploid is developed by creating a cross between *Aegilops tauschii* and durum wheat. Only 0.3% of all CIMMYT lines contained this allele, and none of those lines was further selected. The allele was not observed in the set of global cultivars.

It is a well noted history that since the early days of Green Revolution, the Indo-Gangetic Plains of South Asia has transformed this region from a food deficit to a food surplus region. At present this region produces about 15% of global wheat production and is inhabited by one sixth of the world population. Enough concerns have been expressed about this region being under threat from climate change, mainly heat and water stress, and may get converted to a sub-optimal wheat belt by the year 2050 [3]. A number of challenges in wheat production have been described for India [1] and whole of South Asia [2].

There are indications that useful variation for heat tolerance is available in the wheat gene pool [1,41,42]. The best way to breed for terminal heat tolerance in wheat has been to delay planting so that wheat populations are exposed to high temperatures from heading onwards and thereby there is high chance of selection for only those plants that have high terminal heat tolerance. Likewise, as shown in this study, early sowing can be done to select for lines that have tolerance to early high temperature. However, the reason for superior performance by a good number of lines under early sowing is not fully understood. There is possibility that it is combination of mild vernalization, superior agronomic performance and some unknown genes that might be supporting proper growth and tillering under early heat. In fact, the genetic control of adaptation response and heat tolerance is poorly understood. Major genes underlying adaptation have been identified in recent years, namely those controlling vernalization response, photoperiod sensitivity and development rate "earliness per se" [9–12]. To bridge the gaps in the knowledge of the effects of possible allele combination on adaptation in specific environments was attempted to address (Figure 3 and Table S3). A linear model was fitted with days to heading in different environments as the y-variable and the alleles at the Ppd-D1a, VRN-A1, VRN-B1 and VRN-D1 genes as the x-variables. The effects of the alleles on heading in different environments and the p-values for the test of the significance of the effects were obtained using a two-tailed t-test. The *Vrn* genes determine the control of the spring/winter wheat difference by defining the chilling hours required by the wheat plant to be able to flower, while the *Ppd* genes play an important part in delaying flowering time in the spring after vernalization requirement has been satisfied [43]. The effects of the *Eps* loci may facilitate more subtle manipulation of the life cycle for regional adaptation. QTL studies have described various regions of potential *Eps* genes [13,14,44,45]. Allelic variation for some QTL, for example effects on 4B, 6A, 7D, appear to occur frequently in diverse germplasm. *Vrn*, *Ppd* and *Eps* genes additionally function in epistatic interaction [15]. Large numbers of allele combination will therefore be involved in determining the regulation of growth habit and optimal adaptation to a certain environment [46,47]. Further work will be required to understand the genetic basis of superior performance of wheat lines under early sowing.

Overall, the study resulted in proving the fact that it is possible to obtain wheat lines that can demonstrate significantly superior grain yield and agronomic traits under early sowing conditions of south Asia that has potential to give additional 1 t/ha yield under same agronomic management. Based on the benefits of these trials, Indian Council of Agricultural Research (ICAR), New Delhi initiated coordinated trials for early sowing in India for the first time with the purpose of breeding and encouraging wheat planting in the third week of October. Consequently, in the year 2020, three wheat varieties (DBW187, DBW303 and WH1270) were identified for release for early sown irrigated conditions in the NWPZ of India, the first case of this kind in India (https://www.aicrpwheatbarley.org/wp-content/uploads/2020/08/VIC-proceedings-2020.pdf, accessed on 15 September 2021).

Supplementary Materials: The following are available online at https://www.mdpi.com/article/10.3390/genes12111808/s1, Figure S1: Genotyping results for Ppd-D1, Figure S2: Genotyping results for Vrn-A1, Figure S3: Genotyping results for Vrn-B1, Figure S4: Genotyping results for Vrn-D1, Figure S5: Genotyping results for Vrn-B3 in CIMMYT and country lines, Figure S6: Genotyping results for Ppd-A1, Figure S7: Genotyping of SSR-marker WMS4167 (linked to Ppd-B1) in the CIMMYT lines, Figure S8: Genotyping of SSR-marker WMS4167 (linked to Ppd-B1) in the country lines, Figure S9: Genotyping of SSR-marker WMS291 (linked toVrn-A2) in the CIMMYT lines, Figure S10: Genotyping of SSR-marker WMS291 (linked to Vrn-A2) in the country lines, Figure S11: Genotyping results for Ppd-B1_R36-F31, Figure S12: Biplot and dendrogram for three locations of NWPZ of India for grain yield and yield traits when diverse 50 wheat lines were tested under October sowing of 2013–2014, Figure S13: Biplot for grain yield and yield traits for 50 lines tested under October sowing of 2013–2014 in three locations of NWPZ of India, Figure S14: Biplot and dendrogram for three locations of CPZ of India for grain yield and yield traits when diverse 50 wheat lines were tested under October sowing of 2013–2014, Figure S15: Biplot for grain yield and yield traits for 50 lines tested under October sowing in three locations of CPZ of India in 2013–2014, Figure S16: Biplot and dendrogram for three locations of NWPZ of India for grain yield and yield traits when diverse 30 wheat lines were tested under October sowing of 2014-2015. Figure S17: PCA plot of the traits in the different environments. Figure S18: Line Fit Plot for plant height and grain yield for 3222 lines evaluated under early sown condition. Table S1: Details of lines with pedigree, origin source, selection history, unique GID and other information. Table S2: Type, number and frequency of alleles based on DNA analysis of markers associated with Vrn nd Ppd gene. Table S3: Associations using a two-sided t-test and the p-values for the t-test used to indicate significance of the association among heading and alleles of Ppd and Vrn genes.

Author Contributions: Conceptualization, methodology and supervision: A.K.J., R.P.S. and M.S.R. Formal analysis, data curation, and investigation: J.C., L.C.-H., S.M., J.H.-E., U.K., S.D., A.K.J. and R.P.S. Resources, project administration and funding acquisition: R.P.S., A.K.J. and G.P.S. Writing—original draft: A.K.J., U.K. and R.P.S. Review, editing and visualization: S.D., C.N.M., G.S.M., V.S.S., S.V.S.P., R.N., S.C.M., S.M. and R.P.S. All authors have read and agreed to the published version of the manuscript.

Funding: This research was funded by Deutsche Gesellschaft für Internationale Zusammenarbeit (GIZ), grant number 81141838.

Institutional Review Board Statement: Not applicable.

Informed Consent Statement: Not applicable.

Data Availability Statement: The data will be made available on request. Kindy contact the corresponding author.

Acknowledgments: The support of Pankaj Singh for facilitating trial at BISA Jabalpur and review the manuscript by Philomin Juliana is duly acknowledged. We thank all our co-operators and NARS partners in India for conducting the trials at respective locations. We are grateful to Indian Council of Agricultural Research (ICAR) for fast track release and dissemination of varieties.

Conflicts of Interest: The authors declare no conflict of interest. The funders had no role in the design of the study; in the collection, analyses, or interpretation of data; in the writing of the manuscript, or in the decision to publish the results.

References

1. Joshi, A.K.; Mishra, B.; Chatrath, R.; Ortiz Ferrara, G.; Singh, R.P. Wheat improvement in India: Present status, emerging challenges and future prospects. *Euphytica* **2007**, *157*, 431–446. [CrossRef]
2. Chatrath, R.; Mishra, B.; Ortiz Ferrara, G.; Singh, S.K.; Joshi, A.K. Challenges to wheat production in South Asia. *Proc. Euphytica* **2007**, *157*, 447–456. [CrossRef]
3. Ortiz, R.; Sayre, K.D.; Govaerts, B.; Gupta, R.; Subbarao, G.V.V.; Ban, T.; Hodson, D.; Dixon, J.M.; Iván Ortiz-Monasterio, J.; Reynolds, M. Climate change: Can wheat beat the heat? *Agric. Ecosyst. Environ.* **2008**, *126*, 46–58. [CrossRef]
4. Rodell, M.; Velicogna, I.; Famiglietti, J.S. Satellite-based estimates of groundwater depletion in India. *Nature* **2009**, *460*, 999–1002. [CrossRef] [PubMed]
5. Lobell, D.B.; Burke, M.B.; Tebaldi, C.; Mastrandrea, M.D.; Falcon, W.P.; Naylor, R.L. Prioritizing Climate Change Adaptation Needs for Food Security in 2030. *Science* **2008**, *319*, 607–610. [CrossRef]
6. Bhatt, R.; Kaur, R.; Ghosh, A. Strategies to Practice Climate-Smart Agriculture to Improve the Livelihoods under the Rice-Wheat Cropping System in South Asia. In *Sustainable Management of Soil and Environment*; Springer: Singapore, 2019; pp. 29–71. ISBN 9789811388323.
7. Bhanja, S.N.; Mukherjee, A.; Rangarajan, R.; Scanlon, B.R.; Malakar, P.; Verma, S. Long-term groundwater recharge rates across India by in situ measurements. *Hydrol. Earth Syst. Sci.* **2019**, *23*, 711–722. [CrossRef]
8. Mondal, S.; Singh, R.P.; Crossa, J.; Huerta-Espino, J.; Sharma, I.; Chatrath, R.; Singh, G.P.; Sohu, V.S.; Mavi, G.S.; Sukaru, V.S.P.; et al. Earliness in wheat: A key to adaptation under terminal and continual high temperature stress in South Asia. *Crop. Res.* **2013**, *151*. [CrossRef]
9. Beales, J.; Turner, A.; GriYths, S.; Snape, J.W.; Laurie, D.A. A Pseudo-Response Regulator is misexpressed in the photoperiod insensitive Ppd-D1a mutant of wheat (*Triticum aestivum* L.). *Theor. Appl. Genet.* **2007**, *115*, 721–733. [CrossRef]
10. Yan, L.; Helguera, M.; Kato, K.; Fukuyama, S.; Sherman, J.; Dubcovsky, J. Allelic variation at the VRN-1 promoter region in polyploid wheat. *Theor. Appl. Genet.* **2004**, *109*, 1677–1686. [CrossRef]
11. Fu, D.; ter Szucs, P.; Yan, L.; Helguera, M.; Skinner, J.S.; von Zitzewitz, J.; Hayes, P.M.; Dubcovsky, J. Large deletions within the first intron in VRN-1 are associated with spring growth habit in barley and wheat. *Mol. Gen. Genom.* **2005**, *273*, 54–65. [CrossRef]
12. Bentley, A.R.; Turner, A.S.; Gosman, N.; Leigh, F.J.; Maccaferri, M.; Dreisigacker, S.; Greenland, A.; Laurie, D.A. Frequency of photoperiod-insensitive Ppd-A1a alleles in tetraploid, hexaploid and synthetic hexaploid wheat germplasm. *Plant Breed.* **2011**, *130*, 10–15. [CrossRef]
13. Hanocq, E.; Laperche, A.; Jaminon, O.; Lainé, A.L.; Le Gouis, J. Most significant genome regions involved in the control of earliness traits in bread wheat, as revealed by QTL meta-analysis. *Theor. Appl. Genet.* **2007**, *114*, 569–584. [CrossRef] [PubMed]
14. Griffiths, S.; Simmonds, J.; Leverington, M.; Wang, Y.; Fish, L.; Sayers, L.; Alibert, L.; Orford, S.; Wingen, L.; Herry, L.; et al. Meta-QTL analysis of the genetic control of ear emergence in elite European winter wheat germplasm. *Theor. Appl. Genet.* **2009**, *119*, 383–395. [CrossRef] [PubMed]
15. Whittal, A.; Kaviani, M.; Graf, R.; Humphreys, G.; Navabi, A. Allelic variation of vernalization and photoperiod response genes in a diverse set of North American high latitude winter wheat genotypes. *PLoS ONE* **2018**, *13*, e0203068. [CrossRef]
16. Kiss, T.; Balla, K.; Veisz, O.; Láng, L.; Bedő, Z.; Griffiths, S.; Isaac, P.; Karsai, I. Allele frequencies in the VRN-A1, VRN-B1 and VRN-D1 vernalization response and PPD-B1 and PPD-D1 photoperiod sensitivity genes, and their effects on heading in a diverse set of wheat cultivars (*Triticum aestivum* L.). *Mol. Breed.* **2014**, *34*, 297–310. [CrossRef]
17. Wang, L.; Niu, J.S.; Li, Q.Y.; Qin, Z.; Ni, Y.J.; Xu, H.X. Allelic variance at the vernalization gene locus Vrn-D1 in a group of sister wheat (*Triticum aestivum*) lines and its effects on development. *J. Agric. Sci.* **2015**, *153*, 588–601. [CrossRef]
18. Braun, H.J.; Atlin, G.; Payne, T. Multi-location testing as a tool to identify plant response to global climate change. In *Climate Change and Crop Production*; CABI: Wallingford, UK, 2010; pp. 115–138. ISBN 9781845936334.

19. Crespo-Herrera, L.A.; Crossa, J.; Huerta-Espino, J.; Vargas, M.; Mondal, S.; Velu, G.; Payne, T.S.; Braun, H.; Singh, R.P. Genetic gains for grain yield in cimmyt's semi-arid wheat yield trials grown in suboptimal environments. *Crop Sci.* **2018**, *58*, 1890–1898. [CrossRef]
20. Singh, R.P.; Huerta-Espino, J.; Sharma, R.; Joshi, A.K.; Trethowan, R. High yielding spring bread wheat germplasm for global irrigated and rainfed production systems. *Euphytica* **2007**, *157*, 351–363. [CrossRef]
21. Yan, L.; Loukoianov, A.; Tranquilli, G.; Helguera, M.; Fahima, T.; Dubcovsky, J. Positional cloning of the wheat vernalization gene VRN1. *Proc. Natl. Acad. Sci. USA* **2003**, *100*, 6263–6268. [CrossRef]
22. Zhang, X.K.; Xiao, Y.G.; Zhang, Y.; Xia, X.C.; Dubcovsky, J.; He, Z.H. Allelic variation at the vernalization genes Vrn-A1, Vrn-B1, Vrn-D1, and Vrn-B3 in Chinese wheat cultivars and their association with growth habit. *Crop Sci.* **2008**, *48*, 458–470. [CrossRef]
23. Yan, L.; Fu, D.; Li, C.; Blechl, A.; Tranquilli, G.; Bonafede, M.; Sanchez, A.; Valarik, M.; Yasuda, S.; Dubcovsky, J. The wheat and barley vernalization gene VRN3 is an orthologue of FT. *Proc. Natl. Acad. Sci. USA* **2006**, *103*, 19581–19586. [CrossRef]
24. Distelfeld, A.; Tranquilli, G.; Li, C.; Yan, L.; Dubcovsky, J. Genetic and molecular characterization of the VRN2 loci in tetraploid wheat. *Plant Physiol.* **2009**, *149*, 245–257. [CrossRef]
25. Wilhelm, E.P.; Turner, A.S.; Laurie, D.A. Photoperiod insensitive Ppd-A1a mutations in tetraploid wheat (*Triticum durum* Desf.). *Theor. Appl. Genet.* **2009**, *118*, 285–294. [CrossRef] [PubMed]
26. Zanke, C.; Ling, J.; Plieske, J.; Kollers, S.; Ebmeyer, E.; Korzun, V.; Argillier, O.; Stiewe, G.; Hinze, M.; Beier, S.; et al. Genetic architecture of main effect QTL for heading date in European winter wheat. *Front. Plant Sci.* **2014**, *5*, 217. [CrossRef] [PubMed]
27. Díaz, A.; Zikhali, M.; Turner, A.S.; Isaac, P.; Laurie, D.A. Copy number variation affecting the photoperiod-B1 and vernalization-A1 genes is associated with altered flowering time in wheat (*Triticum aestivum*). *PLoS ONE* **2012**, *7*, e33234. [CrossRef] [PubMed]
28. Crossa, J.; Cornelius, P.L.; Yan, W. Biplots of linear-bilinear models for studying crossover genotype × environment interaction. *Crop Sci.* **2002**, *42*, 619–633. [CrossRef]
29. Evenson, R.; Pray, C.; Rosegrant, M. *Agricultural Research and Productivity Growth in India*; IFPRI Research Report; International Food Policy Research Institute: Washington, DC, USA, 1999.
30. Mondal, S.; Joshi, A.K.; Huerta-Espino, J.; Singh, R.P. Early Maturity in Wheat for Adaptation to High Temperature Stress. In *Advances in Wheat Genetics: From Genome to Field*; Springer: Tokyo, Japan, 2015; pp. 239–245.
31. Mondal, S.; Dutta, S.; Crespo-Herrera, L.; Huerta-Espino, J.; Braun, H.J.; Singh, R.P. Fifty years of semi-dwarf spring wheat breeding at CIMMYT: Grain yield progress in optimum, drought and heat stress environments. *Crop. Res.* **2020**, *250*, 107757. [CrossRef]
32. Sandhu, B.S.; Dhaliwal, N.S.; Sandhu, G.S. Production potential and economics of wheat, Triticum aestivum as influenced by different planting methods in Punjab, India. *J. Appl. Nat. Sci.* **2016**, *8*, 777–781. [CrossRef]
33. Krupnik, T.J.; Schulthess, U.; Ahmed, Z.U.; McDonald, A.J. Sustainable crop intensification through surface water irrigation in Bangladesh? A geospatial assessment of landscape-scale production potential. *Land Use Policy* **2017**, *60*, 206–222. [CrossRef]
34. Reynolds, M.P.; Borlaug, N.E. Impacts of breeding on international collaborative wheat improvement. *J. Agric. Sci.* **2006**, *144*, 3–17. [CrossRef]
35. Stelmakh, A.F. Geographic distribution of Vrn-genes in landraces and improved varieties of spring bread wheat. *Euphytica* **1990**, *45*, 113–118. [CrossRef]
36. Van Beem, J.; Mohler, V.; Lukman, R.; Van Ginkel, M.; William, M.; Crossa, J.; Worland, A.J. Analysis of genetic factors influencing the developmental rate of globally important CIMMYT wheat cultivars. *Crop Sci.* **2005**, *45*, 2113–2119. [CrossRef]
37. Stelmakh, A.F. Genetic effects of Vrn genes on heading date and agronomic traits in bread wheat. *Euphytica* **1992**, *65*, 53–60. [CrossRef]
38. Singh, R.P.; Trethowan, R. Breeding Spring Bread Wheat for Irrigated and Rainfed Production Systems of the Developing World. In *Breeding Major Food Staples*; Blackwell Publishing Ltd.: Hoboken, NJ, USA, 2007; pp. 107–140. ISBN 9780470376447.
39. Farhad, M.; Tripathi, S.B.; Singh, R.P.; Joshi, A.; Bhati, P.; Vishwakarma, M.K.; Mondal, S.; Malik, A.A.; Kumar, U. Multi-trait selection of bread wheat ideotypes for adaptation to early sown condition. *Crop Sci.* **2021**. [CrossRef]
40. Singh, R.P.; Herrera-Foessel, S.; Huerta-Espino, J.; Singh, S.; Bhavani, S.; Lan, C.; Basnet, B.R. Progress Towards Genetics and Breeding for Minor Genes Based Resistance to Ug99 and Other Rusts in CIMMYT High-Yielding Spring Wheat. *J. Integr. Agric.* **2014**, *13*, 255–261. [CrossRef]
41. Witcombe, J.; Hollington, P.; Howarth, C.; Reader, S.; Steele, K. Breeding for abiotic stresses for sustainable agriculture. *Philos. Trans. R. Soc. B Biol. Sci.* **2008**, *363*, 703–716. [CrossRef]
42. Paliwal, R.; Röder, M.S.; Kumar, U.; Srivastava, J.P.; Joshi, A.K. QTL mapping of terminal heat tolerance in hexaploid wheat (*T. aestivum* L.). *Theor. Appl. Genet.* **2012**, *125*, 561–575. [CrossRef]
43. Harris, F.A.J.; Eagles, H.A.; Virgona, J.M.; Martin, P.J.; Condon, J.R.; Angus, J.F. Effect of VRN1 and PPD1 genes on anthesis date and wheat growth. *Crop Pasture Sci.* **2017**, *68*, 195. [CrossRef]
44. Zikhali, M.; Leverington-Waite, M.; Fish, L.; Simmonds, J.; Orford, S.; Wingen, L.U.; Goram, R.; Gosman, N.; Bentley, A.; Griffiths, S. Validation of a 1DL earliness per se (eps) flowering QTL in bread wheat (*Triticum aestivum*). *Mol. Breed.* **2014**, *34*, 1023–1033. [CrossRef]
45. Hill, C.B.; Taylor, J.D.; Edwards, J.; Mather, D.; Langridge, P.; Bacic, A.; Roessner, U. Detection of QTL for metabolic and agronomic traits in wheat with adjustments for variation at genetic loci that affect plant phenology. *Plant Sci.* **2015**, *233*, 143–154. [CrossRef]

46. Chen, C.; Wang, B.; Feng, P.; Xing, H.; Fletcher, A.L.; Lawes, R.A. The shifting influence of future water and temperature stress on the optimal flowering period for wheat in Western Australia. *Sci. Total Environ.* **2020**, *737*, 139707. [CrossRef] [PubMed]
47. Flohr, B.M.; Hunt, J.R.; Kirkegaard, J.A.; Evans, J.R. Water and temperature stress define the optimal flowering period for wheat in south-eastern Australia. *Crop. Res.* **2017**, *209*, 108–119. [CrossRef]

genes

MDPI

Article

Expression Analysis of the *TdDRF1* Gene in Field-Grown Durum Wheat under Full and Reduced Irrigation

Arianna Latini [1], Cristina Cantale [1], Karthikeyan Thiyagarajan [1], Karim Ammar [2] and Patrizia Galeffi [1,*]

[1] Italian National Agency for New Technologies, Energy and Sustainable Economic Development, ENEA, Casaccia Research Center, 00123 Rome, Italy; arianna.latini@enea.it (A.L.); cantalec@gmail.com (C.C.); pltbiotechtkarthi2018@gmail.com (K.T.)

[2] International Maize and Wheat Improvement Centre(CIMMYT), Texcoco 56237, Mexico; k.ammar@cgiar.org

* Correspondence: patrizia.galeffi@enea.it

Abstract: Some of the key genes and regulatory mechanisms controlling drought response in durum wheat have been identified. One of the major challenges for breeders is how to use this knowledge for the achievement of drought stress tolerance. In the present study, we report the expression profiles of the *TdDRF1* gene, at consecutive plant growth stages, from different durum wheat genotypes evaluated in two different field environments. The expression of a possible target gene (*Wdnh13*) of the *TdDRF1* gene was also investigated and analogies with the transcript profiles were found. The results of the qRT-PCR highlighted differences in molecular patterns, thus suggesting a genotype dependency of the *TdDRF1* gene expression in response to the stress induced. Furthermore, a statistical association between the expression of *TdDRF1* transcripts and agronomic traits was also performed and significant differences were found among genotypes, suggesting a relationship. One of the genotypes was found to combine molecular and agronomic characteristics.

Keywords: durum wheat; expression profiles; field trials; qRT-PCR; *TdDRF1* gene; transcription factors; *Wdhn13*

Citation: Latini, A.; Cantale, C.; Thiyagarajan, K.; Ammar, K.; Galeffi, P. Expression Analysis of the *TdDRF1* Gene in Field-Grown Durum Wheat under Full and Reduced Irrigation. *Genes* **2022**, *13*, 555. https://doi.org/10.3390/genes13030555

Academic Editor: Qing Yang

Received: 18 February 2022
Accepted: 18 March 2022
Published: 21 March 2022

1. Introduction

Unmitigated climate change due to increasing greenhouse gas emissions will have an adverse impact on plant growth and crop yield in some areas of the world, including through the more frequent occurrences of drought stress [1,2]. To mitigate against water scarcity and/or irregular availability and to enhance the sustainability of global food production, it is necessary to explore avenues for producing more food with proportionally less water [3]. Cereals are our dominant source of food, with wheat playing a major contribution to human diet and health [4]. Durum wheat (*Triticum turgidum* var. *durum*) is largely cultivated in the Mediterranean basin and other semi-arid and marginal areas, with the milled product being used mainly for making pasta and other staple foods.

Plants deploy complex mechanisms to cope with stresses like dehydration. Several genes have been described that are activated at the transcriptional level, with *cis*- and *trans*-acting factors involved in the expression of dehydration responsive genes [5]. The dehydration responsive element binding (DREB) family of transcription factors (TFs) represents one of the major players involved in abiotic (dehydration, cold, high salinity) stress responses [6]. The DREB proteins interact with the *drought-responsive element* (DRE) motif in the promoter regions of many stress-inducible genes and belong to the larger AP2/ERF (APETALA2/ethylene-responsive factor) family, as the DNA binding and recognition is mediated by the Apetala2 (AP2) domain [6,7]. TFs are therefore good candidates for improving crop tolerance to drought because of their role as master regulators of several clusters of genes [8–10].

A *DREB2*-related gene, namely *TdDRF1* (*Triticum durum dehydration responsive factor 1*), was isolated in durum wheat and reported as producing three forms of transcript through alternative splicing (AS): *TdDRF1.1* and *TdDRF1.3*, encoding putative TFs containing the AP2/EREBP DNA-binding domain and the nuclear localization signal (NLS), and *TdDRF1.2*, encoding a putative abortive protein lacking both the AP2 domain and the NLS. *TdDRF1* gene expression was linked to the plant response to water deficit [11] and its analysis in different genotypes of durum wheat and one triticale cultivar under greenhouse conditions and subjecting plants to a moderate dehydration stress resulted in different genotypic behaviours [12]. Furthermore, a preliminary study of the field expression of *TdDRF1* was also carried out, analysing durum wheat and triticale lines in a short time-course with five sampling points [13]. The above-mentioned studies revealed that *TdDRF1* gene expression had an important genotype-dependence, as also found for other transcription factor genes controlling plant response to abiotic stress [14,15] or key metabolic pathways [16,17]. The relationships among the expression patterns of the three transcripts and the phenotype response to water stress in a complex field environment are still largely unknown, highlighting the need for further research to gain more insight into the gene expression under realistic environmental conditions.

The aim of the present work was to investigate the molecular behaviour of the *TdDRF1* gene in the field during a time-course drought stress experiment in six durum wheat genotypes, and to find whether there was a relationship between the expression of the *TdDRF1* gene and one possible downstream target (*Wdhn13* gene). The association between transcript profile analysis and agronomic performance was also explored.

2. Materials and Methods

2.1. Plant Materials

Six durum wheat genotypes were used in the field experiments (Table 1).

Table 1. Pedigree information, date of release/development, and origin of six durum wheat genotypes (Country of origin: IT, Italy; MX, Mexico).

Genotype Name	Pedigree	Year of Release or Development	Country of Origin
Duilio	Cappelli//Anhinga/Flamingo	1984	IT
Altar C84	Ruff/Free Gallipoli/2/Mexicali75/3/Shwa	1985	MX
AEL *	Gediz//Fgo//Gta/3/Srn_1/4/Totus/ 5/Ente/Mexi_2//Hui/3/Yav_1/Gediz/ 6/Sombra_20/7/Stot//Altar 84/Ald	2005 **	MX
Creso	CpB144×[(Yt54N10B)Cp263Tc3]	1974	IT
Colosseo	Mexa/Creso Mutant	1994	IT
Barnacla	Arment//Srn_3/Nigris_4/3/Canelo_9.1	2003 **	MX

* AEL: Abbreviation used by authors for this advanced experimental line; ** year of development of the breeding line, not yet registered or commercially released.

Duilio, Creso, and Colosseo are Italian commercial varieties. The other three genotypes were developed by the CIMMYT program in Mexico, with Barnacla and AEL being advanced experimental lines and Altar C84 being a high yielding variety commercially released in Mexico and other countries.

2.2. Field Experiments

The field trial was conducted at the CIMMYT experimental station (Campo Experimental Norman Ernest Borlaug, CENEB) near Cd. Obregón (Sonora, Mexico) during the cropping season of 2010. The experiment was arranged in a randomized complete block design with 4 replicates and plots of 3.36 m^2 for each of the two irrigation treatments or testing environments. The two different irrigation conditions were full irrigation (FI), with 550–600 mm of total water supplied by gravity irrigation during the full crop cycle, and reduced irrigation (RI), with 220–250 mm of total water applied through a drip system, all before heading. In both irrigation conditions plots were fertilized optimally as per the site-specific agronomic recommendations using a total of 250 units of nitrogen in the form of urea (50 units at sowing, 100 units at first node, and 100 units at the end of tillering) and phosphorus (50 units applied at sowing). Plots were maintained free of diseases and pests via the uniform application of fungicide and insecticide.

2.3. Agronomic Traits

After mechanical harvest of the whole plots, grain yield and thousand kernel weight were determined and considered in relation with the *TdDRF1* expression profiles.

2.4. RNA Extraction

A time-course experiment was designed with a sampling schedule consisting of 7 collection stages (T1 to T7), as reported in Table 2.

Table 2. Leaf collection schedule implemented during the time-course experiment to establish the expression profile of the *TdDRF1* gene.

Time-Course	Weeks after T1	Month
T1 (1st collection date)	82 days after sowing	End of January
T2 (2nd collection date)	1	Early February
T3 (3rd collection date)	2	February
T4 (4th collection date)	3	February
T5 (5th collection date)	4	End of February
T6 (6th collection date)	5	Early March
T7 (7th collection date)	6	March

This schedule was planned to include the whole growing period, from heading to harvest. For each sampling, ten representative flag leaves were harvested for each plot, pooled together, immediately frozen in liquid nitrogen, and subsequently stored at −80 °C prior to RNA extraction. Total RNA was extracted from the leaves using the TRIzol® Reagent (Invitrogen, Carlsbad, CA, USA) in accordance with the manufacturer's instructions, and lyophilized. Lyophilized RNA samples (each yielding being approximately 25–35 μg of total RNA) were then resuspended in nuclease-free sterile water, qualitatively assessed by agarose gel electrophoresis, and quantified with a NanoDrop ND-1000 Spectrophotometer (NanoDrop Technologies, Wilmington, NC, USA).

2.5. Reverse Transcription, Pre-Amplification, and qRT-PCR

A set of specific primers, designed using the Assay-By-Design software (Applied Biosystems) with a view to obtain three specific and distinguishable fragments corresponding to each *TdDRF1* transcript, were used [11] (Supplementary Figure S1).

Supplementary Figure S2 is a schematic representation of the complete procedure of reverse transcription, pre-amplification, and qRT-PCR. The pre-amplification step was included to optimize the *real-time* reactions. A High-Capacity cDNA Reverse Transcription Kit (Applied Biosystems, Foster City, CA, USA) was used for the reverse transcription reactions. Samples (20 μL) contained 2 μL of 10× RT Buffer, 0.8 μL of 25× dNTP Mix (100 mM), 2 μL of 10× RT Random Primers, 1 μL of RNase Inhibitor (20 U/μL), and 2 μg of total RNA in nuclease-free water. The thermal cycling conditions were 10 min at 25 °C, 2 h at 37 °C, and 5 min at 85 °C and at 4 °C. Pre-amplification reactions (50 μL) contained 2 μL of TaqMan® PreAmp Master Mix 2× (Applied Biosystems), 12.5 μL of pooled assay mix (0.2×, each assay), and 250 ng of cDNA sample in nuclease-free water. Reactions were held at 95 °C for 10 min and then at 95 °C for 15 s and 60 °C for 4 min 14 times. The resulting pre-amplified reactions were then diluted (1:20) in 1× TE buffer and used as the starting material for the subsequent Custom TaqMan® Gene Expression Assays (Applied Biosystems) for the three target transcripts and the endogenous control carried out in the Applied Biosystems 7300 Real-Time PCR System. The final volume (20 μL) of a single PCR reaction contained 10 µL of 2× TaqMan® Universal PCR Master Mix with AmpErase® UNG (Applied Biosystems), 1 µL of 20× Custom TaqMan® Gene Expression Assay (Applied Biosystems), and 2 µL of diluted pre-amplified product as a template. Samples were run in three biological replicates (from three randomized plots) and three technical replicates.

For endogenous control of relative quantification, different wheat genes were tested: *18S rRNA*, *TaSNK1* [18], *actin*, *Ta2291*, and *Ta2776* [19]. The GeNorm algorithm was used to calculate the gene-stability value (*M*) for all reference genes according to:

$$Mj = \frac{\sum_{k=1}^{n} Vjk}{n-1} \quad (1)$$

where *Vjk* represents the arithmetic mean of all pairwise variations [19]. As the gene with the lowest *M* values showed the most stable expression, *TaSNK1* was used in our qRT-PCR assays (Supplementary Table S1).

A relative quantification of the *TdDRF1* transcripts was obtained using the $\Delta\Delta C_T$ method [20] for each sample and results were expressed as normalized relative quantity (NRQ). Furthermore, each expression profile was also calculated as the log2 value of the fold change (FC) (abundance under stress/abundance under control) for each transcript and time.

2.6. Transcripts of the Wdhn13 Gene

Based on the literature, the *Wdhn13* gene was chosen as a putative target of the regulation by the *Td*DRF1 transcription factor [21,22], using the sequence from *Triticum aestivum*, locus AB297677, deposited in NCBI in 2007. A qRT-PCR analysis was performed on RNA samples of AEL and of Barnacla collected from both the FI and RI conditions during the time-course. The following pair of primers was used: FOR 5′-GATGGCAACTACGGGAAGTC-3′ and REV 5′-GCAGCTTGTCCTTGATCTTG-3′, amplifying an 88 bp cDNA fragment, which was cloned for the setup of the standard curve. qRT-PCR reactions were performed using SYBR green technology in accordance with the procedure reported by Vítámvás and colleagues [23].

2.7. Statistical Analyses

All statistical analyses were carried out with IBM SSPS Statistics 23. The analysis of variance (ANOVA) for each parameter was performed at a 95% confidence level and the significant difference between means was tested using Tukey's method when applicable. Furthermore, the significance of contrast between the up-regulated and down-regulated *TdDRF1* transcripts was also calculated using molecular data as a fixed factor and agronomic data (GY and TKW) as the variable ones.

3. Results

3.1. Agronomic Data

The grain yield (GY) and thousand kernel weight (TKW) averaged over four replicates in the two irrigation conditions are reported in Table 3.

Table 3. Grain yield (GY) in *ton/ha* and thousand kernel weight (TKW) in *g* observed for the six durum wheat genotypes, evaluated under full (FI) and reduced (RI) irrigation conditions. Data reported are means ± standard deviations.; * $p < 0.05$; ** $p < 0.01$.

Genotype	FI		RI	
	GY *	TKW	GY **	TKW **
Duilio	6.6 ± 0. 44 (ab)	48.2 ± 2.08 (a)	2.5 ± 0.45 (bc)	41.0 ± 1.0 (bc)
Altar C84	6.6 ± 0. 85 (ab)	41.8 ± 2.89 (a)	1.4 ± 0.08 (a)	43.5 ± 0.50 (c)
AEL	7.1 ± 0. 86 (b)	41.8 ± 1.89 (a)	2.4 ± 0.23 (bc)	34.3 ± 1.44 (a)
Creso	5.3 ± 0. 61 (a)	44.0 ± 4.09 (a)	1.8 ± 0.23 (ab)	46.7 ± 1.53 (d)
Colosseo	5.9 ± 0.01 (ab)	48.8 ± 2.02 (a)	2.4 ± 0.36 (bc)	40.2 ± 1.26 (b)
Barnacla	5.8 ± 0. 68 (ab)	41.8 ± 1.26 (a)	3.1 ± 0.62 (c)	40.0 ± 0.50 (b)

Means with the same letter are not significantly different.

3.2. Expression Profiles of TdDRF1 Gene

For each transcript, ANOVA was carried out using a log transformation of normalized relative quantities (NRQs) of transcript between the two irrigation conditions (FI and RI) for each time. Significant differences are summarized in Table 4. With regard to the putative transcription factors, significant differences were found in the *TdDRF1.3* transcript at T1, in both *TdDRF1.1* and *TdDRF1.3* transcripts at T4, in the *TdDRF1.1* transcript at T5, and in the *TdDRF1.3* transcript at T7.

Table 4. ANOVA summery table. Significant differences in *TdDRF1* transcripts between FI and RI conditions during the time-course are shown.

Time	Transcript	F	*p*
T1	*TdDRF1.2*	F(1,29) = 8.35	<0.01
	TdDRF1.3	F(1,17.47) = 10.78	<0.01
T2	*TdDRF1.2*	F(1,20.88) = 10.5	<0.01
T4	*TdDRF1.1*	F(1,31) = 9.86	<0.01
	TdDRF1.3	F(1,31) = 10.33	<0.01
T5	*TdDRF1.1*	F(1,31) = 4.22	<0.05
	TdDRF1.2	F(1,29) = 11.06	<0.01
T7	*TdDRF1.3*	F(1,31) = 18.75	<0.001

F, sample F statistics; *p*, significance.

For each genotype, NRQs of each transcript calculated during the time-course and rescaled to its own T1 value are shown in Figure 1. Unfortunately, the samples of Altar C84 under the RI condition were lost. As each profile value referred to its own T1, the analyses of transcripts were carried out at a trend level throughout the time-course. Furthermore, the strategy of relative quantification is only suitable for comparing results from the same transcript between treatments, so that results obtained with primer pairs different to each other could not be directly compared [20].

Comparing the genotypes shown in Figure 1, the *TdDRF1.1* transcript under the FI condition showed a slightly variable trend with some exceptions: Altar C84, AEL, Creso, and Colosseo showed a decrease at T3, while Duilio, Altar C84, AEL, and Colosseo showed an increase at T6. On the other hand, under the RI condition, each genotype showed a more distinct behaviour: Duilio showed a clear decrease at each time, AEL, Creso, and Colosseo remained almost constant at their T1 values, with a substantial increase at T7, while Barnacla showed a higher variability during the time-course with two large increases at T2 and T6.

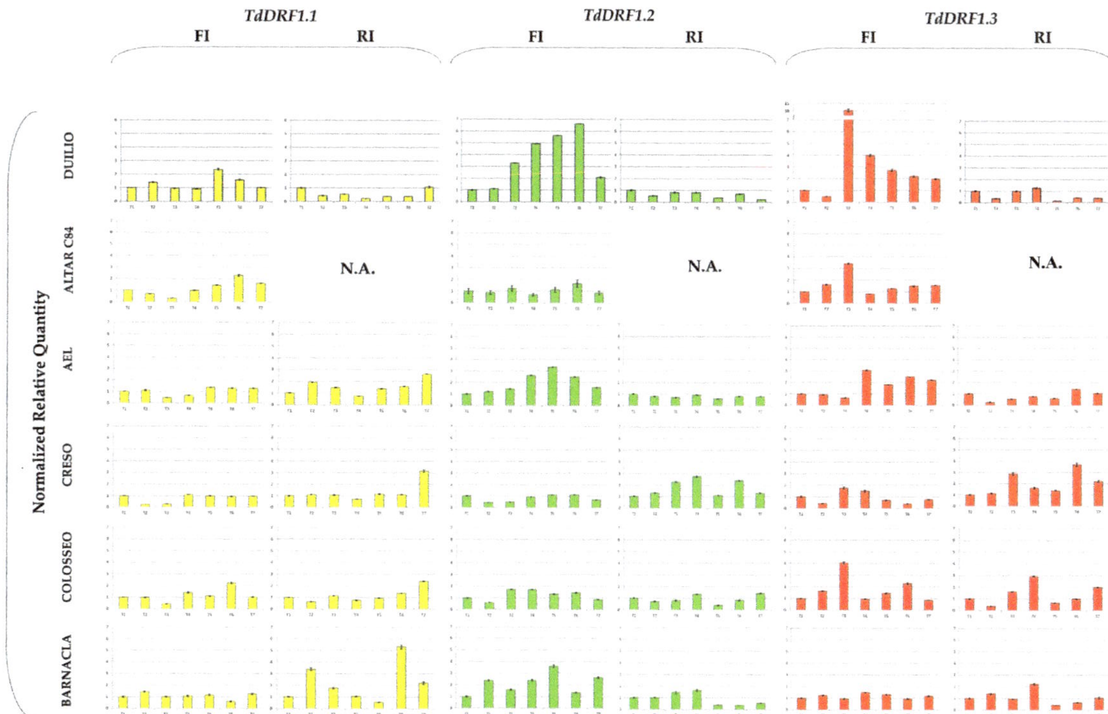

Figure 1. Expression profiles of the *TdDRF1* gene transcripts. Each expression profile is shown as normalized relative quantity rescaled to its own T1 value. Results referred to the six durum wheat genotypes analysed under both FI and RI conditions (*TdDRF1.1* in yellow, *TdDRF1.2* in green and *TdDRF1.3* in red, respectively). The time course consisted of seven points (T1 to T7). N.A., data not available. Data are represented as mean + SE.

With regard to the *TdDRF1.2* transcript under the FI condition, there were differences mainly at T3, T4, and T5. Duilio, AEL, and Barnacla displayed a large increase in comparison with their T1 values, while Altar C84 and Creso remained almost constant and Colosseo showed a slight increase. Under the RI condition, only Creso showed appreciable variations during the time-course.

As regards the *TdDRF1.3* transcript under the FI condition, all genotypes were found to be appreciably variable during the time-course with the exception of Barnacla. In particular, at T3 there was a large increase in Duilio, Altar C84, and Colosseo and, to a lesser extent, Creso, while AEL was late, showing an increase at T4. Under the RI condition, Duilio and AEL were almost constant or less than their initial values, while the other three genotypes showed variations, mainly in Creso and Colosseo and, to a minor extent, in Barnacla.

In addition, a tentative comparison of the trends among the three transcripts was carried out for each genotype. Duilio showed an overall expression of the three transcripts under full irrigation, while under reduced irrigation it seemed to be down-regulated by the water stress. Under full irrigation, Altar C84 showed a complementary activation of the transcription factors (*TdDRF1.1* and *TdDRF1.3*) at different times. AEL showed similar behaviours for the *TdDRF1.2* and *TdDRF1.3* transcripts under full irrigation, while under reduced irrigation neither transcript showed any appreciable changes, with *TdDRF1.1* being variable. Creso showed an almost constant expression of the three transcripts under full irrigation, while under reduced irrigation the *TdDRF1.2* and *TdDRF1.3* transcripts were found to respond to the water stress. On the other hand, Colosseo expressed mainly the *TdDRF1.3* transcript under both conditions, while Barnacla expressed only the *TdDRF1.2* transcript under full irrigation and the *TdDRF1.1* transcript under reduced irrigation with a single peak in the *TdDRF1.3* transcript at T4.

The expression profiles represented as fold change (log2 value of the ratio for abundance under stress and abundance under control) for each transcript are shown in Figure 2. The *TdDRF1.1* transcript turned out to be mostly up-regulated in the genotypes, particularly at T4, T5, and T6, to different extents, with Creso showing the lowest values. The *TdDRF1.2* transcript showed the greatest variability among genotypes, even if all of them were up-regulated at T5. Both AEL and Duilio showed a similar trend, initially down-regulated and then shifting to up-regulated at different time points (T3 and T5, respectively). Creso was mostly down-regulated, while Colosseo showed a swinging trend (up- and down-regulated at different times). Barnacla was largely up-regulated from T5 onwards. The *TdDRF1.3* transcript was largely up-regulated in Barnacla during the whole time-course with a peak at T5 (in fact detectable in all its transcripts). On the contrary, all other genotypes showed a down-regulation throughout the time-course, apart from a light up-regulation in Colosseo and Duilio at T2 and T5, respectively.

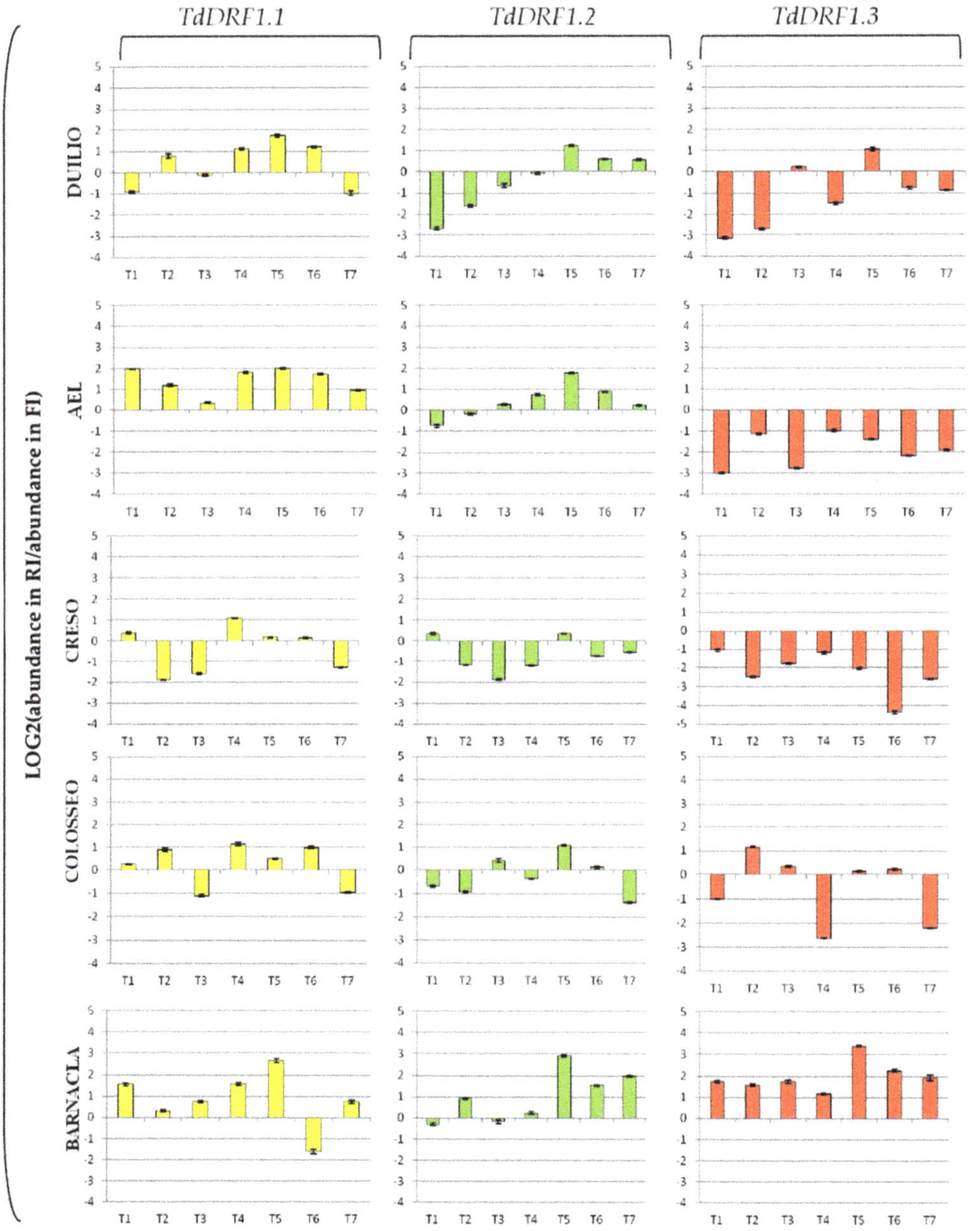

Figure 2. Expression profiles of the three *TdDRF1* transcripts. Each expression profile is shown as fold change (log2 value of the ratio for abundance under stress and abundance under control). Results referred to five durum wheat genotypes analysed under both FI and RI conditions (*TdDRF1.1* in yellow, *TdDRF1.2* in green, and *TdDRF1.3* in red, respectively). The time course consisted of seven points (T1 to T7). Data are represented as mean ± SE.

3.3. Statistical Association between Fold Change of TdDRF1 Transcripts and Traits

With the aim of finding an association between the agronomical data and the fold change in the three transcripts under the reduced irrigation condition, molecular data were represented using a binary code. For each genotype, we assigned a value = 0 representing a down-regulation and a value = 1 representing an up-regulation at each time point, whatever the size (see Figure 2). The analysis was carried out at the most relevant times for each trait analysed, that is, T2 for grain yield (GY) and T6 for thousand kernel weight (TKW), and was extended to neighbour points T1 and T7, respectively. The contrast between the molecular data (up- and down-regulation of transcripts) and the traits analysed is shown in Table 5.

Table 5. Statistical association between molecular and agronomical data under RI condition. Molecular data were represented as a binary code and agronomical traits are referred to as GY and TKW.

Trait	Time	Transcript	Estimated LSM				
			Genotypes	0 Group	Genotypes	1 Group	$p = \|t\|$
GY	T1	*TdDRF1.2*	All others	2.62	Creso	1.79	<0.05
	T1	*TdDRF1.3*	All others	2.28	Barnacla	3.14	<0.01
	T2	*TdDRF1.1*	Creso	1.79	all others	2.62	<0.01
	T2	*TdDRF1.2*	All others	2.28	Barnacla	3.14	<0.01
TKW	T6	*TdDRF1.2*	Creso-Barnacla	43.33	Duilio-AEL-Colosseo	38.50	<0.01
	T7	*TdDRF1.1*	Duilio-Creso-Colosseo	42.61	AEL-Barnacla	37.17	<0.01
	T7	*TdDRF1.2*	Creso-Colosseo	43.42	Duilio-AEL-Barnacla	38.44	<0.05

LSM, Estimated least square means; *p*, significance.

At T2, the *TdDRF1.1* transcript was significantly associated to GY, with $p < 0.01$ and all up-regulated genotypes showing a mean GY significantly higher than Creso (down-regulated). *TdDRF1.2* was also significantly associated ($p < 0.01$) to the yield and Barnacla (up-regulated) showed a GY value significantly higher than the mean GY of all others. On widening the analysis to T1, Barnacla, characterized by up-regulation of the *TdDRF1.3* transcript, again showed a GY value significantly higher than all the others.

Considering the other agronomic trait, TKW, a significant association was found at T6 for *TdDRF1.2*, and the down-regulated genotypes (Creso and Barnacla) showed a TKW value significantly higher than that of the up-regulated group (Duilio, AEL, and Colosseo). On the other hand, widening the analysis to T7, *TdDRF1.1* was also significantly associated to TKW and the group of down-regulated genotypes (Duilio, Creso, and Colosseo) showed a mean TKW value significantly higher than the up-regulated group (AEL and Barnacla).

3.4. Expression Profile of the Wdhn13 Gene

The *Wdhn13* gene encodes for LEA D-11 DHN (dehydrin), which has been reported to respond to water stress in wheat plants [21,24]. The *Wdhn13* gene transcript was analysed in two genotypes, AEL and Barnacla, under both irrigation conditions. Regardless of genotypes, significant differences of the *Wdhn13* gene transcript were found between FI and RI conditions at T1, T2, T4, T6, and T7, as shown in Table 6. The expression profiles of the *Wdhn13* gene transcript obtained analysing the environments separately in two genotypes are reported in Figure 3. Figure 3a refers to the full irrigation condition, while Figure 3b refers to the reduced irrigation condition. No significant differences were found between genotypes under FI, with the exception of T2 ($p < 0.05$), while significant differences were observed under RI (at T1, T3, and T5).

Table 6. ANOVA summery table. Significant differences in *Wdhn13* gene transcript between FI and RI conditions during the time-course.

Time	F	*p*
T1	F(1,10) = 7.13	<0.05
T2	F(1,10) = 14.1	<0.01
T4	F(1,10) = 6.67	<0.05
T6	F(1,9) = 138.06	<0.0001
T7	F(1,10) = 36.86	<0.0001

F, sample F statistics; *p*, significance.

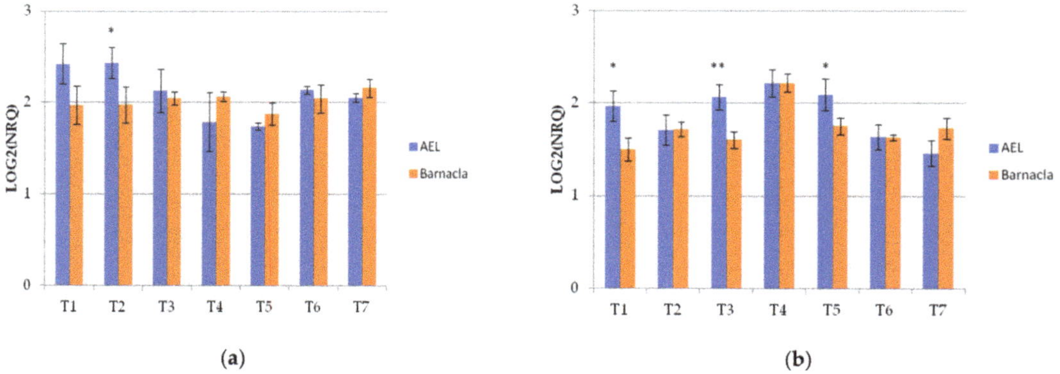

Figure 3. The expression profiles of *Wdhn13* gene transcript in the two genotypes measured under FI and RI conditions. (**a**) The expression profiles of *Wdhn13* transcript in AEL and Barnacla observed under FI condition. (**b**) The expression profiles of *Wdhn13* transcript in AEL and Barnacla observed under RI condition. Data are represented as mean ± SE. Significance codes: ** $p < 0.01$, * $p < 0.05$.

The expression of the *Wdhn13* gene transcript in AEL is shown in Figure 4 and demonstrated significant differences under the two conditions, showing almost opposite concavities crossing at T3 and T6, suggesting up-regulation under reduced irrigation.

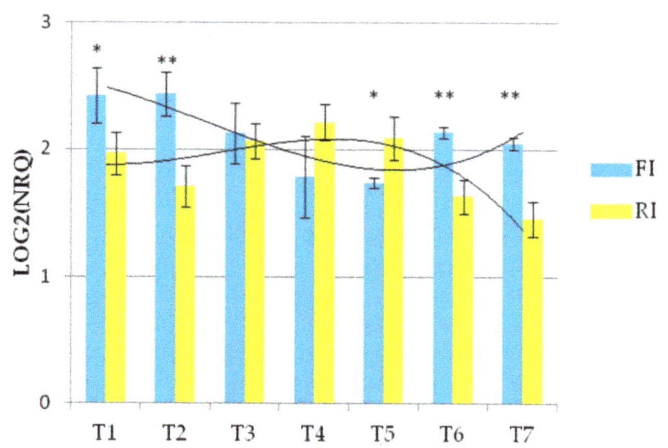

Figure 4. The expression profile of *Wdhn13* gene transcript of AEL under FI and RI conditions. A third-order polynomial trendline was added to highlight the opposite concavities; ** $p < 0.01$, * $p < 0.05$.

The *Wdhn13* gene transcript profiles were also different between the two conditions in Barnacla, showing a larger variability under reduced irrigation, while the expression levels under full irrigation were almost stable (Figure 5).

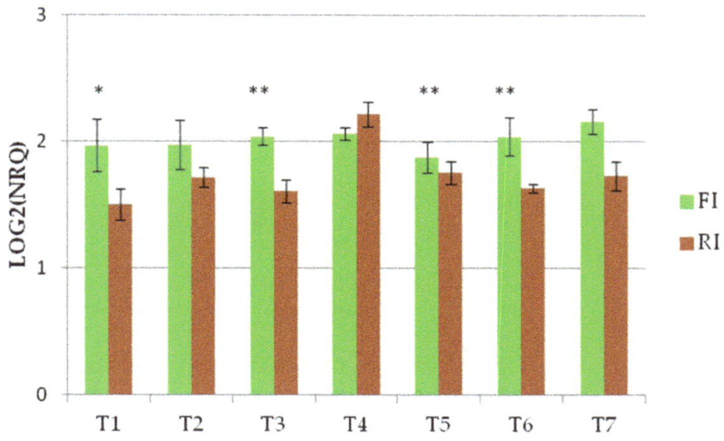

Figure 5. The expression profile of *Wdhn13* transcript of Barnacla under FI and RI conditions ** $p < 0.01$, * $p < 0.05$.

4. Discussion

This study was aimed at analysing and comparing the expression levels of the *TdDRF1* gene in the field under two different irrigation conditions, i.e., full and reduced irrigation (FI and RI, respectively), taking into account the reported involvement of its transcripts under water stress. Six different durum wheat genotypes were studied with the aim of finding any different behaviours among the transcripts. Our results revealed significant differences in the *TdDRF1* transcripts between the two irrigation conditions at different time points (Table 4). Both the *TdDRF1.1* and *TdDRF1.3* transcripts showed significant differences between the two conditions at T4, suggesting their full involvement in drought response, which continued with *TdDRF1.1* at T5 and *TdDRF1.3* at T7, pointing to their possible different role in the *trans*-activation throughout the time. On the other hand, these results could reflect the activity of the alternative splicing mechanism during the time-course, producing and cumulating the three transcripts in different ways as a response to the environment [25]. The modulation of the *TdDRF1.2* transcript is particularly interesting, as the latter is not directly involved in the downstream gene modulation but, with high probability, plays an important role in gene regulation through its expression and degradation via nonsense-mediated mRNA decay or other RNA surveillance mechanisms during transcript maturation, as reported for the abortive forms produced by the alternative splicing mechanism [26,27].

Looking at the trends shown in Figure 1 as a whole, every genotype seemed to have its own transcripts profile, depending mainly on genetics rather than environmental effects, as was previously observed [11,12,28]. The *TdDRF1.1* transcript remained almost steady across the time-course in both conditions, with few exceptions other than the notable behaviour of Barnacla under the reduced irrigation condition. Since the *TdDRF1.1* transcript results from the junction of all four exons (E1-E2-E3-E4) present in the gene sequence, it could be speculated that the quantity of this transcript, as the first product obtained, remains lower than the other ones, taking into account another mechanism very common in plants, known as intron retention [29], which could explain the formation of *TdDRF1.2* (E1-E2-E4) and *TdDRF1.3* (E1-E4) transcripts from an immature *TdDRF1.1* transcript. In addition to a relatively high value for the *TdDRF1.1* transcript, Barnacla also showed an opposite behaviour to the others under reduced irrigation. In the other genotypes, the *TdDRF1.3* transcript under full irrigation increased throughout the time-course, with some exceptions and to different extents, with Duilio up to 10-fold. Under reduced irrigation, Duilio and AEL did not change their initial expression levels, while both Creso and Colosseo showed substantial increases. It is important to underline that all the results are based on a time-course spanning 124 days from sowing with a seven-day window (Table 2); it could be possible that, in some cases and particularly under reduced irrigation, the window was too large to capture the expression modulations based on early and short responses. The *a priori* selection of an experimental design aimed at analysing gene expression in the field is a challenging task, particularly when the study involves transcription factors whose few molecules are sufficient to respond promptly to the stress by producing a large transcriptional burst of stress-responsive gene expression in a short period. Furthermore, in the field, plants may experience several distinct abiotic stresses either concurrently or at different times through the growing season [30], and consequently plant responses to water scarcity are very complex [31,32]. Finally, durum wheat is an allopolyploid (tetraploid) species and, due to its large genome size and the high levels of sequence similarity between the chromosomes duplicated, it is traditionally difficult to analyse [33]. Undeniably, for all the above-mentioned reasons, the number of gene expression studies using plant materials from the field is much less covered by the scientific literature than those under controlled environmental conditions.

With the results represented as the fold change, it is possible to state the expression levels in terms of down- and up-regulation of transcripts. It is worth noting that Duilio and AEL showed similar trends for all transcripts with very few differences: *TdDRF1.1* was always up-regulated, while *TdDRF1.3* was always down-regulated at all times. This molecular behaviour may suggest that the early stress response, aimed at activating *in trans* downstream genes, is mainly based on the 1.1 isoform. On the whole, the three transcripts of *TdDRF1* in Creso were strongly inhibited by water stress, possibly suggesting that this genotype employs other genes to cope with drought stress. Colosseo showed a certain similarity with Duilio and AEL with regard to the *TdDRF1.1* transcript. The molecular behaviour of the *TdDRF1* gene in Barnacla was unique, as all transcripts were largely up-regulated, suggesting a continuous transcription process that cumulates the three transcripts through a very articulated and complex control of alternative splicing [34]. Interestingly, in the contrast between the molecular and agronomic data, the up-regulation of the *TdDRF1.2* and *TdDRF1.3* transcripts, at T2 and T1, respectively, was significantly associated with grain yield, with Barnacla showing a value higher than the mean of all the others. The significant associations found by the statistical analyses are intriguing, but further studies at a molecular level and with a larger panel of different genotypes are necessary to better clarify the role of the transcripts in field stress response and to monitor their possible effects on the different stages of plant development and maturation.

A further aim of this work was to highlight the possible relationship between the *TdDRF1* transcripts and the *Wdhn13* gene, involved in environmental stress tolerance. In a recent study, Mehrabad Pour-Benab and colleagues [35] investigated dehydrin expression in different species of *Triticum* and *Aegilops* under well-watered and drought stress treatments in the greenhouse. Thirty days after applying the water stress, they observed the doubling of the *Wdhn13* expression level in water-stressed plants in the different wheat species. Our analysis in the field showed significant differences in the *Wdhn13* transcript levels between FI and RI conditions, at every time point except T3 and T5. Similarly, the *TdDRF1* transcripts were also found to be significantly different between the two conditions throughout the time-course. It could be speculated that the largest *Wdhn13* differences at T6 and T7 reflect the activity of transcription factors (*TdDRF1.1* at T5 and *TdDRF1.3* at T7) on the promoter of the *Wdhn13* gene, resulting finally in the dehydrin protein, involved in the drought stress tolerance at a cellular level.

Our results showed that the molecular behaviour of specific plant genes is highly dynamic and much more complex than a direct on/off switching. Furthermore, they confirm the main effect of genotype on the *TdDRF1* gene expression profile and, in this regard, the molecular behaviour of Barnacla appears particularly interesting, as it is the only genotype showing both transcription factors active under reduced irrigation during the time-course, even if to different extents. It is particularly intriguing that the grain yield value of Barnacla was the highest and was statistically associated with the up-regulation of two transcripts under reduced irrigation, thus conferring an added value to this genotype.

In conclusion, notwithstanding the intrinsic difficulty of identifying the effect of a single factor in the complex picture of gene expression, this work represents a remarkable contribution to highlighting the role of the *TdDRF1* gene in field conditions. Further experiments are necessary with a larger panel of genotypes and set of genes (transcription factors and target genes) to better understand the molecular plant response to drought, with the final aim of setting up a support tool to assist in orienting breeders' decisions.

Supplementary Materials: The following supporting information can be downloaded at: https://www.mdpi.com/article/10.3390/genes13030555/s1, Figure S1: Specific amplification and detection of the three *TdDRF1* transcripts, Figure S2: Schematic representation of the complete procedure of reverse transcription, pre-amplification, and qRT-PCR, Table S1: Gene-stability values (M) of the analysed reference genes.

Author Contributions: P.G. coordinated the project; A.L., K.A. and P.G. designed the experimental work; K.A. managed the field experiment and elaborated the agronomic data; A.L. and K.T. carried out the molecular analyses; C.C. carried out bioinformatics and the statistical analyses; A.L., C.C. and P.G. analysed the data and interpreted the results; C.C. and P.G. developed the first draft; K.A. and A.L. reviewed and edited the manuscript. All authors have read and agreed to the published version of the manuscript.

Funding: This study was partially supported by the High Relevance Mexico–Italy Project of the Italian Foreign Affairs Ministry and by the Italian RIADE project (Integrated research for applying new technologies and processes for combating desertification). A.L. was supported by Fellowships of the Government of Mexico, "Secretaría de Relaciones Exteriores" for two internships at CIMMYT (Mexico). K.T. was supported by an ENEA International 2-year postdoc Fellowship at ENEA Casaccia Research Centre, Italy.

Institutional Review Board Statement: Not applicable.

Informed Consent Statement: Not applicable.

Data Availability Statement: Not applicable.

Acknowledgments: All authors gave the consent to the acknowledgement. The authors are sincerely grateful to Chiara Rasi, Domenico Di Bianco, and Fabio Felici for their valuable work in the field during the time-course experiment; to the technical staff of CIMMYT experimental station (CENEB) near Cd. Obregón for their care and the valuable field work; to Giulio Marconi (ENEA library service) for his constant assistance; and to Ian Pace, professional mother-tongue reviewer, for the revision of the English text.

Conflicts of Interest: The authors declare no conflict of interest. The funders had no role in the design of the study or in the collection, analyses, and interpretation of data.

Abbreviations

AEL, Advanced experimental line; DRE, *drought-responsive element; DREB,* dehydration responsive element binding gene; FC, fold change; FI, full irrigation; GY, grain yield; NRQ, normalized relative quantity of transcript; RI, reduced irrigation; *TdDRF1*, *Triticum durum* dehydration responsive factor 1 gene; TKW, thousand kernel weight; TF, transcription factor; *Wdhn13*, wheat dehydrin 13 gene.

References

1. Raza, A.; Razzaq, A.; Mehmood, S.S.; Zou, X.; Zhang, X.; Lv, Y.; Xu, J. Impact of climate change on crops adaptation and strategies to tackle its outcome: A review. *Plants* **2019**, *8*, 34. [CrossRef] [PubMed]
2. Pareek, A.; Dhankher, O.P.; Foyer, C.H. Mitigating the impact of climate change on plant productivity and ecosystem sustainability. *J. Exp. Bot.* **2020**, *71*, 451–456. [CrossRef] [PubMed]

3. FAO. *The State of Food and Agriculture 2020. Overcoming Water Challenges in Agriculture*; FAO: Rome, Italy, 2020. [CrossRef]
4. Shewry, P.R.; Hey, S.J. The contribution of wheat to human diet and health. *Food Energy Secur.* **2015**, *4*, 178–202. [CrossRef] [PubMed]
5. Yamaguchi-Shinozaki, K.; Shinozaki, K. Transcriptional regulatory networks in cellular responses and tolerance to dehydration and cold stresses. *Ann. Review Plant Biol.* **2006**, *57*, 781–803. [CrossRef]
6. Agarwal, P.K.; Gupta, K.; Lopato, S.; Agarwal, P. Dehydration responsive element binding transcription factors and their applications for the engineering of stress tolerance. *J. Exp. Bot.* **2017**, *68*, 2135–2148. [CrossRef]
7. Faraji, S.; Filiz, E.; Kazemitabar, S.K.; Vannozzi, A.; Palumbo, F.; Barcaccia, G.; Heidari, P. The *AP2/ERF* gene family in *Triticum durum*: Genome-wide identification and expression analysis under drought and salinity stresses. *Genes* **2020**, *11*, 1464. [CrossRef]
8. Lata, C.; Prasad, M. Role of DREBs in regulation of abiotic stress responses in plants. *J. Exp. Bot.* **2011**, *62*, 4731–4748. [CrossRef]
9. Rabara, R.C.; Tripathi, P.; Rushton, P.J. The potential of transcription factor-based genetic engineering in improving crop tolerance to drought. *OMICS* **2014**, *18*, 601–614. [CrossRef]
10. El-Esawi, M.A.; Al-Ghamdi, A.A.; Ali, H.M.; Ahmad, M. Overexpression of *AtWRKY30* Transcription Factor Enhances Heat and Drought Stress Tolerance in Wheat (*Triticum aestivum* L.). *Genes* **2019**, *10*, 163. [CrossRef]
11. Latini, A.; Rasi, C.; Sperandei, M.; Cantale, C.; Iannetta, M.; Dettori, M.; Ammar, K.; Galeffi, P. Identification of a DREB-related gene in *Triticum durum* and its expression under water stress conditions. *Ann. Appl. Biol.* **2007**, *150*, 187–195. [CrossRef]
12. Latini, A.; Sperandei, M.; Cantale, C.; Arcangeli, C.; Ammar, K.; Galeffi, P. Variability and expression profile of the *DRF1* gene in four cultivars of durum wheat and one triticale under moderate water stress conditions. *Planta* **2013**, *237*, 967–978. [CrossRef] [PubMed]
13. Latini, A.; Sperandei, M.; Sharma, S.; Cantale, C.; Iannetta, M.; Dettori, M.; Ammar, K.; Galeffi, P. Molecular analyses of a DREB-related gene in durum wheat and triticale. In *Biosaline Agriculture and High Salinity Tolerance*; Abdelly, C., Öztürk, M., Ashraf, M., Grignon, C., Eds.; Birkäuser Verlag AG: Basel, Switzerland, 2008; pp. 287–295.
14. Catacchio, C.R.; Alagna, F.; Perniola, R.; Bergamini, C.; Rotunno, S.; Calabrese, F.M.; Crupi, P.; Antonacci, D.; Ventura, M.; Cardone, M.F. Transcriptomic and genomic structural variation analyses on grape cultivars reveal new insights into the genotype-dependent responses to water stress. *Sci. Rep.* **2019**, *9*, 2809. [CrossRef] [PubMed]
15. Baker, C.R.; Stewart, J.J.; Amstutz, C.L.; Ching, L.G.; Johnson, J.D.; Niyogi, K.K.; Adams, W.W., 3rd; Demmig-Adams, B. Genotype-dependent contribution of CBF transcription factors to long-term acclimation to high light and cool temperature. *Plant Cell Environ.* **2021**, *45*, 392–411. [CrossRef] [PubMed]
16. Rodríguez, V.M.; Santiago, R.; Malvar, R.A.; Butrón, A. Inducible maize defense mechanisms against the corn borer *Sesamia nonagrioides*: A transcriptome and biochemical approach. *Mol. Plant Microbe Interact.* **2012**, *25*, 61–68. [CrossRef]
17. Sari, E.; Bhadauria, V.; Vandenberg, A.; Banniza, S. Genotype-dependent interaction of lentil lines with *Ascochyta lentis*. *Front. Plant Sci.* **2017**, *8*, 764. [CrossRef]
18. Kam, J.; Gresshoff, P.; Shorter, R.; Xue, G.-P. Expression analysis of RING zinc finger genes from *Triticum aestivum* and identification of *TaRZF70* that contains four RING-H2 domains and differentially responds to water deficit between leaf and root. *Plant Sci.* **2007**, *173*, 650–659. [CrossRef]
19. Paolacci, A.R.; Tanzarella, O.A.; Porceddu, E.; Ciaffi, M. Identification and validation of reference genes for quantitative RT-PCR normalization in wheat. *BMC Mol. Biol.* **2009**, *10*, 11. [CrossRef]
20. Rieu, I.; Powers, S.J. Real-time quantitative RT-PCR: Design, calculations, and statistics. *Plant Cell* **2009**, *21*, 1031–1033. [CrossRef]
21. Ohno, R.; Takumi, S.; Nakamura, C. Kinetics of transcript and protein accumulation of a low-molecular-weight wheat LEA D-11 dehydrin in response to low temperature. *J. Plant Physiol.* **2003**, *160*, 193–200. [CrossRef]
22. Kobayashi, F.; Takumi, S.; Egawa, C.; Ishibashi, M.; Nakamura, C. Expression patterns of low temperature responsive genes in a dominant ABA-less-sensitive mutant line of common wheat. *Physiol. Plant.* **2006**, *127*, 612–623. [CrossRef]
23. Vítámvás, P.; Kosová, K.; Musilová, J.; Holková, L.; Mařík, P.; Smutná, P.; Klíma, M.; Prášil, I.T. Relationship between dehydrin accumulation and winter survival in winter wheat and barley grown in the field. *Front. Plant Sci.* **2019**, *10*, 7. [CrossRef] [PubMed]
24. Egawa, C.; Kobayashi, F.; Ishibashi, M.; Nakamura, T.; Nakamura, C.; Takumi, S. Differential regulation of transcript accumulation and alternative splicing of a *DREB2* homolog under abiotic stress conditions in common wheat. *Genes Genet. Syst.* **2006**, *81*, 77–91. [CrossRef] [PubMed]
25. Song, L.; Pan, Z.; Chen, L.; Dai, Y.; Wan, J.; Ye, H.; Nguyen, H.T.; Zhang, G.; Chen, H. Analysis of Whole Transcriptome RNA-seq Data Reveals Many Alternative Splicing Events in Soybean Roots under Drought Stress Conditions. *Genes* **2020**, *11*, 1520. [CrossRef] [PubMed]
26. Dubrovina, A.S.; Kiseklev, K.V.; Zhuravlev, Y.N. The role of canonical and noncanonical pre-mRNA splicing in plant stress responses. *Biomed Res Int.* **2013**, 264314. [CrossRef]
27. Göhring, J.; Jacak, J.; Barta, A. Imaging of endogenous messenger RNA splice variants in living cells reveals nuclear retention of transcripts inaccessible to nonsense-mediated decay in Arabidopsis. *Plant Cell* **2014**, *26*, 754–764. [CrossRef]
28. Liu, H.; Able, A.J.; Able, J.A. Integrated analysis of small RNA, transcriptome, and degradome sequencing reveals the water-deficit and heat stress response network in durum wheat. *Int. J. Mol. Sci.* **2020**, *21*, 6017. [CrossRef]
29. Iñiguez, L.P.; Hernández, G. The evolutionary relationship between alternative splicing and gene duplications. *Front. Genet.* **2017**, *8*, 14. [CrossRef]

30. Tester, M.; Bacic, A. Abiotic stress tolerance in grasses. From model plants to crop plants. *Plant Physiol.* **2005**, *137*, 791–793. [CrossRef]
31. Chaves, M.M.; Pereira, J.S.; Maroco, J.; Rodrigues, M.L.; Ricardo, C.P.P.; Osório, M.L.; Carvalho, I.; Faria, T.; Pinheiro, C. How plants cope with water stress in the field? Photosynthesis and growth. *Ann. Bot.* **2002**, *89*, 907–916. [CrossRef]
32. Osakabe, Y.; Osakabe, K.; Shinozaki, K.; Tran, L.-S.P. Response of plants to water stress. *Front. Plant Sci.* **2014**, *5*, 86. [CrossRef]
33. Kuo, T.; Hatakeyama, M.; Tameshige, T.; Shimizu, K.K.; Sese, J. Homeolog expression quantification methods for allopolyploids. *Brief. Bioinform.* **2020**, *21*, 395–407. [CrossRef] [PubMed]
34. Chen, M.; Manley, J.M. Mechanisms of alternative splicing regulation: Insights from molecular and genomics approaches. *Nat. Rev. Mol. Cell Biol.* **2009**, *10*, 741–754. [CrossRef] [PubMed]
35. Mehrabad Pour-Benab, S.; Fabriki-Ourang, S.; Mehrabi, A.-A. Expression of dehydrin and antioxidant genes and enzymatic antioxidant defense under drought stress in wild relatives of wheat. *Biotechnol. Biotechnol. Equip.* **2019**, *33*, 1063–1073. [CrossRef]

genes

MDPI

Article

Genome Wide Association Study Uncovers the QTLome for Osmotic Adjustment and Related Drought Adaptive Traits in Durum Wheat

Giuseppe Emanuele Condorelli [1], Maria Newcomb [2], Eder Licieri Groli [1], Marco Maccaferri [1], Cristian Forestan [1], Ebrahim Babaeian [3], Markus Tuller [3], Jeffrey Westcott White [4], Rick Ward [2], Todd Mockler [5], Nadia Shakoor [5] and Roberto Tuberosa [1,*]

[1] Department of Agricultural and Food Sciences, University of Bologna, 40127 Bologna, Italy; condorelli87@gmail.com (G.E.C.); e.groli@sisonweb.com (E.L.G.); marco.maccaferri@unibo.it (M.M.); cristian.forestan@unibo.it (C.F.)

[2] School of Plant Sciences, College of Agriculture and Life Sciences, Maricopa Agricultural Center, University of Arizona, Maricopa, AZ 85138, USA; maria.newcomb2@usda.gov (M.N.); rickw.ward@gmail.com (R.W.)

[3] Department of Environmental Science, University of Arizona, Tucson, AZ 85721, USA; ebabaeian@arizona.edu (E.B.); mtuller@cals.arizona.edu (M.T.)

[4] US Arid Land Agricultural Research Center, USDA-ARS, Maricopa, AZ 85138, USA; jeff.white.az@gmail.com

[5] Danforth Foundation, St. Louis, MO 63132, USA; tmockler@danforthcenter.org (T.M.); nshakoor@danforthcenter.org (N.S.)

* Correspondence: roberto.tuberosa@unibo.it

Citation: Condorelli, G.E.; Newcomb, M.; Groli, E.L.; Maccaferri, M.; Forestan, C.; Babaeian, E.; Tuller, M.; White, J.W.; Ward, R.; Mockler, T.; et al. Genome Wide Association Study Uncovers the QTLome for Osmotic Adjustment and Related Drought Adaptive Traits in Durum Wheat. *Genes* **2022**, *13*, 293. https://doi.org/10.3390/genes13020293

Academic Editor: Patrizia Galeffi

Received: 11 December 2021
Accepted: 29 January 2022
Published: 2 February 2022

Abstract: Osmotic adjustment (OA) is a major component of drought resistance in crops. The genetic basis of OA in wheat and other crops remains largely unknown. In this study, 248 field-grown durum wheat elite accessions grown under well-watered conditions, underwent a progressively severe drought treatment started at heading. Leaf samples were collected at heading and 17 days later. The following traits were considered: flowering time (FT), leaf relative water content (RWC), osmotic potential (ψs), OA, chlorophyll content (SPAD), and leaf rolling (LR). The high variability (3.89-fold) in OA among drought-stressed accessions resulted in high repeatability of the trait (h^2 = 72.3%). Notably, a high positive correlation (r = 0.78) between OA and RWC was found under severe drought conditions. A genome-wide association study (GWAS) revealed 15 significant QTLs (Quantitative Trait Loci) for OA (global R^2 = 63.6%), as well as eight major QTL hotspots/clusters on chromosome arms 1BL, 2BL, 4AL, 5AL, 6AL, 6BL, and 7BS, where a higher OA capacity was positively associated with RWC and/or SPAD, and negatively with LR, indicating a beneficial effect of OA on the water status of the plant. The comparative analysis with the results of 15 previous field trials conducted under varying water regimes showed concurrent effects of five OA QTL cluster hotspots on normalized difference vegetation index (NDVI), thousand-kernel weight (TKW), and/or grain yield (GY). Gene content analysis of the cluster regions revealed the presence of several candidate genes, including bidirectional sugar transporter SWEET, rhomboid-like protein, and S-adenosyl-L-methionine-dependent methyltransferases superfamily protein, as well as DREB1. Our results support OA as a valuable proxy for marker-assisted selection (MAS) aimed at enhancing drought resistance in wheat.

Keywords: drought; durum wheat; osmotic adjustment; QTL

1. Introduction

Drought is one of the most devastating abiotic stressors limiting crop yield, adaptability, and quality [1,2]. Recent global climate models predict a consistent rainfall reduction in temperate drylands [3–5], hence destabilizing food systems and global food security [6]. The plant reaction to drought is mediated by complex molecular systems linked to the transcriptome [7–10], as well as hormone signaling and metabolism [11–13]. In particular, drought is the major abiotic stress curtailing yield and lowering quality [14,15] in durum wheat (*Triticum turgidum* ssp. *durum*; 2n = 28, AABB), the most cultivated wheat in the Mediterranean regions [16], whose genome sequence was recently assembled *de novo* [17]. Among the strategies adopted by plants to withstand water scarcity [18,19], osmotic adjustment (OA) plays a major role in enhancing drought resistance through an active accumulation of solutes in response to a water potential reduction, thereby preserving cellular turgor [18,20–27]. Active OA maintains relative water content at low leaf water potential in order to sustain plant growth without impairing normal cellular functions [28]. Plants accumulate low-molecular weight organic solutes, such as soluble sugars [24,29] and proline [27,30–32], both of which increase under water stress, hence enhancing OA and contributing to maintain photosynthesis, as well as stomatal conductance, at lower water potentials. To date, the dissection of the genetic basis of OA has received limited attention, mainly due to the difficulty in measuring this trait in more than a limited number of accessions [33], an essential prerequisite to properly map and characterize the effects of the QTLs [34–36] underscoring OA variability. In cereals, the dissection of the OA QTLome has been attempted in rice [37,38] and barley [39,40] based on the evaluation of biparental recombinant inbred lines (RIL) populations, hence surveying only a limited amount of genetic variability as compared to that surveyed in GWAS studies. Herein, we report the results of the first large-scale genetic dissection of the OA QTLome in wheat via GWAS based on the field evaluation of 248 durum wheat elite accessions grown under conditions of progressively increasing drought and previously tested for grain yield in 15 field trials carried out under a broad range of water regimes in Mediterranean countries [41]. Three major QTL clusters were identified, where OA was unrelated to flowering time while being positively associated with the water status of the plant and grain yield as reported in [41], supporting the beneficial role of OA in enhancing drought resistance, most likely through an avoidance strategy. A comparative analysis with the sequence information available for these regions in durum [17] and bread wheat [42] revealed a number of putative candidate genes.

2. Materials and Methods

2.1. Plant Material and Field Management

For this study, 248 durum wheat elite accessions (Durum Panel) were chosen at the University of Bologna (Table S1). Most (189) of these accessions were originally assembled by [43] to represent a large portion of the genetic diversity (Table S2) present in the major improved durum wheat gene pools adapted to Mediterranean environments. The field trial was conducted at the University of Arizona Maricopa Agricultural Center (33.070 °N, 111.974 °W, elevation 360 m) on a Casa Grande Soil (fine-loamy, mixed, superactive, hyperthermic Typic Natrargids) (Figure S1). The Durum Panel was planted on 28 November 2017 according to a row-column experimental design with two replicates. Each accession was evaluated in two-row plots (3.5-m long, 0.76-m row spacing) with an average plant density of 22 plants/m^2. Orita and Tiburon, both representing the Arizona's "Desert-Durum"® wheat, were chosen as border plots. Before planting, granular nitrogen at 112 kg ha^{-1} was incorporated into the soil. Sprinkler irrigation was used to germinate seeds and establish the crop, followed by subsurface drip irrigation matching evapotranspiration for optimal plant growth, once or twice a week as needed. The pressurized subsurface drip irrigation system was installed before planting when one dripline with emitters spaced every 0.30 m was buried at ~0.10 m depth along each seed row. The final irrigation event was on 11 March 11 2018 (i.e., 103 days after sowing, DAS), when ~50% of the accessions had flag leaf sheaths opened (i.e., at Zadoks growth stage 47) [44]. From here on, the whole experiment was subjected to a progressive water deficit until 2–3 April 2018, when plants at the anthesis halfway stage (Zadoks growth stage 65, on average) were harvested to measure total above-ground biomass.

2.2. Meteorological Data and Soil Moisture Monitoring

Daily and hourly meteorological reports for the growing season were obtained from the Arizona Meteorological Network [45]. In addition, high temporal resolution meteorological data, particularly air temperature, relative humidity, and photosynthetic photon flux density (PPFD), for the experimental site were recorded at 5-s intervals with an automated weather station (Clima Sensor US, Adolf Thies GmbH & Co. KG, Göttingen, Germany) and a quantum sensor (SQ-214, Apogee Instruments, Inc., Logan, UT, USA). These data were made available by the TERRA Phenotyping Reference Platform [46]. Vapor pressure deficit (VPD) was calculated as the difference between the saturation and actual vapor pressure [47]. The soil volumetric water content (VWC) was monitored in and between seed rows with time-domain reflectometry (TDR) sensors (True TDR-315, Acclima, Inc., Meridian, ID, USA) installed at three locations within the experiment at 1, 10, and 50 cm depths at each location. Additional soil sensors were installed between rows at 15 cm depth to measure the soil matric potential (Tensiomark, ecoTech Umwelt-Meßsysteme GmbH, Bonn, Germany). All soil sensors recorded data at 15-min intervals throughout the entire growing season. Based on the characterization of the soil hydraulic and physical properties of the experimental site under the TERRA-REF project, the volumetric water contents corresponding to the permanent wilting point (θPWP) and the field capacity (θFC) at 10–15 cm depth were approximately 0.110 and 0.282 m^3/m^3, respectively. The VWC dynamics at the three measurement locations, for the entire growing season, are depicted in Figure S2. The Durum Panel accessions were monitored regularly for above-ground diseases and pests, which remained below threshold levels, hence not requiring control treatments, while growing degree days (GDD) were monitored until harvest (3-4 April 2018) at 125 days after sowing (DAS) (Table S3). Growth stages of each accession were defined based on the basis of the Zadoks scale [44] at 92, 93, 98, 101, 111, and 118 DAS, and flowering time (FT) was recorded when more than 50% of ears in the plot had flowered (anthesis half-way). Plants were harvested 125 days (DAS) to allow for planting the next phenotyping experiment; therefore, biomass data indicate the status at a point in time rather than direct estimates of final yields.

2.3. Evaluation of RWC, ψs, OA, LR, Leaf Chlorophyll Content (SPAD), and Biomass

The entire Durum Panel was evaluated for leaf relative water content (RWC) and osmotic potential (ψs) in well-watered (12 March 2018, 104 DAS) and severe drought (27 March 2018, 119 DAS) conditions. At the first sampling (fully-irrigated conditions) awns were visible on approximately 50% of accessions, while the second sampling was carried out under severe drought conditions when most accessions were at early grain-filling (Figure S3). Fully expanded flag leaves of eight different plants were sampled in each plot (experimental unit) at dawn from 6:00 to 7:00 a.m. Leaves were immediately placed in sealed plastic bags, stored in portable coolers (4 °C) to minimize water loss due to evaporation, and transported to the lab where leaves were removed from the bags. After cutting the leaf tips (5 cm), the remaining leaf portion (average length 15 cm) was cut in the middle to obtain two homogeneous pieces of similar weight, then mixed and stored in Falcon (50 ml) conical centrifuge tubes. One batch was used to measure OA following the "Rehydration method" described in Reference [23]. Leaves were rehydrated for 4 h in distilled water to reach full turgor, then dried, and stored in a freezer (−20 °C). After thawing, the cell sap was collected using a garlic press, and 10 µl were transferred onto a paper sample disc covering the sampling cuvette of a vapor pressure osmometer (Wescor 5520, Logan, UT, US), previously calibrated using the 290, 1000, and 100 mmol kg^{-1} standards. After each measurement, the osmometer cuvette was rinsed using deionized water. Finally, the resulting osmolality (mosmol kg^{-1}) was converted to osmolarity (MPa) using the following formula: ψs (MPa) = −c (mosmol kg^{-1}) × 2.58 × 10^{-3} [48], and OA was measured as the difference between the ψs at full turgor in control and in water-stressed conditions: ψs (control)—ψs (water stress). The other batch was used to measure RWC. Fresh leaves were weighed (FW) then submerged in distilled water in the Falcon tubes and stored at 4 °C for rehydration (10 h). Rehydrated leaves were wiped thoroughly with blotting paper and weighed (turgid weight: TW). Then, leaves were oven-dried at 65 °C for three days prior to measuring the dry weight (DW). In the end, RWC values were computed as follows: [(FW-DW)/(TW-DW)] × 100 [49] (Figure 1). Leaf rolling (LR) was visually estimated at midday (112 DAS) with a 0 (no leaf rolling) to 9 (all leaves severely rolled) score when the majority of the accessions showed a LR > 5. Finally, the chlorophyll content was assessed (114 DAP) based on Soil-Plant Analysis Development (SPAD) estimates obtained with a non-destructive chlorophyll meter SPAD-502Plus (Konica Minolta Sensing, Inc., Osaka, Japan) as an indicator of leaf chlorophyll content and nitrogen (N) status. At the end of the field trial, plants within the entire two-row plots were cut on 3–4 April 2018 with mechanical harvester (Carter Mfg. Co. equipment, Donalsonville, GA, US), while subsamples of 2–3 plants were collected to evaluate moisture content in order to estimate dry biomass.

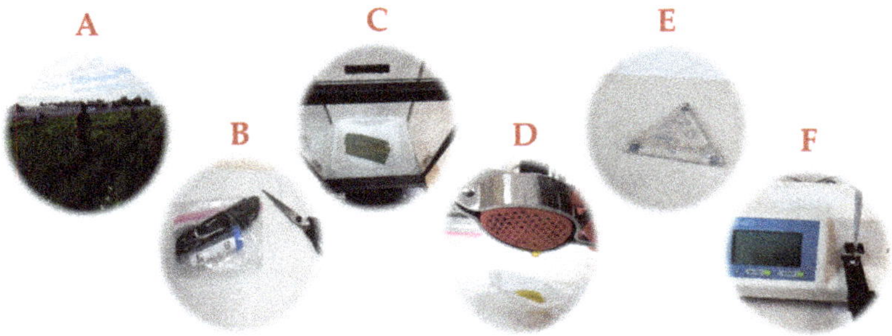

Figure 1. Major "Rehydration method" steps used at the Maricopa Agricultural Center (MAC) to measure osmotic adjustment (OA) and leaf relative water content (RWC): (**A**) Sampling of eight fully expanded homogeneous flag leaves before dawn. (**B**) Stacking the eight leaves and cutting off the tips. The remaining leaf parts (ca. 15 cm long) were cut in the middle to obtain two homogeneous pieces of similar weight, then mixed and inserted in Falcon 50 ml Conical Centrifuge Tubes. (**C**) Weighing of the leaf samples for RWC. (**D**) Collection of leaf cell sap for OA analysis using a garlic press. (**E**) Calibration of the osmometer (Wescor 5520) with sodium chloride solution of increasing concentration. (**F**) Pipetting ca. 10 µl of leaf cell sap onto a paper disc placed on the sampling cuvette of the osmometer (Wescor 5520).

2.4. *Statistical Analysis*

The *lme4* package (*r-project*) and custom R scripts were used to conduct a spatial adjustment analysis considering row and column as random effects, as well as a moving mean of specific size. *R-project* was used to calculate repeatability values (h^2) and Pearson's correlation *r* coefficients among traits. *Minitab 18* software [50] was used to calculate the global percentage of phenotypic variation (global QTL model, R^2%) explained by all QTLs for each trait.

2.5. *SNP Genotyping, Population Structure, and GWAS Model*

Durum panel genomic DNA was extracted using the NucleoSpin®8/96 Plant II Core Kit from Macherey Nagel and sent for SNP genotyping to [51]. The Illumina iSelect 90K wheat SNP assay [52] was used, and genotype calls were acquired as reported in [53]. Markers were assigned on the basis of the tetraploid wheat consensus map reported in [54]. Haploview 4.2 software [55] was used to calculate Linkage Disequilibrium (LD) decay among markers for the A and B genomes, and only Single Nucleotide Polymorphisms (SNPs) with minor allele frequency (MAF) > 0.05 were considered. LD decay pattern based on the consensus genetic distances was inspected considering squared allele frequency correlation (r^2) estimates from all pairwise comparisons among intra-chromosomal SNPs in TASSEL (Trait Analysis by aSSociation, Evolution, and Linkage) *5.2.37*. The Hill and Weir formula [56] was used in R-project to define the confidence interval (CI) for QTLs in accordance with the curve fit and the distance at which LD decays below $r^2 = 0.3$ [57]. Haploview 4.2 tagger function set to $r^2 < 1.0$ was used to calculate a kinship matrix (K) of genetic relationships among individual accessions of the Durum Panel with all non-redundant 7,723 SNPs. Kinship based on Identity-by-State (IBS) among accessions was calculated in TASSEL 5.2.37. In addition, a subset of non-redundant 2,382 SNP markers ($r^2 < 0.5$) was used to evaluate the population structure (Q) in STRUCTURE 2.3.4. software [58] using the corresponding tagger function in Haploview 4.2 software [55]. Numbers of hypothetical subpopulations ranging from k = 2 to 10 were assessed using 50,000 burn-in iterations, followed by 100,000 recorded Markov-Chain iterations, in five independent runs for each k in order to estimate the sampling variance (robustness) of population structure inference. Then, the rate of change in the logarithm of the probability of likelihood (LnP(D)) value between successive k values was considered (Δk statistics) [58], together with the rate of variation (decline) in number of accessions clearly attributed to subpopulations (accessions with Q membership's coefficient ≥ 0.5). Finally, the level of differentiation among subpopulations was measured using the Fixation Index (Fst) among all possible population pairwise combinations [59]. Subsequently, 17,721 SNPs with MAF > 0.05, imputed with LinkImpute (LDkNNi) [60,61], were used for GWAS-Mixed Linear Model [MLM; [62,63] in TASSEL. MLM was specified as follows: $y = X\beta + Zu + e$ [64], where y is the phenotype value, β is the fixed effect due to the marker, and u is a vector of random effects not accounted for by the markers; X and Z are incidence matrices that related y to β and u, while e is the unobserved vector of random residual. In this study, both Kinship matrix (K) and Structure Population (Q) were included as random effects in the model (MLM-Q+K), while flowering time was included as a covariate considering GWAS QQ-plot results (Figures S4 and S5). Then, GWAS p-values and r^2 effects were analyzed, and QTL significance was determined as follows: "highly significant" refers to p-value < 0.0001 and "significant" refers to p-value < 0.001. The QTL confidence interval (CI, in cM) was measured on the basis of the average genetic distance at which LD decayed below r^2 of 0.3 [56], a threshold frequently adopted in GWAS [54,57,65]. Considering a LD of $r^2 = 0.3$, the corresponding inter-marker genetic distance was 3.0 cM [57], and the CI of $\pm$ 3.0 cM based on map positions of QTL tag-SNPs was chosen. Finally, Minitab 18 software [50] was used to calculate the proportion of variance for phenotypic traits explained by selected SNPs.

2.6. Identification of Candidate Genes

The physical position of each QTL was determined by the position of the flanking SNP markers after their alignment on the *Triticum turgidum* ssp. *durum* reference genome of (cv. Svevo) [66]. Genes within the confidence intervals associated with the eight main QTL hotspots were retrieved from the EnsemblPlants database [67], together with their functional annotation and the amino acid sequences of putative proteins. Gene Ontology (GO) term enrichment was determined by comparing the genes included in each QTL to the number of genes annotated in each GO term with g:Profiler web software [68]. Statistical significance of terms for genes in the physical intervals was assessed using the hypergeometric statistic for every term and the g:SCS correction method for multiple testing. Durum wheat GO annotation was retrieved from the Ensembl plant genome database. To identify the most important metabolic pathways associated to eight QTLs, genes within cluster intervals were aligned to KEGG (Kyoto Encyclopedia of Genes and Genomes) database using Reference [69]. Genes annotated within the intervals were compared with their orthologs from *Triticum aestivum* (cv. Chinese Spring; IWGSC RefSeq v1.0) [70]. Identification of candidate genes was further supported by a knowledge network (proteins, biological pathways, phenotypes, and publications) created using the KnetMiner program, using the bread wheat orthologs [71], and by the analysis of temporal and spatial gene expression at the Wheat Expression Browser and ePLANT databases [72], as of September 2021.

3. Results

3.1. Population Structure

The Durum Panel showed a clear population genetic structure with an optimal number of eight (k = 8) subpopulations on the basis of pairwise comparisons among and within subgroups with 155 accessions (62.5%) clearly grouped into one of the eight main gene pools at a Q membership coefficient ≥ 0.5, while the remaining 93 were considered as admixed. The Fixation Index (Fst) and Neighbor Joining tree [73] highlighted a high genetic diversity between the old Italian accessions (S1) and the modern French, North American, Canadian and Australian cultivars (S8), while a considerable admixture among subgroups characterized the ICARDA, CIMMYT, and Italian groups. Subgroups details are reported in Table S2.

3.2. Phenotypic Analysis

Osmotic potential in well-watered (control) conditions (ψs-c; $h^2 = 0.57$) ranged from −1.44 to −0.74 MPa, with an average of −1.13 MPa, while, in water-stressed conditions, (ψs-s; $h^2 = 0.58$) ranged from −2.63 to −1.56 MPa, with an average of −2.00 MPa. The difference between osmotic potential measured at full turgor in well-watered (control: ψs-c) and in water stressed (ψs-s) conditions was considered to compute OA ($h^2 = 0.72$), which showed a normal distribution and ranged from 0.38 to 1.48 MPa, with an average of 0.95 MPa. Figure S3 reports flowering time distribution, while Figures S6 and S7 report box plots and histogram distributions, for OA, ψs-c and ψs-s, RWC-c and RWC-s, LR, and SPAD. RWC-c ($h^2 = 0.29$) ranged from 89.9 to 101.3%, with an average of 95.7%, while RWC-s ($h^2 = 0.78$) ranged from 45.2 to 76.9%, with an average of 62.2% (Figures S6 and S7). Leaf rolling (LR; $h^2 = 0.84$) at 112 DAP ranged from 2.86 to 9.60, with an average of 6.13 (Table 1), while leaf chlorophyll content (SPAD; $h^2 = 0.76$) at 114 DAP ranged from 31.9 to 48.8, with an average of 42.0. The Pearson's correlation coefficient was positive between OA and RWC-s (r = 0.78), while a negative association was found between OA and LR (r = −0.25), RWC-s and ψs-s (r = −0.49), and RWC-s and LR (r = −0.30) (Table 2, Figure 2).

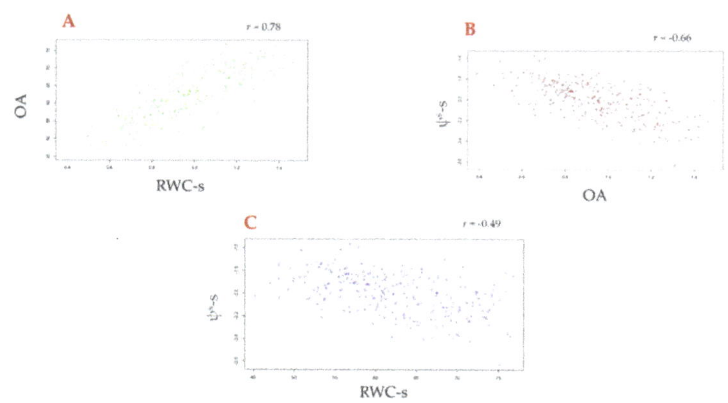

Figure 2. Scatter plot showing Pearson correlation between (**A**) osmotic adjustment (OA) and relative water content under drought (RWC-s), (**B**) OA and osmotic potential under drought (ψs-s), and (**C**) ψs-s and RWC-s. R-project [74].

Table 1. Summary statistics and heritability (%) for osmotic adjustment (OA), osmotic potential under drought stress (ψs-s), osmotic potential under well-watered conditions (ψs-c), relative water content under drought stress (RWC-s), relative water content under well-watered conditions (RWC-c), leaf rolling (LR), and chlorophyll content (SPAD) in a panel of 248 durum wheat elite advanced lines and cultivars from worldwide.

Trait	Min.	Max.	Average	St. Dev.	h^2
OA	0.38 [a]	1.48	0.95	0.22	0.76
ψs-s	−2.63 [a]	−1.56	−2.00	0.18	0.58
ψs-c	−1.45 [a]	−0.75	−1.13	0.12	0.57
RWC-s	45.21 [b]	76.88	62.10	7.11	0.78
RWC-c	89.9 [b]	100.0	95.7	1.56	0.29
LR	2.86	9.60	6.13	1.52	0.84
SPAD	31.9	48.8	42.0	3.20	0.76

[a] megapascal (MPa); [b] % RWC.

Table 2. Pearson's correlation plot among osmotic adjustment (OA), osmotic potential (ψ^s) under full (--c) and deficit irrigation (--s), relative water content (RWC) under full (--c) and deficit irrigation (--s), leaf rolling (LR), and chlorophyll content (SPAD).

Trait	OA	ψ^s-s	ψs-c	RWC-s	RWC-c	LR	SPAD
OA	1	−0.66 ***	0.33 ***	0.78 ***	0.11	−0.25 ***	0.04
ψ^s-s	-	1	0.30 ***	−0.49 ***	−0.16 *	0.13 *	−0.06
ψ^s-c	-	-	1	−0.08	0.02	−0.03	0.03
RWC-s	-	-	-	1	0.13 *	−0.30 ***	−0.02
RWC-c	-	-	-	–	1	−0.08	0.20 **
LR	-	-	-	–	–	1	−0.01
SPAD	-	-	-	–	–	–	1

*** *p*-value < 0.0001, ** 0.0001 < *p*-value < 0.001, * 0.001 < *p*-value < 0.01. *R-project* [74].

3.3. Genetic Analysis

The rate of linkage disequilibrium (LD) decay of the 248 durum wheat elite accessions of the Durum Panel is reported in Figure 3. The average QTL confidence interval (CI) was determined on the basis of the average genetic distance at which LD decayed below r^2 of 0.3 multiplied by 2, corresponding to 2.12 cM (CI = ± 1.06 cM from the QTL tagSNP). Fifteen flowering time QTLs were identified and are reported in Table S4. Major QTLs for flowering time included those on chromosome arms 2AS (*QFT.ubo-2A.1* and *QFT.ubo-2A.2*), on 4AS (*QFT.ubo-2A.1*) and 6BL (*QFT.ubo-6B.1*). Among others, *Ppd-A1* was clearly identified by *QFT.ubo-2A.2* = IWA2526. Using FT as covariate, GWAS analysis (MLM-Q+K) identified 70 significant QTLs (log *p*-value > 3.00) for ψs-c, ψs-s, OA, RWC-s, LR, and/or SPAD, organized into QTL clusters. A larger portion of ψs QTLs were detected under drought (62.5%) as compared to well-watered conditions (37.5%). In particular, two major ψs-s QTLs were observed on chromosomes 1B (*Qψsc.ubo-1B.2*) and 6A (*Qψsc.ubo-6A.1*), with a log *p*-value of 4.68 and 6.04, and with R^2 of 5.84 and 7.88%, respectively. A total of 15 OA QTLs were mapped on chromosome arms 1AL, 1BL, 2AS, 2AL, 2BL, 4AL, 4BS, 5AL, 6AL, 6BL, and 7BS, with the three major ones being those on chromosomes 2B (*QOA.ubo-2B.2*) and 6A (*QOA.ubo-6A.1*and *QOA.ubo-6A.2*) with a log *p*-value of 4.13, 4.01, and 4.45, and R^2 of 4.37, 4.23, and 4.78%, respectively (Table 3). Adopting flowering time as a covariate effectively removed the effects associated to FT on OA, except for *QFT.ubo-2A.2* = *QOA.ubo-2A.1* and *QFT.ubo-6B.1* = *QOA.ubo-6B.1*. Nine RWC-s loci were mapped on five chromosome arms (1BS, 2AS, 4AL, 6AL, and 6BL), and two major QTLs were observed on 4AL (*QRWCs.ubo-4A.2* and *QRWCs.ubo-4A.3*), with a log *p*-value of 4.83 and 4.27, and R^2 of 3.95 and 3.84, respectively (Table 3).

Table 3. Significant GWAS-QTLs for osmotic adjustment (OA) and RWC-s (p-value < 0.001). QTL intervals were defined based on a confidence interval of ± 3.0 cM from the map position of the QTL tagging-SNPs. The rows with grey background indicate the QTLs affecting both OA and RWC-s. Position and peak marker of each QTL region are based on the tetraploid wheat consensus map [54].

QTL	Marker	Chr.	Position (bp)	Position (cM)	Log p-Value	R^2	Allele	Effect
Osmotic Adjustment (OA)								
QOA.ubo-1A.1	IWB27332	1A	508851821	88.3	3.07	3.10	C/T	−1.550
QOA.ubo-1B.1	IWB65251	1B	582533506	93.3	3.17	3.19	C/T	−0.097
QOA.ubo-2A.1	IWB34575	2A	36292525	46.6	3.11	3.11	A/G	0.128
QOA.ubo-2A.2	IWB39807	2A	768563743	206.8	3.08	3.31	C/T	0.090
QOA.ubo-2B.1	IWA2318	2B	656566640	133.0	3.89	4.07	C/T	−0.117
QOA.ubo-2B.2	wPt-0049	2B	781813758	185.8	4.13	4.37	A/T	0.141
QOA.ubo-4A.1	IWB38918	4A	644103100	139.7	3.11	3.12	A/G	−0.165
QOA.ubo-4A.2	IWB34029	4A	717060721	161.7	3.88	4.06	C/T	1.252
QOA.ubo-4B.1	IWB72203	4B	26616372	28.8	3.00	2.48	A/C	0.076
QOA.ubo-5A.1	IWB50381	5A	640718417	198.8	3.24	3.28	A/G	0.159
QOA.ubo-6A.1	wPt-2014	6A	505253000	91.2	4.01	4.23	A/T	0.162
QOA.ubo-6A.2	IWB70454	6A	596626025	117.1	4.45	4.78	C/T	0.181
QOA.ubo-6B.1	IWB33826	6B	437229717	75.3	3.12	3.13	A/G	−0.105
QOA.ubo-6B.2	IWB71722	6B	644758469	114.3	3.21	3.24	A/G	−0.086
QOA.ubo-7B.1	wPt-3147	7B	3571960	3.7	3.13	3.14	A/T	−0.095
Relative Water Content under water stress (RWC-s)								
QTL	**Marker**	**Chr.**	**Position (bp)**	**Position (cM)**	**Log p-Value**	**R^2**	**Allele**	**Effect**
QRWCs.ubo-1B.1	IWB461	1B	628218198	45.3	3.70	3.24	C/T	−4.29
QRWCs.ubo-2A.1	IWB22184	2A	7224905	9.4	3.33	2.86	A/G	−4.25
QRWCs.ubo-4A.1	IWB66212	4A	687621664	140.7	3.02	2.53	A/C	2.74
QRWCs.ubo-4A.2	IWB56811	4A	697055522	147.2	4.83	3.95	C/T	−5.51
QRWCs.ubo-4A.3	IWB55093	4A	707177021	156.9	4.27	3.84	A/G	5.24
QRWCs.ubo-4A.4	IWA3449	4A	720085814	161.7	3.90	3.45	C/T	4.66
QRWCs.ubo-6A.1	IWA4603	6A	597277894	117.7	3.39	2.92	A/G	3.15
QRWCs.ubo-6B.1	IWA7962	6B	454884102	78.8	3.04	2.56	A/G	−6.92
QRWCs.ubo-6B.2	IWB71722	6B	644758469	114.3	3.00	2.44	A/G	−2.46

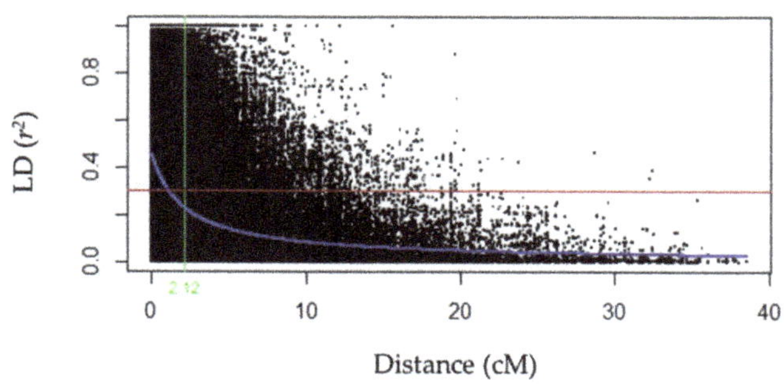

Figure 3. The rate of linkage disequilibrium (LD) decay of the 248 durum wheat elite accessions (Durum Panel). The Hill and Weir formula [56] was used to describe the LD decay of r^2. The LD among SNPs in the Durum Panel was estimated using Haploview 4.2 [54]. The blue curve represents the model fit to LD decay (non-linear regression of r^2 on distance). A confidence interval of 2.12 cM for the QTLs is shown when LD (r^2) is 0.3 (red line). R-project [74].

Nine LR loci were mapped on seven chromosome arms (1BL, 2AL, 3AS, 3AL, 3BL, 6AS, and 6BL), and one major QTL was observed on chromosome arm 2AL (*QLR.ubo-2A.1*), with a log p-value of 4.08, and R^2 of 4.92. As to SPAD, 21 QTLs were mapped on 12 chromosome arms (1AL, 2AL, 2BS, 3BS, 3BL, 4AL, 4BS, 5AL, 6BS, 6BL, 7AL, and 7BS), and three major QTLs were observed on chromosome arms 1AL (*QSPAD.ubo-1A.1*), 4BS (*QSPAD.ubo-4B.1*), and 5AL (*QSPAD.ubo-5A.1*), with a log *p*-value of 6.08, 6.61, and 6.81, and R^2 of 7.87, 8.69, and 8.99, respectively. The global R^2 of the multiple QTL model was 58.0% for ψs-c, 56.5% for ψs-s, 63.6% for OA, 25.7% for RWC-s, 44.1% for LR, and 50.2% for SPAD. On the basis of their concurrent allelic effects on OA and other related traits, eight major QTL clusters were identified: (i) *DR_QTL_cluster_1#* (RWC-s and ψs-s) on 1B, (ii) *DR_QTL_cluster_2#* (OA and ψs-c) on 2B, (iii) DR_QTL_cluster_3# (OA and RWC-s) on 4A, (iv) *DR_QTL_cluster_4#* (OA and SPAD) on 5A, (v) *DR_QTL_cluster_5#* (OA and RWC-s) on 6A, (vi) *DR_QTL_cluster_6#* (OA and RWC-s) on 6B, (vii) *DR_QTL_cluster_7#* (OA, RWC-s, and SPAD) on 6B, and (viii) *DR_QTL_cluster_8#* (OA and SPAD) on 7B (Table 4 and Figure 4).

Table 4. List of GWAS-QTL clusters identified in the Durum Panel and significantly associated with osmotic adjustment (OA), RWC under drought stress (RWC-s), osmotic potential under well-watered conditions (ψs-c) and drought stress (ψs-s), and leaf rolling (LR). The co-localization with previously known normalized difference vegetation index (NDVI), chlorophyll content (SPAD), root growth angle (RGA), thousand kernel weight (TKW), and grain yield (GY) QTLs is reported.

QTL Cluster	Chr.	Position (cM)	Trait	QTLs from Literature
DR_QTL_cluster_1#	1B	45.3	RWC-s, ψ^s-s	UAV-Red-Edge-NDVI [a], TKW [c], GY [d]
DR_QTL_cluster_2#	2B	185.3	OA, ψ^s-c	Tractor-NDVI [a], TKW [c], NDVI [c], Chlorophyll content (SPAD) [c], GY [e],
DR_QTL_cluster_3#	4A	161.7	OA, RWC-s	Tractor-NDVI [a], UAV-Red-Edge NDVI [a], Chlorophyll content (SPAD) [a], GY[f]
DR_QTL_cluster_4#	5A	198.8	OA, SPAD	
DR_QTL_cluster_5#	6A	117.1	OA, RWC-s	UAV-Red-Edge-NDVI [a], RGA/TKW/GY [b],
DR_QTL_cluster_6#	6B	75.3	OA, RWC-s	
DR_QTL_cluster_7#	6B	114.3	OA, RWC-s	GY [f]
DR_QTL_cluster_8#	7B	3.7	OA, SPAD	

[a] Reference [60], [b] Reference [75], [c] Reference [76], [d] Reference [41], [e] Reference [77] and [f] Reference [78].

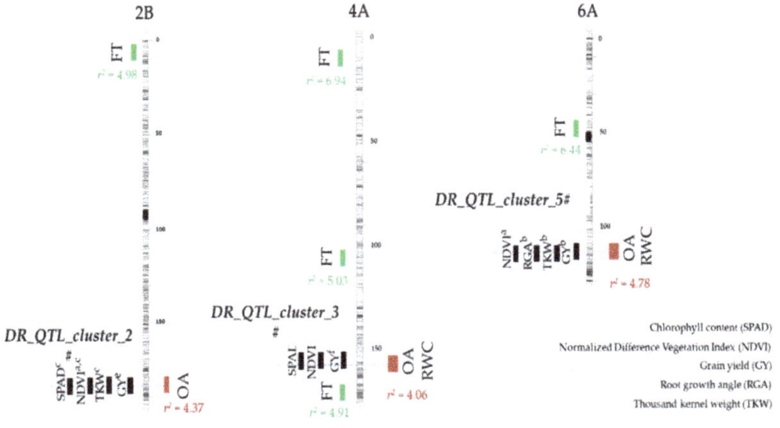

Figure 4. Chromosome position on the durum consensus map [54] and R^2 of the QTL hotspots for osmotic adjustment (OA) and/or relative water content (RWC) on chromosome arms 2BL (*DR_QTL_cluster_2#*), 4AL (*DR_QTL_cluster_3#*), and 6AL (*DR_QTL_cluster_5#*), as well as their overlaps with QTLs previously reported in literature: [a] Reference [60], [b] Reference [75], [c] Reference [76], [d] Reference [41], [e] Reference [77], and [f] Reference [78]. The QTLs for flowering time (FT) are shown in green.

These regions were selected for a more detailed analysis and comparative analysis with previously reported results on grain yield in both durum and bread wheat, as discussed hereafter. Durum wheat genes within the confidence intervals of the eight selected QTL hotspots were retrieved from EnsemblPlants database, together with their functional annotation (Table S5). Gene Ontology (GO) enrichment analysis (Figure 5) and KEGG pathways reconstruction (Table 5) were used to further functionally characterize the genes included in the eight QTL clusters. In parallel, the bread wheat orthologous genes were identified, as well (see Table S5), for the comparison of genes annotated in the syntenic regions of the two wheat species. KnetMiner [71] knowledge networks, constructed using bread wheat orthologs, were integrated to identify putative candidate gene(s) within the confidence interval of each QTL. The confidence interval of *DR_QTL_cluster_1#* on chromosome 1B corresponds to a physical interval of approximately 7.0 Mb with 46 high-confidence (HC) genes in the Svevo genome (Table S5). Among the genes included in the interval, no GO terms were significantly enriched, while KEGG mapping annotated 21 genes to nine functional categories (Table 5). The two most notable candidates in the confidence interval are *TRITD1Bv1G127690*, which encodes a transmembrane protein with transporter activity homologous of the *Arabidopsis* Major facilitator superfamily MEE15, and *TRITD1Bv1G126800*, which encodes for a seven transmembrane MLO-like protein. The confidence interval of the *DR_QTL_cluster_2#* on chromosome 2B corresponds to a 3.2 Mb interval, which contains 63 high-confidence (HC) genes. GO terms associated to "stress response" and "antioxidant activity" were enriched among these genes (Figure 5), due to the presence of 10 peroxidase encoding genes in the QTL interval. KEGG mapping confirmed their annotation in secondary metabolism pathway (Table 5), acting in the phenylpropanoid biosynthesis. In addition to these peroxidase encoding genes that could act in drought-stress response and adaptation, the two most notable candidates in the interval are *TRITD2Bv1G263980*, encoding for a protein kinase and *TRITD2Bv1G264060*, which encodes a DDB1- and CUL4-associated factor-like protein 1. Both genes are located at the confidence interval boundaries and could be functionally related to OA-related aspects (Table S5). Notably, the comparison of the syntenic physical region in *T. aestivum* Chinese Spring evidenced several gaps in the corresponding Svevo region: for 30 HC genes annotated in the Chinese Spring syntenic region, their Svevo orthologs are indeed included in unmapped scaffolds and could, thus, be included among the list of putative candidates. Even if none of these 30 missing genes are drought-stress responsive or OA-related functional annotation, the presence of these gaps in *T. durum* genome assembly and gene annotation could clearly impair candidate gene discovery. The confidence interval of *DR_QTL_cluster_3#* on chromosome 4A corresponds to approximately 3.5 Mb with 33 HC genes (Table S5), mapped to seven KEGG functional categories (Table 5) and enriched in manganese transport-related GO terms (Figure 5). *TRITD4Av1G256080* and *TRITD4Av1G256120* indeed encode for membrane protein of ER body-like proteins, likely working as metal transporters. Additional genes annotated in the QTL interval include *TRITD4Av1G255460*, *TRITD4Av1G255480*, and *TRITD4Av1G255490*, encoding for three Glutathione S-transferases, *TRITD4Av1G255990*, encoding for an RNA-binding family protein with RRM/RBD/RNP motifs, and *TRITD4Av1G256200*, which encodes for a 5′-methylthioadenosine/S-adenosylhomocysteine nucleosidase. In addition, for this QTL hotspot, a gap between *T. durum* and *T. aestivum* chromosome assemblies was found, hence impairing a more proper identification of candidate genes. The confidence interval of *DR_QTL_cluster_4#* on chromosome 5A spans a physical interval of approximately 4.0 Mb, which contains 39 HC genes in the Svevo genome (Table S5). Lyase activity was the unique GO term enriched among the genes within the QTL (Figure 5), while KEGG mapping assigned 17 of them to eight functional groups (Table 5). Interestingly, *TRITD5Av1G246840*, encoding for a putative Phospholipase D potentially implicated in multiple plant stress responses, and *TRITD5Av1G247330*, a putative Lipoxygenase required for jasmonic acid accumulation, were mapped to the "environmental information processing" category. In addition, *TRITD5Av1G247220*, encoding for an UDP-N-acetylglucosamine (UAA) transporter family protein, could be involved in the osmosensory signaling pathway and cell wall organization. The confidence interval of the *DR_QTL_cluster_5#* on chromosome 6A spans approximately 2.0 Mb, with 32 HC genes (Table S5). While the "manganese binding" function was the only GO-enriched term (Figure 5), six of the genes included in the region encode for putative Cinnamoyl CoA reductases and were mapped to the "secondary metabolism"/"phenylpropanoid biosynthesis" pathways (Table 5). The two most notable candidates in the interval are *TRITD6Av1G217800* and *TRITD6Av1G218080*, which encode for two F-box protein PP2, and *TRITD6Av1G217670*, encoding for DREB1, a CRT-binding factor. The confidence interval of *DR_QTL_cluster_6#* on chromosome 6B spans a 5.9 Mb interval, which contains 40 HC genes in the Svevo genome (Table S5). No GO term was found significantly enriched, while KEGG mapping annotated 20 genes to 12 different functional categories (Table 5). In particular, three genes encoding for MYB transcription factors potentially regulating different aspects of stress response were assigned to the "genetic information processing" category. Moreover, *TRITD6Bv1G133070*, orthologs of the CCT motif-containing response regulator protein coding gene of *Arabidopsis*, appears as an even more interesting candidate gene at this locus. The confidence interval of the *DR_QTL_cluster_7#*, located on chromosome 6B, spans approximately 1.2 Mb, with 14 HC genes, including *TRITD6Bv1G207930*, encoding for a protein kinase family protein/WD-40 repeat family protein 3. Both GO-enrichment and KEGG mapping did not identify other genes in this QTL interval (Table 5 and Figure 5). Finally, the confidence interval of *DR_QTL_cluster_8#* on chromosome 7B corresponds to a physical interval of approximately 1.5 Mb, which contains 20 HC genes in the Svevo genome, lacking GO enrichment or predominant KEGG annotations (Table 5 and Figure 5), but including *TRITD7Bv1G002000*, a gene encoding for a photosynthetic NDH subcomplex B3 (Table S5).

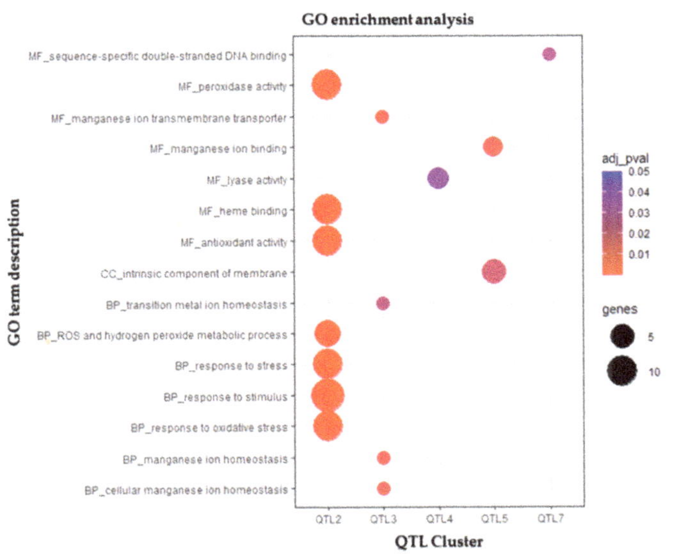

Figure 5. Gene Ontology (GO) enrichment analysis. Dot plot shows GO terms of biological processes (BP), molecular functions (MF) and cellular compartment (CC) identified using g:Profiler [68] to be enriched (adjusted p-value < 0.05) among the genes included in each QTL interval. The size of the dots is based on gene count enriched in the pathway, and the color of the dots represents the adjusted p-values.

Table 5. Summary of KEGG functional pathways mapped for genes included in each QTL interval, grouped based on metabolic activities.

Functional Category	QTL1	QTL2	QTL3	QTL4	QTL5	QTL6	QTL7	QTL8
Carbohydrate metabolism	2			1		1		
Energy metabolism						1	1	
Lipid metabolism	2		2	3		1		1
Nucleotide metabolism						1		
Amino acid metabolism	1	1	4	3	1		1	
Glycan biosynthesis and metabolism						1		
Metabolism of cofactors and vitamins				3				1
Biosynthesis of other secondary metabolites		10			6			
Genetic information processing	1		2			2	1	1
Environmental information processing	1	2	2	2		2	1	
Organismal systems	1					2		
Protein families: metabolism	5	2	1			1	1	
Protein families: genetic information processing	6	2		1	1	5		
Protein families: signaling and cellular processes	2	6	1	1	4	2	1	1
Unclassified: metabolism			1	3		1		
KEGG mapped genes	**21**	**23**	**13**	**17**	**12**	**20**	**6**	**4**
Total genes	**46**	**63**	**33**	**39**	**32**	**40**	**14**	**20**

4. Discussion

A number of authors have proposed OA as an important adaptive mechanism to support higher crop yield under stressful environmental conditions, as reviewed in [24,79]. Notably, grain yield differences have been shown to be positively correlated to OA in cereals [21,30,80–83], hence indicating OA as a valuable proxy to predict grain production [24]. This notwithstanding, the QTLome regulating OA in wheat and other crops remains basically unknown, the main reason being the difficulty to adequately screen the large number (>200) of (i) RILs of the mapping populations and/or (ii) accessions of GWAS mapping panels required for a meaningful QTL discovery. In field conditions, the collection of leaves and their processing must be completed rapidly to minimize the bias introduced by the time of sample collection in an adequately large number of genotypes, an essential prerequisite for identifying and accurately mapping QTLs [19,34]. The QTLome dissection of OA in cereals was first attempted in rice [37,38] and barley [39,40]. In bread wheat, Reference [84] mapped an osmoregulation gene locus [85] located on the short arm on chromosome 7A. However, OA and osmoregulation differ and have different functional meanings. While OA refers to a lowering of osmotic potential (ψs) due to an accumulation of osmolytes in response to a water deficit, osmoregulation refers to the ψs regulation by the addition/removal of osmolytes until the intracellular potential is approximately equal to that of the medium surrounding the cell [20]. The gene described by Morgan regulates turgor pressure and water content by osmotic adjustment [84,85], hence the term osmoregulation. In this study, OA was measured according to the "rehydration method" [23,86]. Although this method was criticized by [85], other authors consider it optimal for screening large populations [20,26,86–89] in view of its merits in terms of labor and cost-effectiveness as compared to the other methods [23]. In our experience, the rehydration of the leaf samples greatly facilitated (i) the cell sap extraction especially in samples collected in water-stressed plants and (ii) the OA screening of the 248 diverse accessions of the Durum Panel. Collectively, this resulted in high OA repeatability (h^2) and a positive and negative correlation with RWC-s and LR, respectively. The positive correlation between OA and RWC clearly indicates an active physiological role of the former to maintain a more favorable water status of the plant, playing a key role for avoiding and mitigating the negative effects of water loss under drought. Overall, our results validate the effectiveness of the "Rehydration method" as an ideal option for handling the large number of samples required for the genetic dissection of the OA QTLome.

4.1. GWAS Mapping and Comparative Analysis with Previous QTL Studies in Durum Wheat

Overall, eight major QTL hotspots were detected on the long arm of chromosomes, 1BL, 2BL, 4AL, 5AL, 6AL, 6BL, and 7BS. The use of flowering time (FT) as covariate for the GWAS analysis reduced the bias caused by the photoperiod-response (*Ppd*) locus and other loci that affect FT, hence allowing a more accurate estimate of QTL effects on a per se basis rather than due to effects related to variability in phenology. Notably, none of the eight major QTL hotspots evidenced by GWAS analysis overlapped with the osmoregulation gene locus described by Reference [85] in bread wheat. *DR_QTL_cluster_1#*, *DR_QTL_cluster_2#*, and *DR_QTL_cluster_5#* overlapped with Normalized Difference Vegetation Index (NDVI) loci identified in 2017 on the same Durum Panel under similar drought conditions using Unmanned Aerial Vehicles (UAV-Sequoia and UAV-Red-Edge), as well as ground-based platforms [60]. Additionally, *DR_QTL_cluster_3#* and *DR_QTL_cluster_5#* overlapped with chlorophyll content (SPAD) loci under drought described in Reference [60]. Both NDVI and SPAD have long been recognized for their ability to estimate crop biomass and predict grain yield [90–94]. *DR_QTL_cluster_2#* overlapped with grain yield, thousand-kernel weight, and NDVI loci previously reported in a durum wheat elite population tested in contrasting thermo-pluviometric conditions [76]. *DR_QTL_cluster_5#* co-mapped with *QRga.ubo-6A.2*, one of the most important loci for root growth angle in durum wheat [75], with thousand-grain weight, particularly under low water availability environments, as well as with grain yield, in the 183 elite accessions of the Durum Panel that were previously evaluated in 15 field trails under a wide range of water regimes [41]. *DR_QTL_cluster_3#* and *DR_QTL_cluster_7#* co-mapped with a major grain yield QTL reported by Reference [78] in an RIL population developed from the hexaploid wheat cross between Chinese Spring × SQ1 evaluated across a broad combination of 24 site × treatment × year combinations. The concurrent effects on grain-yield related traits reported herein fully support the conclusions of Reference [24] on OA being a valuable proxy with a positive effect on crop yield under water-limited conditions and not merely for survival under severe drought. These QTL hotspots will further enhance drought tolerance in durum wheat.

4.2. Candidate Genes

By combining the physical confidence interval position of the QTL hotspots, functional prediction of annotated genes, and biological data mining, we investigated candidate genes from selected QTLs involved in OA and/or drought resistance. Putative drought candidate genes encoding for proteins involved in drought stress responses, as well as grain development, were mapped within the eight major selected QTL hotspots, even if some highlighted gaps in the Svevo genome assembly could hinder candidate gene identification at two QTL hotspots. Among the identified candidates, the seven transmembrane MLO-like protein (*TRITD1Bv1G126800; DR_QTL_cluster_1#*) was shown to act in drought and salt stress responses through signaling of the phytohormone abscisic acid (ABA) [95], with biological knowledge networks analysis strongly supporting its role in oxidative stresses, salt, and drought tolerance (Figure 6A). Similarly, the phospholipase D (PLD) coding gene (*DR_QTL_cluster_4#*) is also involved in ABA responses [96]. Interestingly, genes associated with manganese transport and binding were identified in both *DR_QTL_cluster_3#* and *DR_QTL_cluster_5#*. Moreover, exogenous application of Mn was recently shown to reduce the negative effects caused by drought, harsh temperature, and salinity, increasing ROS detoxification and secondary metabolism [97,98]. Despite the clear enrichment for stress response-related GO terms among genes at *DR_QTL_cluster_2#*, the identification of 10 tandemly duplicated peroxidase encoding genes acting in the phenylpropanoid biosynthetic make the construction of a knowledge network and putative candidate gene more complex and less reliable. Conversely, of considerable interest is, instead, the dehydration-responsive element-binding protein DREB (*DR_QTL_cluster_5#*; Figure 6B), that belongs to a family of plant-specific transcription factors that can specifically bind to DRE/CRT elements in the response to abiotic stresses, such as drought, salt, and low temperature [99,100], reviewed in [101]. In addition, *TRITD6Bv1G133070* (*DR_QTL_cluster_6#*), orthologs of the *Arabidopsis* CCT motif-containing response regulator protein, was shown to be involved in both the generation of circadian rhythms and long-term drought adaptation [102]. Finally, *TRITD7Bv1G002000* (*DR_QTL_cluster_8#*, affecting both OA and SPAD) encodes for chloroplast NAD(P)H dehydrogenase complex, involved in cyclic electron flow around photosystem I to produce ATP [103].

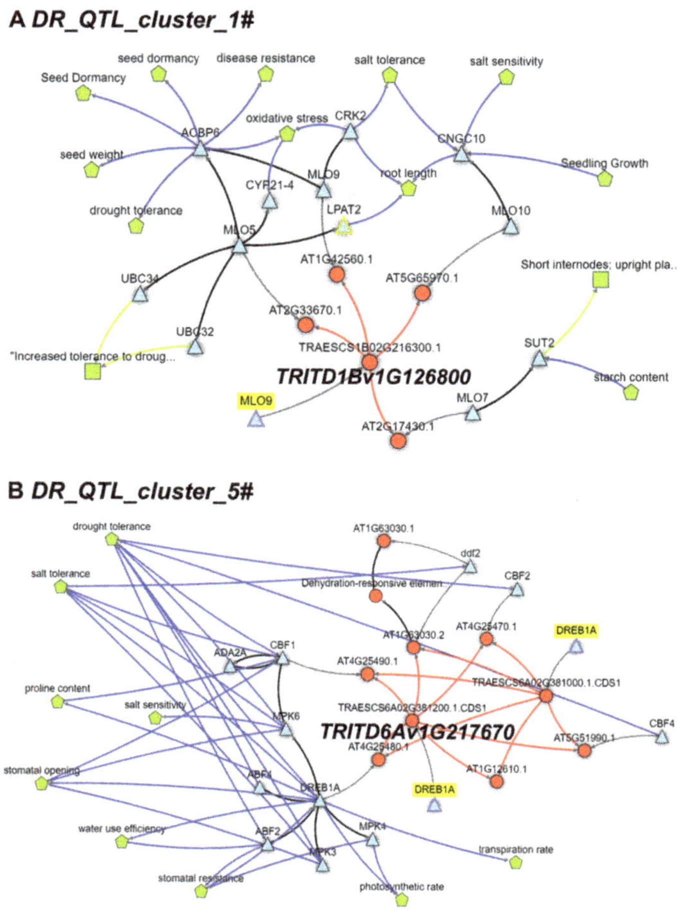

Figure 6. KnetMiner network views displaying knowledge networks of selected candidate genes Table S5. *TRITD1Bv1G126800* (**A**), encoding for a seven transmembrane MLO-like protein (*DR_QTL_cluster_1#*), and *TRITD6Av1G217670* (**B**), which encodes for a dehydration-responsive element-binding protein DREB (*DR_QTL_cluster_5#*). Networks were constructed using the *Triticum aestivum* orthologous genes *TraesCS1B02G216300* and *TraesCS6A02G381200*, respectively, and can be accessed using the following links: https://knetminer.com/beta/knetspace/network/528cbd3a-52d5-40b5-91be-f59323db55a3 (accessed on 1 April 2021) and https://knetminer.com/beta/knetspace/network/44b31582-bfaa-495f-9272-87a4f06d40a6 (accessed on 1 April 2021).

5. Conclusions

This study is the first to report QTLs for OA via GWAS mapping in wheat. From a methodological standpoint, our results support the validity of the "Rehydration method" as the fastest and most effective protocol for large-scale screening of OA under well-watered and drought conditions. The genetic variants within the Durum Panel evaluated herein allowed for the detection of significant loci for OA, ψs, RWC, LR, and SPAD, with eight multiple concurrent QTL hotspots, all unrelated to phenology, hence being more valuable from a breeding standpoint. Importantly, five of these clusters (*DR_QTL_cluster_1#,DR_QTL_cluster_2#, DR_QTL_cluster_3#, DR_QTL_cluster_5#*, and *DR_QTL _cluster _7#*) co-located with QTLs for grain yield and/or grain yield-related traits described in previous multi-environmental studies and, in one case (*DR_QTL_cluster_5#*), co-located with a major QTL controlling root growth angle which has been demonstrated to play a relevant role in maintaining the water access in deep soils during wheat terminal drought [104]. The candidate genes identified within the confidence intervals of selected drought response-specific QTL hotspots provide useful markers for future breeding schemes. Our results support the role of OA as an important drought-stress adaptive mechanism with beneficial effects on the plant water status in durum wheat.

Supplementary Materials: The following are available online at https://www.mdpi.com/article/10.3390/genes13020293/s1, Figure S1: field trial experiment, Figure S2: volumetric water content analysis, Figure S3: Zadoks growth stage intervals & flowering time, Figure S4: Q-Q (quantile-quantile) of the GWAS analysis for osmotic adjustment (OA), Figure S5: Manhattan plot of the GWAS analysis for osmotic adjustment (OA), Figure S6: box plots for (A) relative water content (RWC) and (B) osmotic potential (ψs) under full and deficit irrigation, Figure S7: histograms, Table S1: list of the 248 durum wheat accessions, Table S2: genetic population structure of the Durum Panel through model-based clustering in STRUCTURE 2.3.4 software, Table S3: growing degree days, Table S4: list of QTLs for flowering time, Table S5: candidate genes.

Author Contributions: E.B., J.W.W., M.M., M.N., M.T., N.S., R.T., R.W. and T.M. designed the experiment. M.N., E.B., J.W.W. and R.W. managed the field experiment, G.E.C., E.L.G. and M.N. collected the leaf samples in the field-grown durum panel under well-irrigated and water-stressed conditions. G.E.C. and E.L.G. conducted the laboratory measurements. G.E.C., E.L.G. and M.M. produced the genotypic data. C.F. and M.M. conducted the candidate gene analysis. G.E.C., C.F., M.M. and R.T. analyzed the data, interpreted the results, and developed the first draft. All authors have read and agreed to the published version of the manuscript.

Funding: This study is part of the TERRA REF experiment, funded by the Advanced Research Projects Agency-Energy (ARPA-E), U.S. Department of Energy, under Award Number DE-AR0000594. This study was supported in part by USDA-ARS agreement no. 58-2020-6-019 from the USDA Agricultural Research Service.

Institutional Review Board Statement: Only Plants, No animals or humans were tested, so it can be eliminated.

Informed Consent Statement: Only Plants, No animals or humans were tested, so it can be eliminated.

Data Availability Statement: Not applicable.

Acknowledgments: This study is the result of the collaborative project among the Department of Agricultural and Food Sciences (DISTAL) of the University of Bologna (Italy), Maricopa Agricultural Center (MAC) of the University of Arizona (USA), and the U.S. Arid-Land Agricultural Research Center of USDA ARS (USA). The views and opinions of authors expressed herein do not necessarily state or reflect those of the United States Government or any agency thereof. The authors are thankful for all the support and technical assistance provided by John T. Heun, Pedro Andrade-Sanchez, and Zbigniew S. Kolber. The authors are grateful to the late Abraham Blum for his encouragement to pursue this study and for his valuable suggestions.

Conflicts of Interest: The authors declare that the research was conducted in the absence of any commercial or financial relationships that could be construed as a potential conflict of interest.

Abbreviations

CI, confidence interval; DAS, days after sowing; DW, dry weight; Fst, fixation index; FT, flowering time; FW, fresh weight; GDD, growing degree days; GO, Gene Ontology; GWAS, genome-wide association study; GY, grain yield; K, kinship matrix; LD, linkage disequilibrium; LR, leaf rolling; MAF, minor allele frequency; MAS, marker-assisted selection; MLM, mixed linear model; NDVI, normalized difference vegetation index; OA, osmotic adjustment; PPFD, photosynthetic photon flux density; Q, structure population; QTL, quantitative trait locus, RGA, root growth angle; RIL, recombinant inbred lines; RWC, relative water content; SNP, single nucleotide polymorphism; SPAD, soil plant analysis development; TDR, time-domain re-flectometry; TKW, thousand kernel weight; TW, turgid weight; VPD, vapor pressure deficit; VWC, volumetric water content; ψs, osmotic potential; θPWP, permanent wilting point; θFC, field capacity.

References

1. Wang, W.-S.; Pan, Y.-J.; Zhao, X.-Q.; Dwivedi, D.; Zhu, L.-H.; Ali, J.; Fu, B.-Y.; Li, Z.-K. Drought-induced site-specific DNA methylation and its association with drought tolerance in rice (*Oryza sativa* L.). *J. Exp. Bot.* **2010**, *62*, 1951–1960. [CrossRef] [PubMed]
2. Hasanuzzaman, M.; Mahmud, J.A. Drought Stress Tolerance in Wheat: Omics Approaches in Understanding and Enhancing Antioxidant Defense. In *Abiotic Stress-Mediated Sensing and Signaling in Plants: An Omics Perspective*; Zargar, S.M., Ed.; Springer: New York, NY, USA, 2018; pp. 267–307.
3. Kelley, C.P.; Mohtadi, S.; Cane, M.A.; Seager, R.; Kushnir, Y. Climate change in the Fertile Crescent and implications of the recent Syrian drought. *Proc. Natl. Acad. Sci. USA* **2015**, *112*, 3241–3246. [CrossRef] [PubMed]
4. Schlaepfer, D.R.; Bradford, J.; Lauenroth, W.K.; Munson, S.M.; Tietjen, B.; Hall, S.; Wilson, S.D.; Duniway, M.C.; Jia, G.; Pyke, D.A.; et al. Climate change reduces extent of temperate drylands and intensifies drought in deep soils. *Nat. Commun.* **2017**, *8*, 14196. [CrossRef] [PubMed]
5. Kyratzis, A.C.; Skarlatos, D.P.; Menexes, G.C.; Vamvakousis, V.F.; Katsiotis, A. Assessment of Vegetation Indices Derived by UAV Imagery for Durum Wheat Phenotyping under a Water Limited and Heat Stressed Mediterranean Environment. *Front. Plant Sci.* **2017**, *8*, 1114. [CrossRef]
6. Lesk, C.; Rowhani, P.; Ramankutty, N. Influence of extreme weather disasters on global crop production. *Nature* **2016**, *529*, 84–87. [CrossRef]
7. Habash, D.; Baudo, M.; Hindle, M.; Powers, S.J.; Defoin-Platel, M.; Mitchell, R.; Saqi, M.; Rawlings, C.; Latiri, K.; Araus, J.L.; et al. Systems Responses to Progressive Water Stress in Durum Wheat. *PLoS ONE* **2014**, *9*, e108431. [CrossRef]
8. Budak, H.; Hussain, B.; Khan, Z.; Ozturk, N.Z.; Ullah, N. From Genetics to Functional Genomics: Improvement in Drought Signaling and Tolerance in Wheat. *Front. Plant Sci.* **2015**, *6*, 1012. [CrossRef]
9. Medina, S.; Vicente, R.; Nieto-Taladriz, M.T.; Aparicio, N.; Chairi, F.; Diaz, O.V.; Araus, J.L. The Plant-Transpiration Response to Vapor Pressure Deficit (VPD) in Durum Wheat Is Associated With Differential Yield Performance and Specific Expression of Genes Involved in Primary Metabolism and Water Transport. *Front. Plant Sci.* **2019**, *9*, 1994. [CrossRef]
10. Chu, C.; Wang, S.; Paetzold, L.; Wang, Z.; Hui, K.; Rudd, J.C.; Xue, Q.; Ibrahim, A.M.H.; Metz, R.; Johnson C.D.; et al. RNA-seq analysis reveals different drought tolerance mechanisms in two broadly adapted wheat cultivars 'TAM 111'and 'TAM 112'. *Sci. Rep.* **2021**, *11*, 4301. [CrossRef]
11. Yoshida, T.; Mogami, J.; Yamaguchi-Shinozaki, K. ABA-dependent and ABA-independent signaling in response to osmotic stress in plants. *Curr. Opin. Plant Biol.* **2014**, *21*, 133–139. [CrossRef]
12. Urano, K.; Maruyama, K.; Jikumaru, Y.; Kamiya, Y.; Yamaguchi-Shinozaki, K.; Shinozaki, K. Analysis of plant hormone profiles in response to moderate dehydration stress. *Plant J.* **2017**, *90*, 17–36. [CrossRef]
13. Kumar, M.; Patel, M.K.; Kumar, N.; Bajpai, A.B.; Siddique, K.H.M. Metabolomics and Molecular Approaches Reveal Drought Stress Tolerance in Plants. *Int. J. Mol. Sci.* **2021**, *22*, 9108. [CrossRef]
14. Guzmán, C.; Autrique, J.E.; Mondal, S.; Singh, R.P.; Govindan, V.; Morales-Dorantes, A.; Posadas-Romano, G.; Crossa, J.; Ammar, K.; Peña, R.J. Response to drought and heat stress on wheat quality, with special emphasis on bread-making quality, in durum wheat. *Field Crop. Res.* **2016**, *186*, 157–165. [CrossRef]
15. Slama, A.; Mallek-Maalej, E.; Ben Mohamed, H.; Rhim, T.; Radhouane, L. A return to the genetic heritage of durum wheat to cope with drought heightened by climate change. *PLoS ONE* **2018**, *13*, e0196873. [CrossRef]
16. Ren, J.; Sun, D. Genetic diversity revealed by single nucleotide polymorphism markers in a worldwide germplasm collec-tion of durum wheat. *Int. J. Mol. Sci.* **2013**, *14*, 7061–7088. [CrossRef]
17. Maccaferri, M.; Harris, N.S.; Twardziok, S.O.; Pasam, R.K.; Gundlach, H.; Spannagl, M.; Ormanbekova, D.; Lux, T.; Prade, V.M.; Milner, S.G.; et al. Durum wheat genome highlights past domestication signatures and future improvement targets. *Nat. Genet.* **2019**, *51*, 885–895. [CrossRef]
18. Blum, A. Osmotic Adjustment and Growth of Barley Genotypes under Drought Stress. *Crop Sci.* **1989**, *29*, 230–233. [CrossRef]
19. Ludlow, M.M.; Muchow, R.C. A Critical Evaluation of Traits for Improving Crop Yields in Water-Limited Environments. *Adv. Agron.* **1990**, *43*, 107–153. [CrossRef]

20. Turner, N.C.; Jones, H.G. Turgor maintenance by osmotic adjustment: A review and evaluation. In *Adaptation of Plants to Water and High Temperature Stress*; Turner, N.C., Kramer, P.J., Eds.; Wiley Interscience: New York, NY, USA, 1980; pp. 87–103.
21. Morgan, J.M. Osmoregulation and water stress in higher plants. *Annu. Rev. Plant Physiol.* **1984**, *35*, 299–319. [CrossRef]
22. Kramer, P.J.; Boyer, J.S. *Water Relations of Plants and Soils*; Academic Press: San Diego, CA, USA, 1995.
23. Babu, R.C.; Pathan, M.S.; Blum, A.; Nguyen, H.T. Comparison of measurement methods of osmotic adjustment in different rice cultivars. *Crop Sci.* **1998**, *39*, 150–158. [CrossRef]
24. Blum, A. Osmotic adjustment is a prime drought stress adaptive engine in support of plant production. *Plant Cell Environ.* **2017**, *40*, 4–10. [CrossRef]
25. Manavalan, L.P.; Nguyen, H.T. Drought tolerance in crops: Physiology to genomics. In *Plant Stress Physiology*, 2nd ed.; Sha-bala, S., Ed.; CAB International: Boston, MA, USA, 2017; Volume 1, pp. 1–23.
26. Turner, N.C. Turgor maintenance by osmotic adjustment, an adaptive mechanism for coping with plant water deficits. *Plant Cell Environ.* **2016**, *40*, 1–3. [CrossRef]
27. Qayyum, A.; Al Ayoubi, S.; Sher, A.; Bibi, Y.; Ahmad, S.; Shen, Z.; Jenks, M.A. Improvement in drought tolerance in bread wheat is related to an improvement in osmolyte production, antioxidant enzyme activities, and gaseous exchange. *Saudi J. Biol. Sci.* **2021**, *28*, 5238–5249. [CrossRef]
28. Chen, H.; Jiang, J.G. Osmotic adjustment and plant adaptation to environmental changes related to drought and salinity. *Environ. Rev.* **2010**, *18*, 309–319. [CrossRef]
29. Munns, R.; Weir, R. Contribution of Sugars to Osmotic Adjustment in Elongating and Expanded Zones of Wheat Leaves during Moderate Water Deficits at Two Light Levels. *Funct. Plant Biol.* **1981**, *8*, 93–105. [CrossRef]
30. Johnson, R.C.; Nguyen, H.T.; Croy, L.I. Osmotic adjustment and solute accumulation in two wheat genotypes differing in drought re-sistance. *Crop Sci.* **1984**, *24*, 957–962. [CrossRef]
31. Mattioni, C.; Lacerenza, N.G.; Troccoli, A.; De Leonardis, A.M.; Di Fonzo, N. Water and salt stress-induced alterations in proline metabolism of Triticum durum seedlings. *Physiol. Plant.* **1997**, *101*, 787–792. [CrossRef]
32. Liang, W.; Ma, X.; Wan, P.; Liu, L. Plant salt-tolerance mechanism: A review. *Biochem. Biophys. Res. Commun.* **2018**, *495*, 286–291. [CrossRef]
33. Turner, N.C.; O'Toole, J.C.; Cruz, R.; Yambao, E.; Ahmad, S.; Namuco, O.; Dingkuhn, M. Responses of seven diverse rice cultivars to water deficits II. Osmotic adjustment, leaf elasticity, leaf extension, leaf death, stomatal conductance and photosynthesis. *Field Crop. Res.* **1986**, *13*, 273–286. [CrossRef]
34. Tuberosa, R. Phenotyping for drought tolerance of crops in the genomics era. *Front. Physiol.* **2012**, *3*, 347. [CrossRef]
35. Gálvez, S.; Iwgsc, T.; Mérida-García, R.; Camino, C.; Borrill, P.; Abrouk, M.; Ramírez-González, R.H.; Biyiklioglu, S.; Amil-Ruiz, F.; Dorado, G.; et al. Hotspots in the genomic architecture of field drought responses in wheat as breeding targets. *Funct. Integr. Genom.* **2019**, *19*, 295–309. [CrossRef] [PubMed]
36. Verma, A.; Niranjana, M.; Jha, S.K.; Mallick, N.; Agarwal, P. QTL detection and putative candidate gene prediction for leaf rolling under moisture stress condi-tion in wheat. *Sci. Rep.* **2020**, *10*, 18696. [CrossRef] [PubMed]
37. Lilley, J.; Ludlow, M.M.; McCouch, S.R.; O'Toole, J.C. Locating QTL for osmotic adjustment and dehydration tolerance in rice. *J. Exp. Bot.* **1996**, *47*, 1427–1436. [CrossRef]
38. Robin, S.; Pathan, M.S.; Courtois, B.; Lafitte, R.; Carandang, S.; Lanceras, S.; Amante, M.; Nguyen, H.T.; Li, Z. Mapping osmotic adjustment in an advanced back-cross inbred population of rice. *Theor. Appl. Genet.* **2003**, *107*, 1288–1296. [CrossRef]
39. Teulat, B.; This, D.; Khairallah, M.; Borries, C.; Ragot, C.; Sourdille, P.; Leroy, P.; Monneveux, P.; Charrier, A. Several QTLs involved in osmotic adjustment trait variation in barley. *Theor. Appl. Genet.* **1998**, *96*, 688–698. [CrossRef]
40. Teulat, B.; Merah, O.; Souyris, I.; This, D. QTLs for agronomic traits from a Mediterranean barley progeny grown in several environments. *Theor. Appl. Genet.* **2001**, *103*, 774–787. [CrossRef]
41. Maccaferri, M.; Sanguineti, M.C.; Demontis, A.; El-Ahmed, A.; Moral, L.G.D.; Maalouf, F.; Nachit, M.; Nserallah, N.; Ouabbou, H.; Rhouma, S.; et al. Association mapping in durum wheat grown across a broad range of water regimes. *J. Exp. Bot.* **2011**, *62*, 409–438. [CrossRef]
42. International Wheat Genome Sequencing Consortium. A chromosome-based draft sequence of the hexaploid bread wheat (*Triticum aestivum*) genome. *Science* **2014**, *345*, 1251788. [CrossRef]
43. Maccaferri, M.; Sanguineti, M.C.; Natoli, V.; Ortega, J.L.A.; Salem, M.B.; Bort, J.; Chenenaoui, C.; Ambrogio, E.D.; Moral, L.G.D.; Montis, A.D.; et al. A panel of elite accessions of durum wheat (*Triticum durum* Desf.) suitable for association mapping studies. *Plant Genet. Resour.* **2006**, *4*, 79–85. [CrossRef]
44. Zadoks, J.C.; Chang, T.T.; Konzak, C.F. A decimal code for the growth stages of cereals. *Weed Res.* **1974**, *14*, 415–421. [CrossRef]
45. AZMET. Available online: https://cals.arizona.edu/azmet/06.htm (accessed on 6 December 2021).
46. TERRA-REF. Available online: https://terraref.org/ (accessed on 6 December 2021).
47. Allen, R.G.; Pereira, L.S. Crop evapotranspiration: Guidelines for computing crop water requirements-FAO Irrigation and drainage paper 56. *Fao* **1998**, *300*, D05109.
48. Bajji, M.; Lutts, S.; Kinet, J.M. Water deficit effects on solute contribution to osmotic adjustment as a function of leaf ageing in three du-rum wheat (Triticum durum Desf.) cultivars performing differently in arid conditions. *Plant. Sci.* **2001**, *160*, 669–681. [CrossRef]

49. Barrs, H.D. Determination of water deficits in plant tissue. In *Water Deficits and Plant Growth*; Kozolvski, T.T., Ed.; Academic Press: New York, NY, USA; London, UK, 1968; Volume 1, pp. 235–368.
50. Minitab Statistical Software release 9. Available online: https://www.minitab.com/en-us/support/minitab/minitab-software-updates/ (accessed on 6 December 2021).
51. TraitGenetics. Available online: http://www.traitgenetics.com/en/ (accessed on 6 December 2021).
52. Wang, S.; Wong, D.; Forrest, K.; Allen, A.; Chao, S.; Huang, B.E.; Maccaferri, M.; Salvi, S.; Milner, S.G.; Cattivelli, L.; et al. Characterization of polyploid wheat genomic diversity using a high-density 90 000 single nucleotide polymorphism array. *Plant Biotechnol. J.* **2014**, *12*, 787–796. [CrossRef]
53. Maccaferri, M.; Zhang, J.; Bulli, P.; Abate, Z.; Chao, S.; Cantu, D.; Bossolini, E.; Chen, X.; Pumphrey, M.; Dubcovsky, J. A Genome-Wide Association Study of Resistance to Stripe Rust (*Puccinia striiformis* f. sp. *tritici*) in a Worldwide Collection of Hexaploid Spring Wheat (*Triticum aestivum* L.). *G3 Genes Genomes Genet.* **2015**, *5*, 449–465. [CrossRef]
54. Maccaferri, M.; Ricci, A.; Salvi, S.; Milner, S.G.; Noli, E.; Martelli, P.L.; Casadio, R.; Akhunov, E.; Scalabrin, S.; Vendramin, V.; et al. A high-density, SNP-based consensus map of tetraploid wheat as a bridge to integrate durum and bread wheat genomics and breeding. *Plant Biotechnol. J.* **2015**, *13*, 648–663. [CrossRef]
55. Barrett, J.C.; Fry, B. Haploview: Analysis and visualization of LD and haplotype maps. *Bioinformatics* **2005**, *21*, 263–265. [CrossRef]
56. Hill, W.; Weir, B. Variances and covariances of squared linkage disequilibria in finite populations. *Theor. Popul. Biol.* **1988**, *33*, 54–78. [CrossRef]
57. Liu, W.; Maccaferri, M.; Bulli, P.; Rynearson, S.; Tuberosa, R.; Chen, X.; Pumphrey, M. Genome-wide association mapping for seedling and field resistance to *Puccinia striiformis* f. sp. *tritici* in elite durum wheat. *Theor. Appl. Genet.* **2016**, *130*, 649–667. [CrossRef]
58. Pritchard, J.K.; Stephens, M.; Donnelly, P. Inference of population structure using multilocus genotype data. *Genetics* **2000**, *155*, 945–959. [CrossRef]
59. Evanno, G.; Regnaut, S.; Goudet, J. Detecting the number of clusters of individuals using the software structure: A simulation study. *Mol. Ecol.* **2005**, *14*, 2611–2620. [CrossRef]
60. Condorelli, G.E.; Maccaferri, M.; Newcomb, M.; Andrade-Sanchez, P.; White, J.W.; French, A.N.; Sciara, G.; Ward, R.; Tuberosa, R. Comparative Aerial and Ground Based High Throughput Phenotyping for the Genetic Dissection of NDVI as a Proxy for Drought Adaptive Traits in Durum Wheat. *Front. Plant Sci.* **2018**, *9*, 893. [CrossRef]
61. Money, D.; Gardner, K.; Migicovsky, Z.; Schwaninger, H.; Zhong, G.-Y.; Myles, S. LinkImpute: Fast and Accurate Genotype Imputation for Nonmodel Organisms. *G3 Genes Genomes Genet.* **2015**, *5*, 2383–2390. [CrossRef]
62. Yu, J.; Pressoir, G.; Briggs, W.H.; Bi, I.V.; Yamasaki, M.; Doebley, J.F.; McMullen, M.D.; Gaut, B.S.; Nielsen, D.M.; Holland, J.B.; et al. A unified mixed-model method for association mapping that accounts for multiple levels of relatedness. *Nat. Genet.* **2005**, *38*, 203–208. [CrossRef]
63. Bradbury, P.J.; Zhang, Z.; Kroon, D.E.; Casstevens, T.M.; Ramdoss, Y.; Buckler, E.S. TASSEL: Software for association mapping of complex traits in diverse samples. *Bioinformatics* **2007**, *23*, 2633–2635. [CrossRef]
64. Zhang, Z.; Ersoz, E.; Lai, C.-Q.; Todhunter, R.J.; Tiwari, H.K.; Gore, M.A.; Bradbury, P.J.; Yu, J.; Arnett, D.K.; Ordovas, J.M.; et al. Mixed linear model approach adapted for genome-wide association studies. *Nat. Genet.* **2010**, *42*, 355–360. [CrossRef]
65. Berger, G.L.; Liu, S.; Hall, M.D.; Brooks, W.S.; Chao, S.; Muehlbauer, G.J.; Baik, B.-K.; Steffenson, B.; Griffey, C.A. Marker-trait associations in Virginia Tech winter barley identified using genome-wide mapping. *Theor. Appl. Genet.* **2013**, *126*, 693–710. [CrossRef]
66. Alemu, A.; Feyissa, T.; Maccaferri, M.; Sciara, G.; Tuberosa, R.; Ammar, K.; Acevedo, M.; Letta, D.T.; Abeyo, B. Genome-wide association analysis unveils novel QTLs for seminal root system architecture traits in Ethiopian durum wheat. *BMC Genom.* **2021**, *22*, 1–16. [CrossRef] [PubMed]
67. EnsemblPlants Database, Release 50. Kyoto Encyclopedia of Genes and Genomes Database. Available online: https://plants.ensembl.org/Triticum_turgidum/Info/Index (accessed on 6 December 2021).
68. Raudvere, U.; Kolberg, L.; Kuzmin, I.; Arak, T.; Adler, P.; Peterson, H.; Vilo, J. g:Profiler: A web server for functional enrichment analysis and conversions of gene lists (2019 update). *Nucleic Acids Res.* **2019**, *47*, W191–W198. [CrossRef]
69. Available online: https://www.genome.jp/kegg/ (accessed on 20 June 2020).
70. IWGSC RefSeq v1.0. Available online: https://plants.ensembl.org/Triticum_aestivum/Info/Index (accessed on 6 December 2021).
71. Hassani-Pak, K.; Singh, A. KnetMiner: A comprehensive approach for supporting evidence-based gene discovery and com-plex trait analysis across species. *Plant Biotechnol. J.* **2021**, *19*, 1670–1678. [CrossRef]
72. ePLANT databases. Available online: http://bar.utoronto.ca/eplant_wheat/ (accessed on 6 December 2021).
73. Saitou, N.; Nei, M. The neighbour-joining method: A new method for reconstructing phylogenetic trees. *Mol. Biol. Evol.* **1987**, *4*, 406–426.
74. R Core Team. *R: A Language and Environment for Statistical Computing*; R Foundation for Statistical Computing: Vienna, Austria, 2016; Available online: https://www.R-project.org/ (accessed on 17 July 2019).
75. Maccaferri, M.; El-Feki, W.; Nazemi, G.; Salvi, S.; Canè, M.A.; Colalongo, M.C.; Stefanelli, S.; Tuberosa, R. Prioritizing quantitative trait loci for root system architecture in tetraploid wheat. *J. Exp. Bot.* **2016**, *67*, 1161–1178. [CrossRef]

76. Graziani, M.; Maccaferri, M.; Royo, C.; Salvatorelli, F.; Tuberosa, R. QTL dissection of yield components and morpho-physiological traits in a durum wheat elite population tested in contrasting thermo-pluviometric conditions. *Crop Pasture Sci.* **2014**, *65*, 80–95. [CrossRef]
77. Sanguineti, M.C.; Li, S.; Maccaferri, M.; Corneti, S.; Rotondo, F.; Chiari, T.; Tuberosa, R. Genetic dissection of seminal root architecture in elite durum wheat germplasm. *Ann. Appl. Biol.* **2007**, *151*, 291–305. [CrossRef]
78. Quarrie, S.A.; Steed, A.; Calestani, C.; Semikhodskii, A.; Lebreton, C.; Chinoy, C.; Steele, N.; Pljevljakusić, D.; Waterman, E.; Weyen, J.; et al. A high-density genetic map of hexaploid wheat (*Triticum aestivum* L.) from the cross Chinese Spring × SQ1 and its use to compare QTLs for grain yield across a range of environments. *Theor. Appl. Genet.* **2005**, *110*, 865–880. [CrossRef] [PubMed]
79. Abdelrahman, M.; Burritt, D.J.; Tran, L.-S.P. The use of metabolomic quantitative trait locus mapping and osmotic adjustment traits for the improvement of crop yields under environmental stresses. *Semin. Cell Dev. Biol.* **2017**, *83*, 86–94. [CrossRef]
80. Morgan, J.; Condon, A. Water Use, Grain Yield, and Osmoregulation in Wheat. *Funct. Plant Biol.* **1986**, *13*, 523–532. [CrossRef]
81. Blum, A.; Pnuel, Y. Physiological attributes associated with drought resistance of wheat cultivars in a Mediterranean envi-ronment. *Crop Pasture Sci.* **1990**, *41*, 799–810. [CrossRef]
82. Blum, A.; Zhang, J.X. Consistent differences among wheat cultivars in osmotic adjustment and their relationship to plant production. *Field Crops Res.* **1999**, *64*, 287–291. [CrossRef]
83. Blum, A. Drought resistance, water-use efficiency, and yield potential—Are they compatible, dissonant, or mutually exclu-sive. *Aust. J. Agric. Res.* **2005**, *56*, 1159–1168. [CrossRef]
84. Morgan, J. A Gene Controlling Differences in Osmoregulation in Wheat. *Funct. Plant Biol.* **1991**, *18*, 249–257. [CrossRef]
85. Morgan, J.; Tan, M. Chromosomal Location of a Wheat Osmoregulation Gene Using RFLP Analysis. *Funct. Plant Biol.* **1996**, *23*, 803–806. [CrossRef]
86. Wilson, J.; Ludlow, M.; Fisher, M.; Schulze, E. Adaptation to Water Stress of the Leaf Water Relations of Four Tropical Forage Species. *Funct. Plant Biol.* **1980**, *7*, 207–220. [CrossRef]
87. Blum, A. *Plant Breeding for Stress Environments*, 1st ed.; Blum, A., Ed.; CRC Press: Boca Raton, FL, USA, 1988; p. 231.
88. Zhang, J.; Nguyen, H.T.; Blum, A. Genetic analysis of osmotic adjustment in crop plants. *J. Exp. Bot.* **1999**, *50*, 291–302. [CrossRef]
89. Mart, K.B.; Veneklaas, E.J.; Bramley, H. Osmotic potential at full turgor: An easily measurable trait to help breeders select for drought tolerance in wheat. *Plant Breed.* **2016**, *135*, 279–285. [CrossRef]
90. Lewis, J.E.; Rowland, J.; Nadeau, A. Estimating maize production in Kenya using NDVI: Some statistical considerations. *Int. J. Remote Sens.* **1998**, *19*, 2609–2617. [CrossRef]
91. Araus, J.L.; Casadesus, J. Recent tools for the screening of physiological traits determining yield. In *Application of Physiology in Wheat Breeding*; Reynolds, M.P., Ortiz-Monasterio, J.I., Eds.; CIMMYT: Batan, Mexico, 2001; pp. 59–77.
92. Salinero, E.C. Analisis de Imagenes: Extraccion de Informacion Tematica. In *Teledetectión Ambiental La Observación de la Tierra Desde el Espacio*, 3rd ed.; Salinero, E.C., Ed.; Ariel: Barcelona, Spain, 2002.
93. Bail, M.L.; Jeuffroy, M.H. Is it possible to forecast grain protein content and yield of several varieties from chlorophyll meter measurements? *Eur. J. Agron.* **2005**, *23*, 379–391. [CrossRef]
94. Labus, M.P.; Nielsen, G.A.; Lawrence, R.L.; Engel, R.; Long, D.S. Wheat yield estimates using multi-temporal NDVI satellite imagery. *Int. J. Remote Sens.* **2002**, *23*, 4169–4180. [CrossRef]
95. Acevedo-Garcia, J.; Kusch, S.; Panstruga, R. Magical mystery tour: MLO proteins in plant immunity and beyond. *New Phytol.* **2014**, *204*, 273–281. [CrossRef]
96. Bargmann, B.; Munnik, T. The role of phospholipase D in plant stress responses. *Curr. Opin. Plant Biol.* **2006**, *9*, 515–522. [CrossRef]
97. Rahman, A.; Hossain, M.S. Manganese-induced salt stress tolerance in rice seedlings: Regulation of ion homeostasis, antiox-idant defense and glyoxalase systems. *Physiol. Mol. Biol. Plants* **2016**, *22*, 291–306. [CrossRef]
98. Ye, Y.; Medina-Velo, I.A. Can abiotic stresses in plants be alleviated by manganese nanoparticles or compounds? *Ecotoxicol. Environ. Saf.* **2019**, *184*, 109671. [CrossRef]
99. Liu, Q.; Kasuga, M. Two transcription factors, DREB1 and DREB2, with an EREBP/AP2 DNA binding domain separate two cellular signal transduction pathways in drought- and low-temperature-responsive gene expression, respectively, in Ara-bidopsis. *Plant Cell* **1998**, *10*, 1391–1406. [CrossRef]
100. Cantale, C.; Di Bianco, D.; Thiyagarajan, K.; Ammar, K.; Galeffi, P. B genome specific polymorphism in the *TdDRF1* gene is in relationship with grain yield. *Planta* **2018**, *247*, 459–469. [CrossRef]
101. Feng, K.; Hou, X.L. Advances in AP2/ERF super-family transcription factors in plant. *Crit. Rev. Biotechnol.* **2020**, *40*, 750–776. [CrossRef]
102. Simon, N.M.L.; Graham, C.A.; Comben, N.E.; Hetherington, A.M.; Dodd, A.N. The Circadian Clock Influences the Long-Term Water Use Efficiency of Arabidopsis. *Plant Physiol.* **2020**, *183*, 317–330. [CrossRef]
103. Ifuku, K.; Endo, T.; Shikanai, T.; Aro, E.-M. Structure of the Chloroplast NADH Dehydrogenase-Like Complex: Nomenclature for Nuclear-Encoded Subunits. *Plant Cell Physiol.* **2011**, *52*, 1560–1568. [CrossRef]
104. Alahmad, S.; Hassouni, K.E.; Bassi, F.M.; Dinglasan, E.; Youssef, C.; Quarry, G.; Aksoy, A.; Mazzucotelli, E.; Juhasz, A.; Able, J.A.; et al. A Major Root Architecture QTL Responding to Water Limitation in Durum Wheat. *Front. Plant Sci.* **2019**, *10*, 436. [CrossRef]

genes

Article

Transcriptome Analysis Reveals Differentially Expressed Genes That Regulate Biosynthesis of the Active Compounds with Methyl Jasmonate in Rosemary Suspension Cells

Deheng Yao, Zihao Zhang, Yukun Chen, Yuling Lin, Xuhan Xu and Zhongxiong Lai *

Institute of Horticultural Biotechnology, Fujian Agriculture and Forestry University, Fuzhou 350002, China; yaodh337@126.com (D.Y.); zhangzihao863@163.com (Z.Z.); cyk68@163.com (Y.C.); buliang84@163.com (Y.L.); xxuhan@163.com (X.X.)

* Correspondence: laizx01@163.com

Abstract: To study the effects of Methyl jasmonates (MeJA) on rosemary suspension cells, the antioxidant enzymes' change of activities under different concentrations of MeJA, including 0 (CK), 10 (M10), 50 (M50) and 100 μM MeJA (M100). The results demonstrated that MeJA treatments increased the activities of phenylalanine ammonla-lyase (PAL), superoxide dismutase (SOD), peroxidase (POD), catalase (CAT) and polyphenol oxidase (PPO) and reduced the contents of hydrogen peroxide (H_2O_2) and malondialdehyde (MDA), thus accelerating the ROS scavenging. Comparative transcriptome analysis of different concentrations of MeJA showed that a total of 7836, 6797 and 8310 genes were differentially expressed in the comparisons of CKvsM10, CKvsM50, CKvsM100, respectively. The analysis of differentially expressed genes (DEGs) showed phenylpropanoid biosynthesis, vitamin B6, ascorbate and aldarate metabolism-related genes were significantly enriched. The transcripts of flavonoid and terpenoid metabolism pathways and plant hormone signal transduction, especially the jasmonic acid (JA) signal-related genes, were differentially expressed in CKvsM50 and CKvsM100 comparisons. In addition, the transcription factors (TFs), e.g., *MYC2*, *DELLA*, *MYB111* played a key role in rosemary suspension cells under MeJA treatments. qRT-PCR of eleven DEGs showed a high correlation between the RNA-seq and the qRT-PCR result. Taken together, MeJA alleviated peroxidative damage of the rosemary suspension cells in a wide concentration range via concentration-dependent differential expression patterns. This study provided a transcriptome sequence resource responding to MeJA and a valuable resource for the genetic and genomic studies of the active compounds engineering in rosemary.

Keywords: *Rosmarinus officinalis* Lour.; suspension cells; MeJA; antioxidant enzymes; RNA-seq; qRT-PCR; transcription factors

Citation: Yao, D.; Zhang, Z.; Chen, Y.; Lin, Y.; Xu, X.; Lai, Z. Transcriptome Analysis Reveals Differentially Expressed Genes That Regulate Biosynthesis of the Active Compounds with Methyl Jasmonate in Rosemary Suspension Cells. *Genes* **2022**, *13*, 67. https://doi.org/10.3390/genes13010067

Academic Editor: Patrizia Galeffi

Received: 31 October 2021
Accepted: 23 December 2021
Published: 27 December 2021

1. Introduction

Rosemary (*Rosmarinus officinalis* Lour.) is a famous ornamental and medicinal homologous plant. Rosemary, as an excellent natural antioxidant and preservative, has been used in various industries widely [1]. Studies have shown that the main components of the active functions of rosemary include terpenoids, phenols, for example a new flavonoid 6″-O-(E)-feruloylhomoplantaginin, and the phenolic diterpene antioxidants (PDAs), carnosic acid and carnosol and rosmarinic acid [2–4]. The active ingredients of rosemary are widely used in anti-tumor, anti-cancer, anti-despondency, anti-virus, anti-inflammatory activity, regulating the immune system and other activities. Carnosic acid could alleviate H_2O_2 induced hepatocyte damage through the SIRT1 pathway and rosemary extract, and significantly up regulate the expression of *Nrf2* in colon cells and inhibit an HCT116 xenograft tumor formation in mice. Moreover, rosemary extract shows a higher antioxidant potential and increases the oxidative stability of oil by more than 30% compared to conventional synthetic antioxidants [5–11]. Researchers had previously attempted to regulate the synthesis of functional metabolites using various methods. MeJA as a signal molecule showed extensive regulations to the secondary metabolism, of which regulation was considered particularly important in changing the synthesis of plant functional metabolites in cells [12,13]. In addition, plant tissue and cell culture techniques were the most efficient methods for obtaining functional metabolites. Our group has established a rosemary suspension cells culture system to study the influence of MeJA treatment on functional metabolites.

MeJA had been used to elicit defense responses in many species through enhancing the secondary metabolites production [14,15], such as volatile terpenoids in *Amomum villosum*, triterpene in *Euphorbia pekinensis* and tropane alkaloids in *Hyoscyamus niger* [16–18]. An analysis of transcriptome after MeJA treatment could find key genes involved in the biosynthesis of active compounds and be used to unveil the relation between genes and metabolism [19,20]. In *Catharanthus roseus*, the MeJA-responsive expression of terpenoid indole alkaloids biosynthesis genes was controlled by protein *CrMYC2*'s regulation of *ORCA* gene expression, regulating a series of terpenoid indole alkaloids biosynthesis genes [21]. The TFs AP2/ERF and bHLH cooperatively mediate jasmonate-elicited nicotine biosynthesis, which via the JA induced signaling cascade leads to increased nicotine biosynthesis in *Tobacco* [22,23]. In *Artemisia annua* suspension cells, exogenous MeJA induced the expression of *CYP71AV1* and promoted the accumulation of artemisinin [24].

MeJA treatment can induce the biosynthesis of many secondary metabolites (terpenoids, phenylpropanoids) and acts as an elicitor of secondary metabolite production across the plant kingdom [25]. *SlMYC1* acts synergistically with *SlEOT1* in the transactivation of the *SlTPS5* promoter to induce the biosynthesis of terpene in *Solanum lycopersicum* with JA treatment [26]. There were 13 predicted genes that could participate in the biosynthesis of flavonoids under MeJA treatment in *E. breviscapus* [27]. Transcript analysis suggested that MeJA up-regulated transcripts of terpenoids and flavonoids, *ObAS1* and *ObAS2* were identified and characterized in *Sweet Basil* [28]. MeJA treatment revealed differential expression of genes involved in phenylpropanoid biosynthesis (*IiPAL, IiC4H, Ii4CL*), and lignin biosynthesis (*IiCAD, IiC3H, IiCCR, IiDIR* and *IiPLR*), and 112 putative AP2/ERF TFs in *Isatis indigotica* [29]. Based on the gene annotation of the transcriptome, 104 unigenes were identified and their responses to MeJA induction were investigated involved in the biosynthesis of indole, terpenoid and phenylpropanoid [30]. The results showed that transcriptional levels of *SgHMGR, SgSQS, SgCS* and *SgCYP450* were up-regulated and their responses in the presence of MeJA were related to the concentration and timing of MeJA treatment in *Siraitia grosvenorii* [31]. MeJA-regulated rubber biosynthesis, based on a differential expression analysis, showed that the expression of *HMGCR, FPPS, IDI, GGPPS, REF/SRPP* and transcription factors (bHLH, MYB, AP2/EREBP and WRKY) increased with MeJA treatment in TKS [32]. Based on the date of the transcriptome in *Taxus*, there were 18 genes showing increased transcript abundance following elicitation of MeJA, which was involved in the biosynthesis of terpenoid backbone, and then multiple candidates for the unknown steps in paclitaxel biosynthesis were identified [33]. A total of 40,952 unigenes and 19 coumarin compounds, 7 cytochrome-P450, 8 multidrug resistance transporter unigenes and 8 marker compounds were obtained, involved in coumarins biosynthesis and transport pathway with a parallel analysis of transcript and metabolic profiles in *Peucedanum praeruptorum* [34]. The secondary metabolites are the main components of their active functions in plant. Therefore, transcriptome technology was widely used as an effective means to research the biosynthesis of active compounds and the mining of key enzyme genes. MeJA could promote the biosynthesis of terpenoids and phenylpropanoids, through revealing differential expression of the genes involved in biosynthesis and plant hormone signal transduction and TFs in plants.

At present, no study has shown the transcriptome of rosemary responding to MeJA. In our study, we investigated the transcriptome of rosemary suspension cells responding to different concentrations of MeJA using high-throughput sequencing technologies. Putative gene expression profiles of rosemary suspension cells were investigated and DEGs were classified under different concentrations of MeJA. By comparing and analyzing the sequencing data of control and illuminated groups, the genes involved in the regulation of primary and secondary metabolism and their regulatory networks were established. These experiments reveal dynamic gene expression changes in responding to different concentrations of MeJA and provide new insights into the genetic and genomic regulation of plant functional metabolites.

2. Materials and Methods

2.1. Plant Material and MeJA Treatments

Rosemary callus was obtained by the following methods: treating rosemary leaves with 0.1% mercury bichloride solution, then cutting the leaves into small pieces about 0.5 cm × 0.5 cm, using the inoculation method that the back of leaves contacts with the medium, on Murashige and Skoog (MS) medium (30 g/L sucrose, pH 5.8) with 0.5 mg/L 1-Naphthaleneacetic acid (NAA) and 4.0 mg/L N-(Phenylmethyl)-9H-purin-6-amine (6-BA). After the suspension culture for several generations in the MS liquid medium supplemented with 1.0 mg/L 2,4-dichlorophenoxyacetic acid (2,4-D), which lasted for 8 days at 25 ± 0.5 °C with shaker speed 120 rpm in the dark, it could derive rosemary suspension cell lines with high cell viability and stable growth. The culture was performed by transferring 4 g·FW/20 mL of 6-day-old culture (cells plus medium) to 80 mL of the fresh growth medium, which lasted for 8 days. MeJA were sterilized and added to the medium on day 6 of the culture process. Baes on the pre-experiment, 10, 50 and 100 μM MeJA treatments could promote the accumulation of rosmarinic acid, carnosic acid and flavonoids in rosemary suspension cells, so we chose the concentration of MeJA solution was 0, 10, 50 and 100 μM. Each test was repeated three times. After 48 h treatment, all materials were stored at −80 °C for later use.

2.2. RNA-Seq Library Construction

Total RNA was isolated from rosemary suspension cells (three replicates) with the RNAprep Pure Plant Kit (TIANGEN, Beijing, China). The integrity and concentration of the RNA samples were further measured using an Agilent 2100 Bioanalyzer (Agilent, CA, USA) and the purity of the RNA samples was assessed using the NanoPhotometer®spectrophotometer (NP80, IMPLEN, Munich, Germany). RNA libraries were prepared using the True-seq RNA sample preparation kit according to the manufacturer's instructions. The constructed library was tested on the Agilent 2100 Bioanalyzer and the ABI StepOnePlus Real-Time PCR System. Finally, mRNA libraries were sequenced on an Illumina HiSeq 4000 platform (Shenzhen, China).

2.3. Sequencing, Assembly and Annotation of the Transcriptome

The cDNA libraries of four samples were sequenced by 2 × 100 bp paired-end sequencing on an Illumine HiSeq platform according to the manufacture's instructions. We used Trinity to assemble clean reads by de novo, while removing PCR duplication to improve assembly efficiency. Then the assembled transcripts were clustered by tgicl to remove redundancy and obtain UniGene. Trinity consists of three independent modules: inchworm, chrysalis and butterfly, which process a large number of reads in turn. The assembled UniGene will be annotated with seven functional databases (KEGG (Kyoto Encyclopedia of Genes and Genomes), GO (Gene Ontology), NR, NT, Swissprot, Pfam and KOG).

2.4. Quantification of Gene Expression Levels

We used bowtie2 to align clean reads to the genome sequence and RESM (http://deweylab.biostat.wisc.edu/rsem/rsem-calculate-expression.html, Access date: 30 January 2019, RESM Version: v1.2.8; RESM Parameter: default;) to calculate the gene expression level of each sample. RESM is a software package for RNA-seq reads to calculate the expression of genes and transcript subtypes.

2.5. Differential Expression Genes Analysis

Degseq method is based on Poisson distribution. In this project, DEG detection is carried out according to the method described by Wang et al. 2010 [35]. Differential expression analysis of the four treatments was performed using the DEGSeq R package.

p-Values were adjusted using the Benjamini and Hochberg method. A corrected P-value (false discovery rate, FDR) of 0.001 and a fold change of 2 were set as the default threshold for defining significant differential expression.

2.6. GO and KEGG Enrichment Analyses of Differentially Expressed Genes

According to the GO (http://www.geneontology.org/, Access date: 30 January 2019) and KEGG (http://www.genome.jp/kegg/, Access date: 30 January 2019) annotation results and the official classification, DEGs were classified, and the enrichment factors were analyzed using the phyper function in R software (https://en.wikipedia.org/wiki/Hypergeometric_distribution, Access date: 30 January 2019).

2.7. Validation of the DEGs by qRT-PCR

To validate the RNA-Seq results, 11 DEGs were subjected to qRT-PCR analysis performed on the LightCycler480 real-time PCR system (Roche, Basel, Switzerland) in a 20 μL final volume containing 10 μL of 2 × SYBR Premix Ex TaqTM (Takara, Shanghai, China), 1 μL of 10 × diluted cDNA, and 0.8 μL specific primer pairs, and 7.4 μL of ddH_2O. The changes in mRNA expression were calculated using the comparative $2^{-\Delta\Delta Ct}$ method. Specific primers were designed using DNAMAN V6.0; the primer pair sequences are listed in Table S1. All treatments were analyzed in three biological replicates.

2.8. Measurement of Antioxidant Enzymes and Non-Enzymatic Antioxidants

Briefly, 0.1 g of rosemary suspension cells (fresh weight) were taken from each of the treatment groups and rapidly frozen with liquid nitrogen. The samples were maintained at 2–8 °C after melting. Then, 1 mL PBS (pH 7.4) was added, followed by homogenization by hand or grinders and centrifugation for 20 min at the speed of 2000–3000 rpm. Supernatant was removed. The activities of PAL, SOD, POD, CAT, PPO and the contents of MDA, H_2O_2, proline (pro) were assayed using ELISA Kit (Weilan, Shanghai, China) and a micro-plate reader (Rayto RT-6100) according to the manufacturer's instructions. The principle of the assay: the kit assay plant PAL, SOD, POD, CAT, PPO, MDA, H_2O_2 and proline level in the sample, use purified plant PAL, SOD, POD, CAT, PPO, MDA, H_2O_2 and proline antibody to coat microtiter plate wells, make solid-phase antibody, then add PAL, SOD, POD, CAT, PPO, MDA, H_2O_2 and proline to the wells. Combined antibody labeled with HRP became an antibody-antigen-enzyme-antibody complex. After washing completely, TMB substrate solution was added. TMB substrate became blue. At HRP catalyzed enzyme, the reaction is terminated by the addition of a sulphuric acid solution and the color change is measured spectrophotometrically at a wavelength of 450 nm. The concentration of PAL, SOD, POD, CAT, PPO, MDA, H_2O_2 and proline in the samples is then determined by comparing the OD of the samples to the standard curve. Assay procedure: add standard, set standard wells, test sample wells, add standard 50 μL to standard well; add sample, set blank wells separately, test sample well, add sample dilution 40 μL to test sample well, then add testing sample 10 μL (sample final dilution is 5-fold), add sample to wells, do not touch the well wall as far as possible, and gently mix; add enzyme, add HRP-conjugate reagent 100 μL to each well, except blank well; incubate, after closing the plate with closure plate membrane, incubate for 60 min at 37 °C; configurate liquid, 20-fold wash solution diluted 20-fold with distilled water and reserve; wash, uncover the closure plate membrane, discard liquid, dry by swinging, add washing buffer to every well, still for 30 s then drain, repeat 5 times, dry by patting; color, add chromogen solution A 50 uL and B to each well, evade the light preservation for 15 min at 37 °C; stop the reaction, add stop solution 50 μL to each well, stop the reaction (blue color changes to yellow); assay, take blank well as zero, read absorbance at 450 nm after adding stopping solution within 15 min.

2.9. Statistical Analysis

Quantitative results for rosemary metabolite content, enzyme activity and gene expression analyses are presented as the means ± standard deviations (SDs) of at least three biological replicates. The effects of MeJA conditions on metabolite contents and gene expression were analyzed by one-way analysis of variance (ANOVA) followed by Duncan's test using SPSS version 19.0. Figures were prepared using GraphPad Prism 8.0 and Excel 2016 software.

3. Results

3.1. Physiological and Biochemical Indexes of Rosemary Suspension Cells under Different Concentrations of MeJA

In our study, measuring the physiological and biochemical indicators in rosemary suspension cells is helpful for understanding how the synthesis of important metabolites is promoted by MeJA. After different concentrations of MeJA treatment for 48 h in rosemary suspension cells, the activities of PAL were the highest in the 100 μM MeJA treatment group, followed by the 50 and 10 μM MeJA treatment groups, they were higher than the CK treatment (Figure 1A). Similarly, the activities of SOD, POD, CAT and PPO were the highest in the 100 μM MeJA treatment group, followed by the 50 and 10 μM MeJA treatment groups, they were higher than the CK treatment (Figure 1B–E). The concentration of MDA and H_2O_2 were the lowest in the 100 μM MeJA treatment group, the highest of the concentration was the CK treatment (Figure 1F,G). For the proline contents, the highest was the 50 μM MeJA treatment group and the lowest was the 100 μM MeJA treatment group (Figure 1H). The result indicating that the MeJA activated the rosemary suspension cells enzyme antioxidant system accelerate the ROS scavenging, the antioxidant enzyme activity increased with the increase of MeJA concentration and had a key role in rosemary suspension cells.

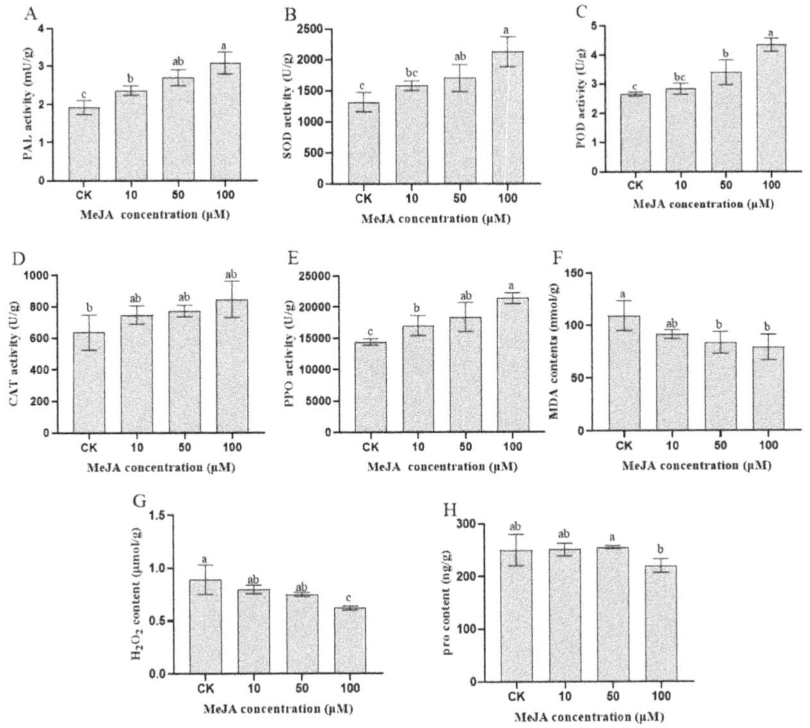

Figure 1. Physiological and biochemical indicators of rosemary suspension cells under different concentrations of MeJA. (**A**) changes in PAL activities. (**B**) changes in SOD activities. (**C**) changes in POD activities. (**D**) changes in CAT activities. (**E**) changes in PPO activities. (**F**) changes in MDA contents. (**G**) changes in H_2O_2 contents. (**H**) changes in proline contents. Values represent means ± SDs of three replicates. Different lower-case letters indicate statistically significant differences at the 0.05 level by one-way ANOVA with Duncan's test.

3.2. RNA-Seq Analysis of Rosemary Suspension Cells

To study the effects of different concentrations of MeJA on rosemary suspension cells at a transcriptional level, four mRNA libraries were constructed and sequenced from the four concentrations of MeJA treatment (CK_(1,2,3), M10_(1,2,3), M50_(1,2,3) and M100_(1,2,3)). After removing the linker sequence, the RNA-Seq data of the twelve rose-mary suspension cells libraries produced 66,314,608 to 71,191,500 reads, respectively, due to differences in concentrations of MeJA. The values of Q20 were higher than 97% in all the samples, indicating the high reliability of the rosemary suspension cells transcriptome sequencing data (Tables S2–S4).

In addition, unigenes were exposed to Nr, Nt, SwissPort, Pfam, KEGG, GO and KOG databases using BLAST analysis (E-value < 0.00001) (Table S5). The unigenes annotated in the Nr database were counted, the top five species were annotated: *Sesamum indicum* L. (65.02%), *Erythranthe guttata* L. (17.59%), *Dorcoceras hygrometricum* L. (2.08%), *Salvia miltiorrhiza* L. (2.08%) and *Ipomoea nil* L. (0.75%). The 17.59% and 2.08% of the total number of Nr annotations accounted for, respectively, 2.08% and 0.75% (Figure S1). The distribution of gene functions in GO was grouped into biological processes, cellular component and molecular function, cellular process (14,755 unigenes), cell (14,796 unigenes), binding (25,721 unigenes) and catalytic activity (23,776 unigenes) were dominant subcategories (Figure S2). Similarly, KOG functional classification showed that the uppermost classification was general function prediction only (15,177 unigenes) followed by signal transduction mechanisms (9078 genes). In addition, 1205 unigenes were annotated for secondary metabolites biosynthesis, transport and catabolism (Figure S3). At last, unigenes were also annotated against KEGG database for understanding advanced-level utilities and functions of the biological structure. Among 80,961 unigenes were annotated in 136 pathways, metabolism (102 pathways, 40,920 unigenes) was the most significant category, a substantial number of genes were related to carbohydrate metabolism (6277 genes) and amino acid metabolism (3835 gens) (Figure S4).

3.3. Global Analysis of Gene Expression across the Four Distinct Samples under Different Concentrations of MeJA

In our study, there were 107,512, 102,762, 107,366 and 107,646 expressed genes in CK, M10, M50 and M100 sample libraries. Among these, 103,386 expressed genes were present in all four sample libraries. However, only 572, 621, 531 and 571 genes were uniquely present in CK, M10, M50 and M100 sample libraries, respectively (Figure 2A), which suggested that distinct spatial transcriptional patterns were present in the sample libraries. In order to evaluate the differences in molecular response among four samples, gene expression was normalized to FPKM. After filtering with FPKM> 10, a total of 14,104 (13.11%), 13,880 (12.87%), 13,981 (12.97%) and 13,940 (12.95%) genes were expressed in CK, M10, M50 and M100, respectively (Table S6). The top 20 most enriched (FPKM) genes ranged from 682 to 3814, 781 to 3294, 663 to 3844 and 709 to 4294, respectively. The top 20 most expressed genes from the four libraries were shown as *defensin-like cystein-rich peptide, extracellular ribonuclease LE, PREDICTED: titin-like, pathogen-related protein STH-2, hypothetical protein SELMODRAFT_431225, aquaporin-like protein, partial, major pollen allergen Lol p 11-like and extensin-3*—highly expressed in CK, M10, M50 or M100 sample libraries (Table S7).

To further understand the changes in the rosemary suspension cells transcriptome under different concentrations of MeJA, Poisson D was used to calculate the expression level of each gene. We filtered the DEGs with $|\log2\text{fold change}| \geq 1$ and FDR < 0.001 between these six pairs the comparisons were as follows: CKvsM10, CKvsM50, CKvsM100, M10vsM50, M10vsM100, M50vsM100, among these contained 7836, 6797, 8310, 10,240, 11,890 and 5260 DEGs, respectively, which included 3596, 325, 4528, 5070, 6537 and 3017 up-regulated genes and 4240, 3772, 3782, 5170, 5353 and 2243 down-regulated genes (Figure 2B). The comparative analysis of the CKvsM10, CKvsM50, and CKvsM100 by Venn diagram showed that 1581 genes were commonly differentially expressed, and 4188, 2409 and 3663 genes were unique to each comparison, respectively; there were 749 genes commonly up-regulated, while 1672, 822 and 2110 genes in each comparison were uniquely up-regulated; similarly, there were 556 genes commonly down-regulated, while 2516, 1587 and 1553 genes in each comparison were uniquely down-regulated, respectively (Figure 2C–E). These DEGs' analysis results revealed that rosemary suspension cells transcriptome undergoes significantly dynamic changes under different concentrations of MeJA. Therefore, the transcriptome datasets of rosemary suspension cells under different concentrations of MeJA might serve as a valuable molecular resource for future studies.

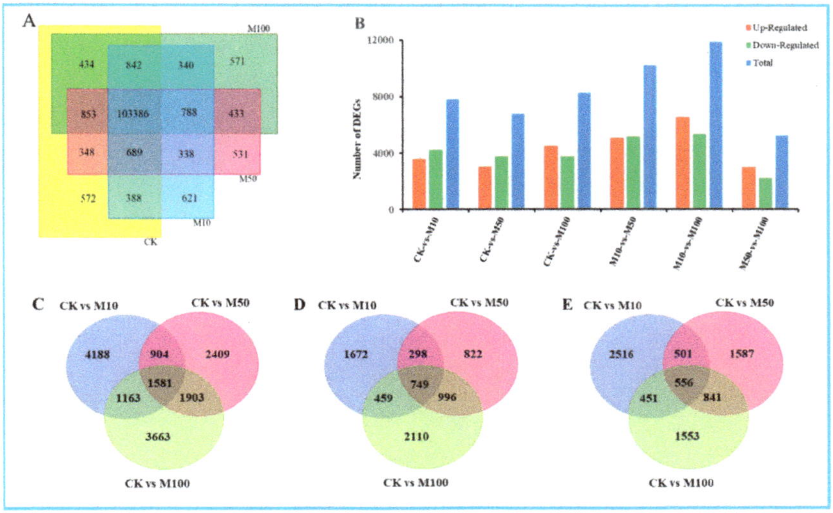

Figure 2. Statistical analysis of differentially expressed unigenes in rosemary suspension cells under different concentrations of MeJA. (**A**) The Venn diagram of expressed genes in four MeJA treatments. (**B**) statistic of up/down-regulated genes in pairwise comparisons. (**C**) Venn diagram of DEGs under MeJA treatment. (**D**) Venn diagram of the unique and common regulated DEGs up-regulated of DEGs under MeJA treatments. (**E**) Venn diagram of the unique and common regulated DEGs down-regulated of DEGs under MeJA treatments.

3.4. GO Enrichment Analysis of DEGs in Rosemary Suspension Cells

To further understand the potential functions of the DEGs under different concentrations of MeJA, GO terms assignment to classify the functions of DEGs was performed in pairwise comparisons under three GO main categories: biological process, cellular component and molecular function (Table S8). The results of GO enrichment analysis in all comparisons showed the enrichment of most biological processes in the CKvsM10 and CKvsM50 combinations were significantly higher than other combinations, DEGs were mainly related to xyloglucan metabolic process and response to glucose in the CKvsM10, CKvsM50 and CKvsM100 comparisons (Figure 3A). The enrichment of most cellular components in the CKvsM50 combination was significantly higher than other combinations, DEGs were mainly related to apoplast and cell wall in the CKvsM10, CKvsM50 and CKvsM100 comparisons (Figure 3B). The enrichment of most molecular functions in the CKvsM10 combination was significantly higher than other combinations, DEGs were mainly related to hydrolase activity, alpha-L-arabinofuranosidase activity and oxidoreductase activity in the CKvsM10, CKvsM50 and CKvsM100 comparisons (Figure 3C). In summary, indicating that the effect of different concentrations of MeJA on rosemary suspension cells was particularly obvious and had substantially different responses to different concentrations of MeJA.

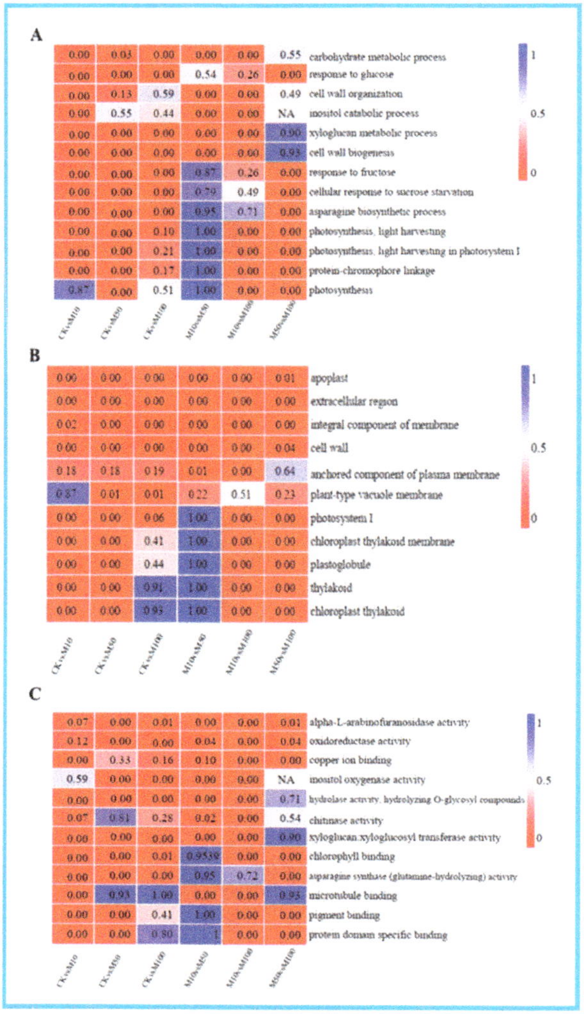

Figure 3. GO-terms enrichment analysis. From the red to blue corresponds to the numerical value of corrected Q and significant enriched GO-terms from the low to the high. (**A**) biological process; (**B**) cellular component; (**C**) molecular function.

3.5. KEGG Enrichment and Mapman Analysis of DEGs in Rosemary Suspension Cells

To investigate the metabolic pathways of the DEGs, we used the KEGG database to classify the DEGs function and mapped the top 20 KEGG pathways enriched among the DEGs according to the enrichment factors identified (Figures S4 and S5). The first 20 enrichment pathways of the six combinations (CKvsM10, CKvsM50, CKvsM100, M10vsM50, M10vsM100, M50vsM100) were compared and analyzed (Figure 4). Phenylpropanoid biosynthesis was enriched in the top 5 in the CKvsM10, CKvsM50, CKvsM100, M10vsM50 and M10vsM100 comparisons. Phenylpropanoid biosynthesis is closely related to the biosynthesis of flavonoids, and phenylalanine is closely related to the synthesis of rosmarinic acid, indicating that there were significant differences in the biosynthesis of flavonoids and rosmarinic acid in rosemary suspension cells by different concentrations of MeJA.

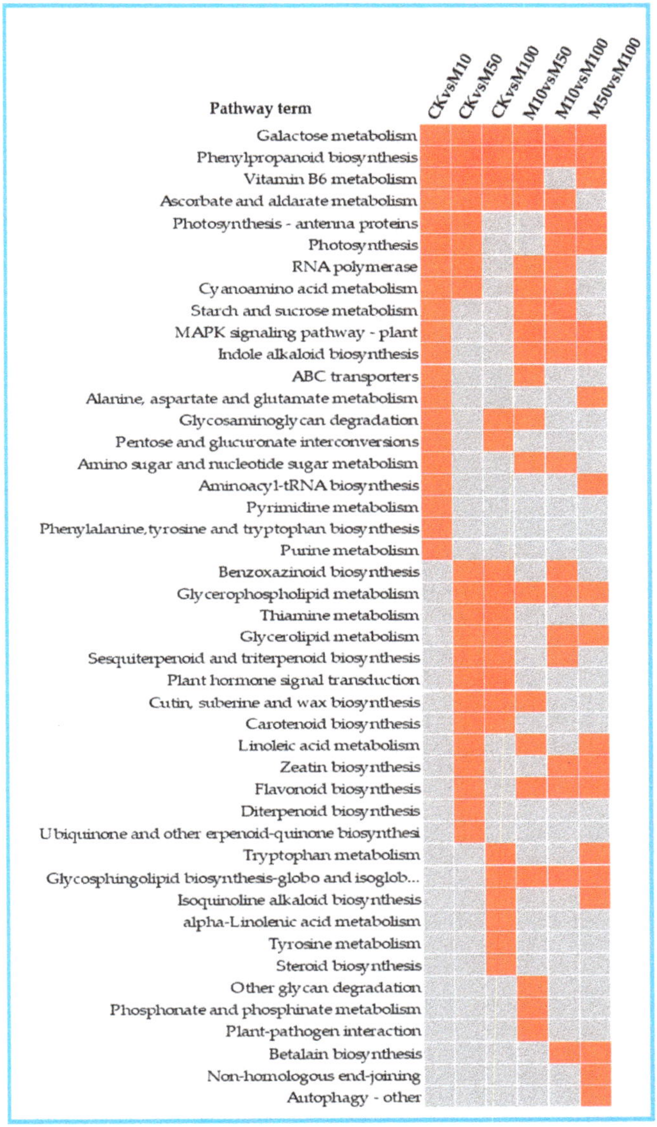

Figure 4. The top 20 KEGG pathways enriched by DEGs in in the comparisons. Red indicates significant enrichment and gray indicates no significant enrichment.

The first 20 enrichment pathways of the three combinations (CKvsM10, CKvsM50, CKvsM100) were compared and analyzed (Figure 4). Galactose metabolism, phenylpropanoid biosynthesis, vitamin B6 metabolism, ascorbate and aldarate metabolism were enriched in the three comparisons among the top 20 pathways (Figure 4). This indicated that these pathways differed significantly under different MeJA concentrations treatments. Some pathways were enriched in the top 20 in two combinations, photosynthesis-antenna proteins, photosynthesis and RNA polymerase were enriched in the top 20 in CKvsM10 and CKvsM50 (Figure 4). This indicated that these pathways might play important roles in responding to 10 and 50 μM MeJA treatments. Glycosaminoglycan degradation and pentose and glucuronate interconversions were enriched in the top 20 in CKvsM10 and

CKvsM100 (Figure 4), suggesting that 10 and 100 μM MeJA were more effective for them. Benzoxazinoid biosynthesis, sesquiterpenoid and triterpenoid biosynthesis, plant hormone signal transduction and carotenoid biosynthesis were enriched in the top 20 in CKvsM50 and CKvsM100 (Figure 4), suggesting that 50 and 100 μM MeJA were more effective for these pathways. Some pathways were only enriched in the top 20 for one combination (Figure 4). MAPK signaling pathway-plant, indole alkaloid biosynthesis, ABC transporters, and phenylalanine, tyrosine and tryptophan biosynthesis were only enriched in the top 20 in CKvsM10. Linoleic acid metabolism, zeatin biosynthesis, flavonoid biosynthesis, diterpenoid biosynthesis and ubiquinone and other terpenoid-quinone biosynthesis were only enriched in the top 20 in CKvsM50. Tryptophan metabolism, isoquinoline alkaloid biosynthesis, alpha-Linolenic acid metabolism, tyrosine metabolism and steroid biosynthesis were only enriched in the top 20 in CKvsM100, indicating that 50 and 100 μM MeJA were more effective than 10 μM MeJA for metabolites such as terpenoids and flavonoids.

Mapman analysis of the DEGs was distributed in the cell wall, lipids, ascorbate and glutathione, sucrose and amino acid pathways (Figure S6A–C). More up-regulated DEGs were identified in the CKvsM100 comparison than the CKvsM10 and CKvsM50 comparisons in lipids, ascorbate and glutathione and amino acid pathways. More DEGs were identified in the CKvsM50 and CKvsM100 than the CKvsM100 comparison in the flavonoid metabolism pathway and more numbers of up-regulated DEGs were identified in the CKvsM50 and CKvsM100 comparisons. More DEGs were identified in the CKvsM10 and CKvsM100 than the CKvsM50 comparison in Non-MVA, MVA and phenylpropanoids pathways (Figure S6D–F). The above results indicate that 100 μM MeJA affected the metabolism including amino acids lipids, ascorbate, glutathione terpenoids and total flavonoids significantly more than 10 and 50 μM MeJA in rosemary suspension cells.

GO and KEGG enrichment and Mapman analyses of the DEGs showed that MeJA affected the synthesis of rosemary suspension cells' metabolites via multiple pathways, including plant hormone signal transduction, reactive oxygen species (ROS) clearance, osmotic balance, phenylpropanoid biosynthesis and many other metabolism pathways that may play important regulatory roles.

3.6. Differential Expression Analysis of Plant Hormone Signal Transduction Related Genes during Rosemary Suspension Cells under Different Concentrations of MeJA

To reveal the potential key genetic factors in rosemary suspension cells under different concentrations of MeJA, we found a total of 16 DEGs were assigned to JA biosynthesis, including 1 *LOX2S*, 2 *AOS*, 1 *AOC*, 2 *OPR*, 1 *OPCL1*, 7 *ACX*, 1 *MEP2* and 1 *J-O-MT*. *LOX2S*, *OPCL1* and *ACX* (Unigene8348_All, Unigene10587_All) were up-regulated from 0 to 100 μM MeJA treatments, *OPR* (Unigene16468_All) and *ACX* (Unigene3995_All, Unigene5819_All) were down-regulated from 0 to 100 μM MeJA treatments. *AOS* was down-regulated under 10 μM MeJA treatment but up-regulated under 100 μM MeJA treatment. Other DEGs showed different expression patterns in the three combinations (Figure S7A).

Meanwhile, we found a total of 25 DEGs were assigned to JA signal transduction, including 2 *JAR1*, 1 *COI1*, 9 *JAZ* and 13 *MYC2*. Only 1 *JAR1* (Unigene18877_All) was up-regulated under 10 μM MeJA treatment, others were down-regulated. *COI1* was drastically up-regulated from 0 to 100 μM MeJA treatment. *JAZ* (CL9927.Contig1_All) was down-regulated under 10 μM MeJA treatment, others were up-regulated. 1 *MYC2* (CL4564.Contig5_All) was down-regulated from 0 to 100 μM MeJA treatment, 6 *MYC2* (CL2762.Contig2_All, CL4726.Contig3_All, CL7026.Contig2_All, CL9744.Contig1_All, CL12032.Contig4_All, Unigene30434_All) were down-regulated from 0 to 100 μM MeJA treatments, the number of up-regulated DEGs under 100 μM MeJA treatment were the most, secondly was 50 μM MeJA treatment, the least was 10 μM MeJA treatment (Figure S7B). The result showed that 100 μM MeJA treatment can significantly induce the DEGs expression of JA biosynthesis and signal transduction pathway compared to10 and 50 μM MeJA, indicating that 100 μM MeJA could significantly affect JA biosynthesis and signal transduction in rosemary suspension cells.

In our study, based on the KEGG and other annotation, plant hormone signal transduction was the representative pathway. There were 248 significant DEGs were assigned to the plant hormone signal transduction pathway involved in auxin (59 DEGs), cytokinine (56 DEGs), gibberellin (GA, 34 DEGs), abscisic acid (ABA, 15 DEGs), ethylene (ET, 44 DEGs), brassinosteroid (BR, 32 DEGs), and salicylic acid (SA, 8 DEGs) (Figure S7C–I). The result of the analysis of these DEGs showed that 69 DEGs were up-regulated under different concentrations of MeJA, including 4 *AUX1*, 3 *TIR1*, 14 *AUX-IAA*, 5 *ARF* and 8 *TF*, 87 DEGs were down-regulated, for example, 1 *SAUR*, 10 *TF*, 22 *B-ARR*, 4 *A-ARR*, 3 *ABF*, 6 *ETR* and 3 *CTR1*. The result also showed that 50 and 100 μM MeJA could significantly inhibit the DEGs expression of the salicylic acid signal transduction pathway (Figure S7C). There were more numbers of up-regulated DEGs in auxin and ABA signal transduction under different concentrations of MeJA (Figure S7D,E), but more numbers of down-regulated DEGs in other hormone signals transduction pathways under different concentrations of MeJA (Figure S7F–I).

3.7. Phenylpropanoid Biosynthesis Related Genes Were Differential Expressed during Rosemary Suspension Cells under Different Concentrations of MeJA

Phenylpropanoid biosynthesis was the representative KEGG pathway, we found 138, 104 and 138 DEGs assigned to phenylpropanoid biosynthesis in CKvsM10, CKvsM50 and CKvsM100 combinations, respectively (Figure 5B). These DEGs include 5 *PAL*, 4 *4CL*, 8 *CCR*, 6 *HCT*, 8 *C3′H* (*CYP98A*), 8 *F6H1*, 3 *CSE*, 1 *caffeoyl-CoA-O-methyltransferase*, 21 *β-glucosidase*, 27 *cinnamyl-alcohol dehydrogenase*, 66 *peroxidase*. There were more numbers of up-regulated DEGs than down-regulated DEGs under 10 μM MeJA treatments, but there were more numbers of down-regulated DEGs under 50 and μM MeJA treatment. Among these DEGs, *PAL* (CL3764.Contig6_All), *4CL* (CL5102.Contig4_All), *CCR* (Unigene974_All), *HCT* (CL10076.Contig1_All), *F6H1* (CL352.Contig1_All), peroxidase (Unigene16028_All), *β-glucosidase* (CL41.Contig21_All), *caffeoyl-CoA-O-methyltransferase* (CL11417.Contig1_All), *cinnamyl-alcohol dehydrogenase* (CL5952.Contig4_All) were drastically up-regulated from 0 to 100 μM MeJA treatment (Figures 5A and S8). The result showed that 10 μM MeJA treatment can significantly induce the DEGs expression of phenylpropanoid biosynthesis compared to 50 and 100 μM MeJA in rosemary suspension cells.

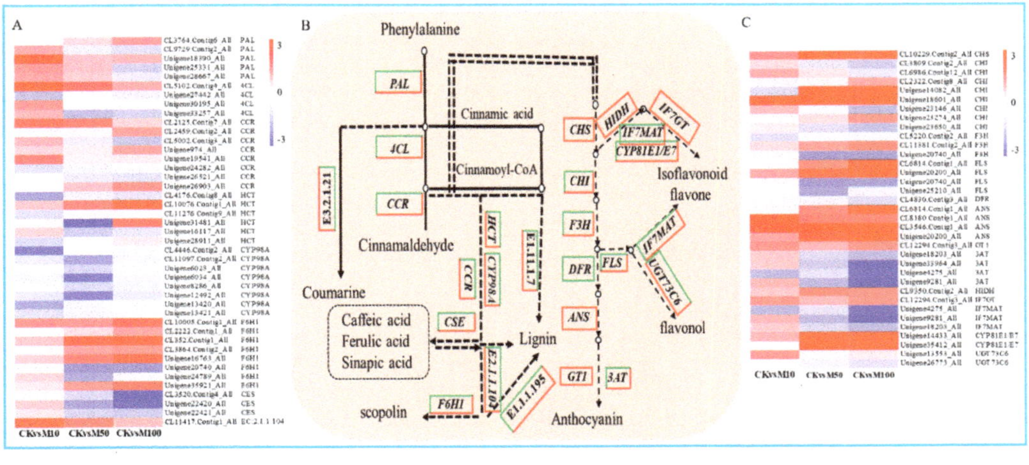

Figure 5. DEGs assigned to phenylpropanoid and flavonoid biosynthesis pathways under MeJA treatments. (**A**) heat map of the expression of DEGs in phenylpropanoid biosynthesis. (**B**) simplified diagram of phenylpropanoid and flavonoid biosynthetic pathway. (**C**) heat map of the expression of DEGs in flavonoid biosynthesis.

To reveal the potential key genetic factors in rosemary suspension cells under different concentrations of MeJA, we found a total of 35 DEGs were assigned to flavonoids biosynthesis (Figure 5B). Among these DEGs, these were 14 DEGs drastically up-regulated under different concentrations of MeJA treatments, include 1 *CHS*, 1 *CHI*, 1 *F3H*, 2 *FLS*, 4 *ANS*, 1 *GT1*, 1 *HIDH*, 1 *IF7GT*, 1 *CYP81E1/E7*, 1 *UGT73C6*. Instead, these were 5 DEGs drastically down-regulated from 0 to 100 μM MeJA treatment, include 1 *F3H*, 1 *DFR*, 1 *3AT*, 1 *IF7MAT* and 1 *UGT73C6*. The other 16 DEGs showed different differential expression patterns under different concentrations of MeJA, most of the DEGs were up-regulated under 10 μM MeJA treatment, but down-regulated under 50 and 100 μM MeJA treatment, for example, *CHI* (Unigene14082_All), which was down-regulated more significant under 100 than 50 μM MeJA treatment (Figure 5C). Therefore, indicating that different concentrations of MeJA could significantly affect the accumulation of flavonoids in rosemary suspension cells.

3.8. Terpenoid Biosynthesis Related Genes Were Differential Expressed during Rosemary Suspension Cells under Different Concentrations of MeJA

In our study, the results showed that a large number of DEGs were found in the terpenoid biosynthesis and terpenoids biosynthesis pathways (Figure 6). There were 6 significant DEGs assigned to the MVA pathway, including 1 *HMGCS*, 3 *HMGCR*, 1 *MVK*, and 1 *PMVK*. Among these DEGs except *HMGCS* (CL375.Contig4_All) and *HMGCR* (Unigene29235_All, CL6936.Contig10_All) were drastically up-regulated under different concentrations of MeJA treatments. There were 3 significant DEGs that were assigned to the MEP pathway, including 2 *DXS* and 1 *ispH*. *DXS* were down-regulated under 10 μM MeJA treatment, but up-regulated under 50 and 100 μM MeJA treatments, ispH was drastically down-regulated under different concentrations of MeJA treatments. The result showed that 100 μM MeJA treatment could significantly induce the DEGs expression of MVA biosynthesis compared to 50 and 100 μM MeJA, 10 μM MeJA treatment could significantly inhibit the DEGs expression of MVA biosynthesis in rosemary suspension cells. In addition, the DEGs *FPS* (*FDPS*) and *GGPS*, which directly act on terpenoids to synthesize precursors geranyl diphosphate (GPP), farnesyl pyrophosphate (FPP) and geranylgeranyl pyrophosphate (GGPP) and other DEGs in the terpenoid biosynthesis pathway showed differential expression patterns under different concentrations of MeJA, for example *FDPS* (CL1142.Contig2_All, Unigene14392_All), *GGPS*, *FACE2*, *ICMT*, *PCME* (Unigene7748_All) and *FOLK* were up-regulated from 0 to 100 μM MeJA treatment. We also found a large number of DEGs were assigned to ubiquinone and other terpenoid-quinone biosynthesis, carotenoid biosynthesis, sesquiterpenoid and triterpenoid biosynthesis, zeatin biosynthesis, steroid biosynthesis, monoterpenoid biosynthesis, diterpenoid biosynthesis and N-Glycan biosynthesis pathways (Figure S9). Therefore, different concentrations of MeJA could significantly affect the accumulation of terpenoids in rosemary suspension cells.

3.9. Transcription Factors Are Important in Rosemary Suspension Cells under Different Concentrations of MeJA

In this study, a total of 551 TFs were annotated, belonging to 43 TFs families in rosemary suspension cells (Figure 7A). Under the MeJA treatments, many differentially expressed TFs were found (Figure 7B,C). For example, MYB (15.97%), AP2-EREBP (14.88%), bHLH (8.17%), NAC (7.62%), WRKY (5.63%), G2-like (4.54%), C2H2 (4.54%), MADS (3.45%), Tify (3.09%), C3H (2.90%) and LOB (2.90%) which frequency were more than 2.90%, indicating that these TFs might play important regulatory roles in the synthesis of rosemary suspension cells important metabolites and the stress-response.

In order to understand the differentially expressed TFs in rosemary suspension cells, there were 320, 305 and 434 differentially expressed TFs in the CKvsM10, CKvsM50 and CKvsM100 comparisons, respectively (Figure 7D–F). The classification of differentially expressed TFs by family showed that MYB (48, 15.00%), AP2-EREBP (38, 11.88%), WRKY (28, 8.75%), G2-like (20, 6.25%), bHLH (20, 6.25%) in the CKvsM10 comparison; MYB (48, 15.74%), AP2-EREBP (42, 13.77%), NAC (30, 9.84%), WRKY (21, 6.89%), bHLH (21, 6.89%) in the CKvsM50 comparison; MYB (61, 15.72%), AP2-EREBP (57, 14.69%), bHLH (38, 9.79%), NAC (35, 9.02%), WRKY (22, 5.67%) in the CKvsM100 comparison. There were more differentially expressed TFs in the CKvsM100 than the CKvsM10 and CKvsM50 comparisons. Indicating that, MYB was the most significantly differentially expressed TFs in rosemary suspension cells responding to MeJA, AP2-EREBP, bHLH and WRKY might play an important role in rosemary suspension cells responding to different concentrations of MeJA through differential expression or specific expression.

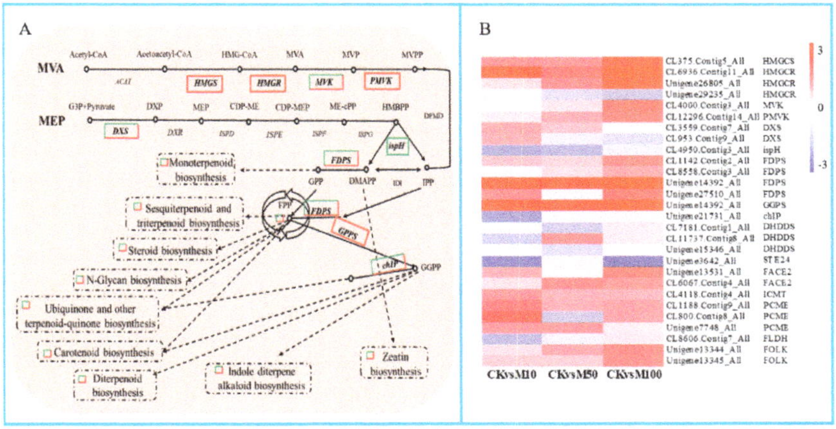

Figure 6. DEGs assigned to terpenoid biosynthesis pathway under MeJA treatments. (**A**) simplified diagram of the terpenoid biosynthesis pathway. (**B**) heat map of the expression of DEGs related to terpenoid backbone biosynthesis pathway.

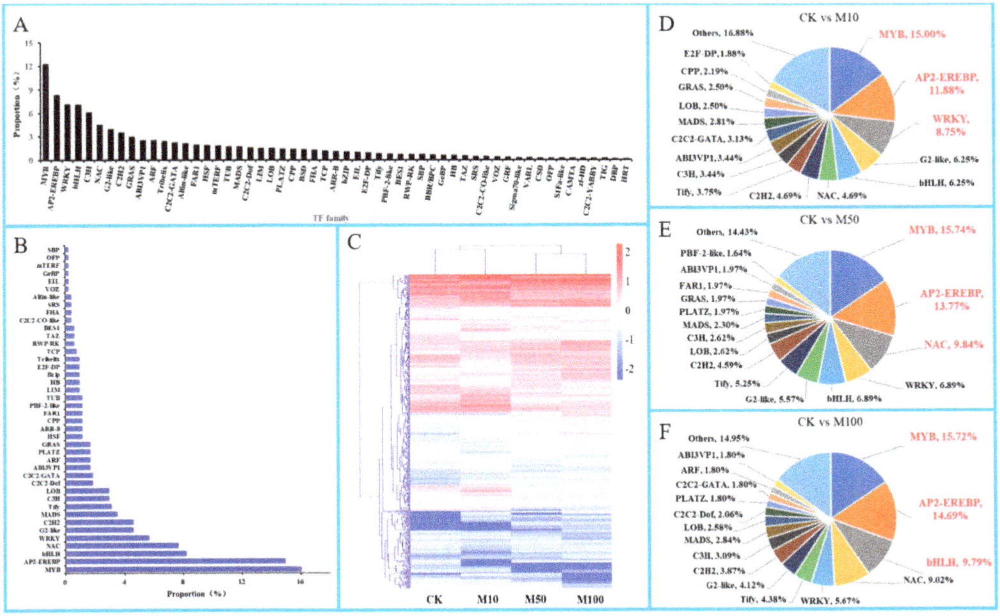

Figure 7. TFs in rosemary suspension cells under MeJA treatments. (**A**) classification statistics of TFs. (**B**) the Frequency distribution of differentially expressed TFs. (**C**) cluster analysis of expression profiles of TFs. (**D**) statistic of differentially TFs in the comparisons of CKvsM10. (**E**) statistic of differentially expressed TFs in the comparisons of CKvsM50. (**F**) statistic of differentially expressed TFs in the comparisons of CKvsM100.

3.10. qRT-PCR Verification of DEGs Related to MeJA

In order to further verify the reliability of gene expression profiles obtained by RNA-seq, the expression pattern of 11 DEGs was estimated by qRT-PCR in rosemary suspension cells under different concentrations of MeJA (Figure 8). The expression of *LOX2S*, *AOS*, *JAZ*, *MYC2*, *PAL*, *CHS*, *ANS*, *FPS* and *GPS* were the highest in the 100 μM treatment, only the expression of *AOS* was the lower in the 10 μM than the 0 μM MeJA (CK) treatment, while the expression of *4CL* was the highest in the 10 μM treatment. The expression pattern of *DFR* was the highest in 0 μM MeJA, followed by 50, 10 and 100 μM MeJA treatments. In summary, the qPCR verification results showed that the relative expression of the genes was similar to the transcriptome, affirming the consistency of RNA-seq data and our results.

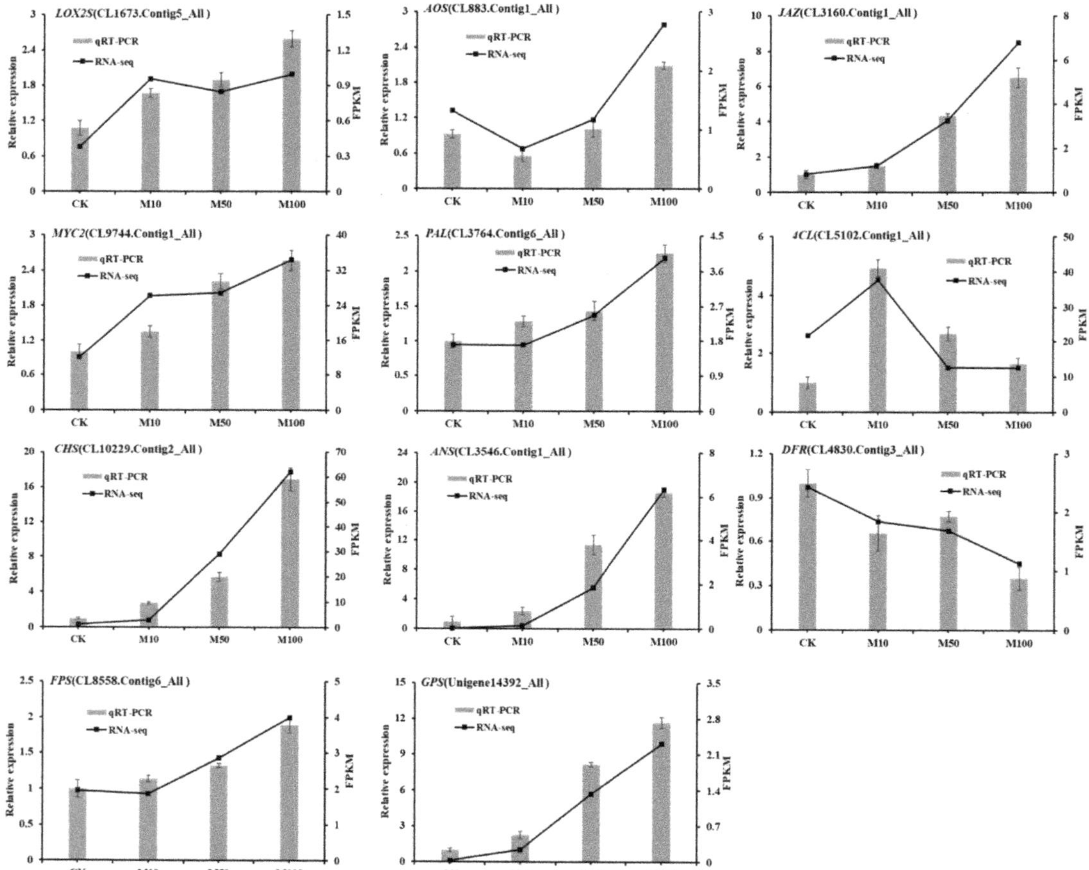

Figure 8. qRT-PCR verification of DEGs in rosemary suspension cells responding to MeJA.

4. Discussion

4.1. Phenylpropanoid and Terpenoid Pathway Were Closely Related to the Synthesis of Important Metabolite in Rosemary Suspension Cells

In our study, phenylpropanoid and terpenoid backbone biosynthesis pathways were significantly enriched. Phenylpropanoid biosynthesis was closely related to the biosynthesis of flavonoids and rosmarinic acid. Exogenous MeJA regulated phytoalexin and polyphenol biosynthesis by up-regulating *PAL* in *Arabidopsis* [36]. PAL was the key rate limiting enzyme, its activity was enhanced after plant stress [37]. In plants, MVA and MEP pathway provided premises for terpenoids biosynthesis [38]. During the process of rosemary suspension cells under different concentrations of MeJA, a total of 157, 35 and 143 DEGs involved in the phenylpropane, flavonoids and terpenoids biosynthesis, respectively, mostly were up-regulated. The key synthase genes *FDPS* and *GGPS* of precursors as GPP were up-regulated in rosemary suspension cells under 100 μM MeJA, which was similar to the results of *Taxus* and *Chrysanthemum indicum* [39,40]. Therefore, MeJA might enhance phenylpropane and terpenoids biosynthesis by regulating the DEGs in the phenylpropanoid and terpenoids biosynthesis pathways. In summary, those DEGs might involve in regulating secondary metabolites, such as flavonoids, rosmarinic acid and carnosic acid.

4.2. ROS Scavenging Systems Improved Tolerance of Rosemary Suspension Cells Responding to MeJA

After the external environment changed, the balance was destroyed and ROS was over-produced to damage the plants, which could induce the synthesis of phytoalexin [41,42]. Plant cells mainly respond to stress through enzymatic and non-enzymatic antioxidant systems [43–45]. In *Arabidopsis*, 177 PLP enzymes might regulate the expression of genes of hormone synthesis and signal transduction, *AtPDX2* could be induced under light, drought and low-temperature stress [46–48]. In *Arabidopsis vtc1* mutant, the content of Asc was 70% lower than the wild type, while the content of ABA and the expression of *NCED* were significantly increased [49]. Asc played an important role in protecting the body and avoiding normal metabolism from oxidative stress [50–52]. After PEG-6000 and H_2O_2 treatments, the antioxidant enzyme system and secondary metabolites synergistically eliminated excess ROS in *S.baicalensis* [53,54]. Proline confers tolerance to Cd-stress in tobacco *BY-2* cells by different mechanisms [55]. Different type of stresses stimulated plants to secrete flavonoids [56,57]. In our study, the activities of SOD, CAT, POD, PPO and PAL were significantly increased, the contents of H_2O_2 and MDA were reduced in rosemary suspension cells under different concentrations of MeJA. These phenomena improved the repair capacity of oxidative damage and ROS scavenging ability in plant cells and thereby enhancing the ability to cope with external factors in plant [58]. The changes of enzymatic and non-enzymatic antioxidants by MeJA could prevent oxidative damage, and promote the accumulation of secondary metabolites, such as flavonoids in rosemary suspension cells.

4.3. Plant Hormone Signal Transduction Played a Key Role in Rosemary Suspension Cells Responding to MeJA

Plant hormones play an important role in the synthesis of secondary metabolites, and the regulatory process involves a variety of signal transduction and interaction factors [59]. In *Arabidopsis arf6* and *arf8* single mutants and sesquimutants, *ARF6* and *ARF8* gene dosage affected the accumulation of JA and the expression of *MYB* [60]. Over-expression *miR393* could stabilize *ARF1* and *ARF9* to increase glucosinolate and decrease camalexin in *Arabidopsis* [61]. Exogenous cytokinin and ethylene up-regulated the alkaloid production through independent pathways in periwinkle suspension cells [62]. Exogenous ABA, GA and ethylene increased the levels of phenolic acids by activating the PAL and TAT in *Salvia miltiorrhiza* hairy roots [63]. DELLA would inhibit the transcriptional activation of downstream target genes by binding to PIF3 and PIF4 [64–66]. In *Tripterygium wilfordii* suspension cells, the contents of triptolide and triptolide increased significantly with exogenous ABA after 10 days [67], and the expression of *AACT*, *MCT*, *CMK* and *CYP450* in terpenoids biosynthesis increased significantly by exogenous MeJA [68]. The content of total essential oil was significantly increased by exogenous BR in *Mentha canadensis* [69]. Under MeJA treatment, the expression of *ODC*, *ADC*, *PMT*, *QPRT* and *bHLH* in nicotine biosynthesis were up-regulated, which affected the accumulation of nicotine and other pyridine organisms in *Nicotiana tabacum* [70]. *JAZ8* participated in the biosynthesis of phenolic acids in *Salvia miltiorrhiza* under MeJA treatment [71]. Over-expression *ORCA3* enhanced the expression of *Tdc*, *Str*, and *D4h* in terpenoid biosynthetic genes increased the accumulation of terpenoid indole alkaloids in *Catharanthus roseus* [72,73]. MeJA induced the expression of JA biosynthesis and signaling pathway genes, and the transcription factor *MYC2* was released from JAZ to transcribe and activate the down-stream response genes in *Arabidopsis* [74–76], *Hevea brasiliensis* [77], *Nicotiana tabacum* [78], *Catharanthus roseus* [79], *Chrysanthemum indicum* [80] and *Vitis vinifera* [81]. In our study, many genes showed differential expressions under MeJA treatment, e.g., *ARF, DELLA, EIN3, JAZ, MYC2*, further study of these genes is required. Thus, 100 μM MeJA could significantly induce the genes of JA biosynthesis and signal transduction and inhibit certain hormone signal transduction, especially SA signal transduction. MeJA could regulate the biosynthesis of secondary metabolites via the complex network of plant hormone signal transduction in rosemary suspension cells, such as terpenoids, flavonoids.

4.4. Transcription Factors Played an Important Role in Rosemary Suspension Cells Responding to MeJA

TFs played a key role in plant growth, development, secondary metabolism and resistance to stress [82]. In *Salvia miltiorrhiza*, *PAP1* could induce the expression of *PAL*, *C4H* and *CHS* to promote the accumulation of anthocyanins [83], *MYB39* and *MYB4* could negatively regulate the synthesis of rosmarinic acid [84,85], *ERF115* had different regulatory effects on salvianolic acid and tanshinone in different tissues and organs [86]. *MYC2* mediated plant responding to stress, regulated the genes in the biosynthesis of many secondary metabolites [87]. AP2/ERF, bHLH, MYB, NAC, WRKY and bZIP TFs families were related to the metabolism of terpenoids, MYB and bHLH could induce plant anthocyanin accumulation [88,89]. In *Arabidopsis*, several R2R3-MYB were involved in the regulation of flavonoid biosynthesis [90–92], *MYB4* negative controlled sinapate ester biosynthesis through down-regulated *C4H* in a UV-dependent manner [93], *MYB11, -12, -111* regulated flavonol biosynthesis by up-regulated *CHS, CHI, F3H, F3′H* and *FLS* [90,94], *MYB75, -90, -113, -114* controlled anthocyanin biosynthesis in vegetative [95], *MYB123* controlled the biosynthesis of proanthocyanidins in the seed coat [96]. *MYB5, -14* played a key role in seed coat polymer biosynthesis in *Medicago truncatula* [97]. *OsWRKY13* was an important TF to resist the infection of rice blast fungus by directly or indirectly regulating the genes of SA and JA signaling pathway [98]. Over-expression *li049*, the expressions of genes related to lignan biosynthesis were significantly up-regulated and the content was increased in *Isatis indigotica* hairy root [99]. *TcDRREB* regulated the key gene of the paclitaxel synthesis pathway to increase the content of paclitaxel increased with JA treatment in *Taxus* suspension cells [100]. DELLA could interact with JAZ1, reduce the inhibition of *JAZ1* to *MYC2* in *Arabidopsis* [101]. *RIM1* was a negative regulator for JA signaling in *Rice* [102]. In our study, the classification of them showed that MYB was most related to MeJA. R2R3-MYB were differentially expressed under MeJA treatment, *MYB111* was significantly up-regulated under 50 and 100 μM MeJA, and down-regulated under 10 μM MeJA. Therefore, MeJA might play a regulatory role in the synthesis of rosmarinic acid, terpenoids, flavonoids and other metabolites by including the MYB, AP2-EREBP, *MYC2* and *DELLA* in rosemary suspension cells.

In conclusion, MeJA increased the activities of PAL, SOD, POD, CAT and PPO, and reduced the contents of H_2O_2 and MDA, thus accelerated the ROS scavenging. A comparative analysis of global gene expression patterns provided subsets of DEGs in rosemary suspension cells under MeJA treatment. Our study revealed the expression profiles of genes involved in plant hormones signaling pathway, flavonoids, terpenoid backbone and phenylpropanoid biosynthesis pathway, TFs, indicating their regulatory role in the synthesis of the active compounds in rosemary suspension cells under MeJA treatment. We suggested a feasible working model based on the result (Figure 9). These transcriptomic data provided new insights into future functional studies, as a means of studying the molecular mechanisms on the biosynthesis of active compounds and the mining of key enzyme genes in rosemary suspension cells.

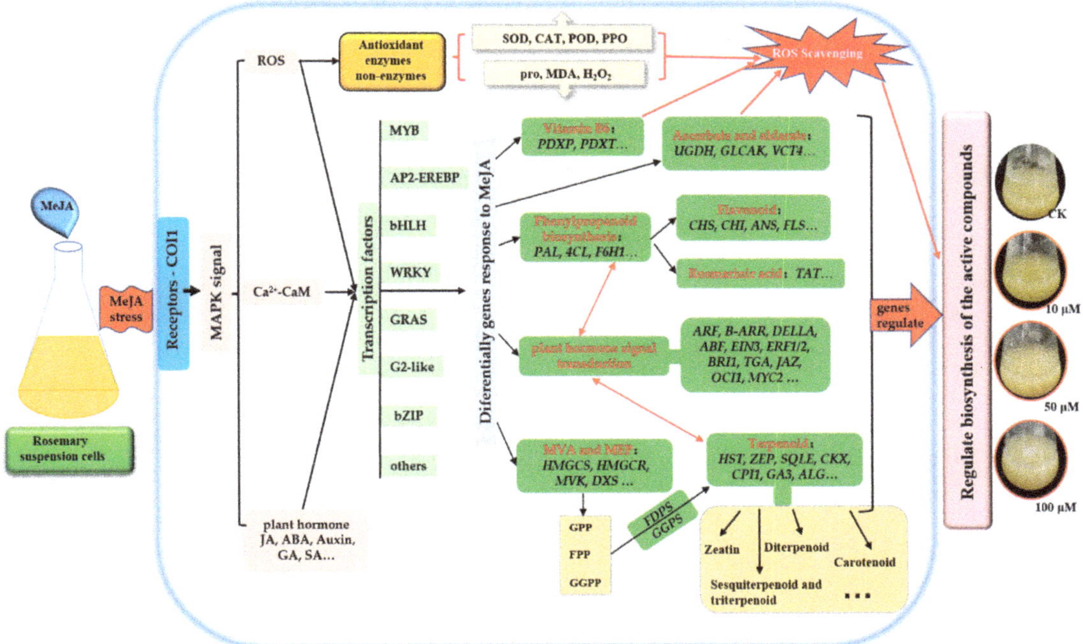

Figure 9. Schematic representation of rosemary suspension cells responding to MeJA. Antioxidant enzymes and non-enzymes, DEGs of many biosynthesis pathways played an essential beneficial in rosemary suspension cells responding to MeJA. Red indicates the key biosynthesis and signals transduction pathways, italics indicate key DEGs.

Supplementary Materials: The following Supplementary Materials are available online at www.mdpi.com/xxx/s1, Figure S1: Species distribution of the Nr database, Figure S2: GO Functional classification, Figure S3: KOG Functional classification, Figure S4: KEGG assignment of Unigenes, Figure S5: Top 20 enriched and regulated KEGG pathways, Figure S6: Rosemary suspension cells primary and secondary metabolic pathways under different concentrations of MeJA, Figure S7: DEGs assigned to plant hormone signal transduction pathway under MeJA treatments, Figure S8: DEGs assigned to Phenylpropanoid biosynthesis pathway under MeJA treatment, Figure S9: DEGs assigned to Terpenoid biosynthesis pathway under MeJA treatment, Table S1: Primers used for real-time quantitative PCR, Table S2: Summary of the sequencing data in each sample, Table S3: All_Unigene length distribution statistics, Table S4: The statistical resscriptome assembly, Table S5: The statistical resscriptome assembly, Table S6. Gene expression values for by FPKM, Table S7. The top 20 most expressed genes (FPKM) form CK, M10, M50, M100 library, Table S8: Statistical analysis of differential gene annotations to GOs in different comparison Groups.

Author Contributions: D.Y. performed the experiments, collected and analyzed the data, prepared the original draft. Z.L. and D.Y. designed the work. Z.L., Y.L., Y.C. and X.X. revised the manuscript carefully. Z.Z. provided assistance and collected the data. All authors have read and agreed to the published version of the manuscript.

Funding: This research was funded by the Science and Technology Plan Major Projects of Fujian Province (2015NZ0002-1), and the Construction of Plateau Discipline of Fujian Province (102/71201801101).

Institutional Review Board Statement: Not applicable.

Informed Consent Statement: Not applicable.

Data Availability Statement: Not applicable.

Conflicts of Interest: The authors declare no conflict of interest.

References

1. Borrás-Linares, I.; Stojanović, Z.; Quirantes-Piné, R.; Arráez-Román, D.; Švarc-Gajić, J.; Fernández-Gutiérrez, A.; Segura-Carretero, A. Rosmarinus Officinalis Leaves as a Natural Source of Bioactive Compounds. *Int. J. Mol. Sci.* **2014**, *15*, 20585–20606. [CrossRef] [PubMed]
2. Bai, N.; He, K.; Roller, M.; Lai, C.-S.; Shao, X.; Pan, M.-H.; Ho, C.-T. Flavonoids and Phenolic Compounds from *Rosmarinus officinalis*. *J. Agric. Food Chem.* **2010**, *58*, 5363–5367. [CrossRef]
3. Bicchi, C.; Binello, A.; Rubiolo, P. Determination of phenolic diterpene antioxidants in rosemary (*Rosmarinus officinalis* L.) with different methods of extraction and analysis. *Phytochem. Anal.* **2000**, *11*, 236–242. [CrossRef]
4. del Baño, M.J.; Lorente, J.; Castillo, J.; Benavente-García, O.; Marín, M.P.; Del Río, J.A.; Ortuño, A.; Ibarra, I. Flavonoid Distribution during the Development of Leaves, Flowers, Stems, and Roots of *Rosmarinus officinalis*. Postulation of a Biosynthetic Pathway. *J. Agric. Food Chem.* **2004**, *52*, 4987–4992. [CrossRef]
5. da Silva Bomfim, N.; Nakassugi, L.P.; Oliveira, J.F.; Kohiyama, C.Y.; Mossini, S.A.; Grespan, R.; Nerilo, S.B.; Mallmann, C.A.; Abreu Filho, B.A.; Machinski, M., Jr. Antifungal activity and inhibition of fumonisin production by *Rosmarinus officinalis* L. essential oil in Fusarium verticillioides (Sacc.) Nirenberg—ScienceDirect. *Food Chem.* **2015**, *166*, 330–336. [CrossRef]
6. Badreddine, B.S.; Olfa, E.; Samir, D.; Hnia, C.; Lahbib, B.J.M. Chemical composition of Rosmarinus and Lavandula essential oils and their insecticidal effects on Orgyia trigotephras (Lepidoptera, Lymantriidae). *Asian Pac. J. Trop. Med.* **2015**, *8*, 98–103. [CrossRef]
7. Kiran, C.R.; Sasidharan, I.; Kumar, D.R.S.; Sundaresan, A. Influence of natural and synthetic antioxidants on the degradation of Soybean oil at frying temperature. *J. Food Sci. Technol.* **2015**, *52*, 5370–5375. [CrossRef]
8. Dias, L.S.; Menis, M.E.C.; Jorge, N. Effect of rosemary (*Rosmarinus officinalis*) extracts on the oxidative stability and sensory acceptability of soybean oil. *J. Sci. Food Agric.* **2015**, *95*, 2021–2027. [CrossRef] [PubMed]
9. Hu, Y.; Zhang, N.; Fan, Q.; Lin, M.; Zhang, C.; Fan, G.; Zhai, X.; Zhang, F.; Chen, Z.; Yao, J. Protective efficacy of carnosic acid against hydrogen peroxide induced oxidative injury in HepG2 cells through the SIRT1 pathway. *Can. J. Physiol. Pharmacol.* **2015**, *93*, 625–631. [CrossRef] [PubMed]
10. Sedighi, R.; Zhao, Y.; Yerke, A.; Sang, S. Preventive and protective properties of rosemary (*Rosmarinus officinalis* L.) in obesity and diabetes mellitus of metabolic disorders: A brief review. *Curr. Opin. Food Sci.* **2015**, *2*, 58–70. [CrossRef]
11. Yan, M.; Li, G.; Petiwala, S.M.; Householter, E.; Johnson, J.J. Standardized rosemary (*Rosmarinus officinalis*) extract induces Nrf2/sestrin-2 pathway in colon cancer cells. *J. Funct. Foods* **2015**, *13*, 137–147. [CrossRef]
12. Bonfill, M.; Mangas, S.; Moyano, E.; Cusido, R.M.; Palazon, J. Production of centellosides and phytosterols in cell suspension cultures of Centella asiatica. *Plant Cell, Tissue Organ Cult.* **2011**, *104*, 61–67. [CrossRef]
13. Mai, V.C.; Drzewiecka, K.; Jeleń, H.; Narożna, D.; Rucińska-Sobkowiak, R.; Kęsy, J.; Floryszak-Wieczorek, J.; Gabryś, B.; Morkunas, I. Differential induction of Pisum sativum defense signaling molecules in response to pea aphid infestation. *Plant Sci.* **2014**, *221*, 1–12. [CrossRef]
14. Qiqing, C.; Yunfei, H.; Geng, L.; Yujia, L.; Wei, G.; Luqi, H. Effects of Combined Elicitors on Tanshinone Metabolic Profiling and SmCPS Expression in Salvia miltiorrhiza Hairy Root Cultures. *Molecules* **2013**, *18*, 7473–7485.
15. Zhang, S.; Yan, Y.; Wang, B.; Liang, Z.; Liu, Y.; Liu, F.; Qi, Z. Selective responses of enzymes in the two parallel pathways of rosmarinic acid biosynthetic pathway to elicitors in Salvia miltiorrhiza hairy root cultures. *J. Biosci. Bioeng.* **2014**, *117*, 645–651. [CrossRef] [PubMed]
16. Wang, H.; Yang, J.; De Ng, K.; He, X.; Zhan, R.; Liang, T. Methyl Jasmonate Affects Metabolism and Gene Transcription of Volatile Terpenoids from *Amomum villosum* Lour. *Mod. Tradit. Chin. Med. Mater. Med.-World Sci. Technol.* **2014**, *16*, 1528–1536.
17. Zhang, W.J.; Cao, X.Y.; Jiang, J.H. Triterpene biosynthesis in Euphorbia pekinensis induced by methyl jasmonate. *Guihaia* **2015**, *35*, 590–596.
18. Zhang, L.; Yang, B.; Lu, B.; Kai, G.; Wang, Z.; Xia, Y.; Ding, R.; Zhang, H.; Sun, X.; Chen, W.; et al. Tropane alkaloids production in transgenic Hyoscyamus niger hairy root cultures over-expressing Putrescine N-methyltransferase is methyl jasmonate-dependent. *Planta* **2007**, *225*, 887–896. [CrossRef]
19. Xiao, Y.; Gao, S.; Di, P.; Chen, J.; Zhang, L. Methyl jasmonate dramatically enhances the accumulation of phenolic acids in Salvia miltiorrhiza hairy root cultures. *Physiol. Plant.* **2010**, *137*, 1–9. [CrossRef]
20. Adio, A.M.; Casteel, C.L.; Vos, M.D.; Kim, J.H.; Joshi, V.; Li, B.; Juéry, C.; Daron, J.; Jander, K.G. Biosynthesis and Defensive Function of Nd-Acetylornithine, a Jasmonate-Induced Arabidopsis Metabolite C W. *Plant Cell* **2011**, *23*, 3303–3318. [CrossRef]
21. Zhang, H.; Hedhili, S.; Montiel, G.; Zhang, Y.; Chatel, G.; Pré, M.; Gantet, P.; Memelink, J. The basic helix-loop-helix transcription factor CrMYC2 controls the jasmonate-responsive expression of the ORCA genes that regulate alkaloid biosynthesis in Catharanthus roseus. *Plant J.* **2011**, *67*, 61–71. [CrossRef]
22. Boer, K.D.; Tilleman, S.; Pauwels, L.; Bossche, R.V.; Sutter, V.D.; Vanderhaeghen, R.; Hilson, P.; Hamill, J.D.; Goossens, A. APETALA2/ETHYLENE RESPONSE FACTOR and basic helix-loop-helix tobacco transcription factors cooperatively mediate jasmonate-elicited nicotine biosynthesis. *Plant J.* **2011**, *66*, 1053–1065. [CrossRef] [PubMed]
23. Shoji, T.; Hashimoto, K.T. Clustered Transcription Factor Genes Regulate Nicotine Biosynthesis in Tobacco. *Plant Cell* **2010**, *22*, 3390–3409. [CrossRef]

24. Caretto, S.; Quarta, A.; Durante, M.; Nisi, R.; De Paolis, A.; Blando, F.; Mita, G. Methyl jasmonate and miconazole differently affect arteminisin production and gene expression in Artemisia annua suspension cultures. *Plant Biol.* **2011**, *13*, 51–58. [CrossRef] [PubMed]
25. De Geyter, N.; Gholami, A.; Goormachtig, S.; Goossens, A. Transcriptional machineries in jasmonate-elicited plant secondary metabolism. *Trends Plant Sci.* **2012**, *17*, 349–359. [CrossRef]
26. Spyropoulou, E.A.; Haring, M.A.; Schuurink, R.C. RNA sequencing on Solanum lycopersicum trichomes identifies transcription factors that activate terpene synthase promoters. *BMC Genom.* **2014**, *15*, 1–16. [CrossRef]
27. Chen, R.-B.; Liu, J.-H.; Xiao, Y.; Zhang, F.; Chen, J.-F.; Ji, Q.; Tan, H.-X.; Huang, X.; Feng, H.; Huang, B.-K.; et al. Deep Sequencing Reveals the Effect of MeJA on Scutellarin Biosynthesis in Erigeron breviscapus. *PLoS ONE* **2015**, *10*, e0143881. [CrossRef]
28. Misra, R.C.; Maiti, P.; Chanotiya, C.; Shanker, K.; Ghosh, S.; Petit, J.; Bres, C.; Just, D.; Garcia, V.; Mauxion, J.-P.; et al. Methyl Jasmonate-Elicited Transcriptional Responses and Pentacyclic Triterpene Biosynthesis in Sweet Basil. *Plant Physiol.* **2014**, *164*, 1028–1044. [CrossRef]
29. Chen, R.; Li, Q.; Tan, H.; Chen, J.; Ying, X.; Ma, R.; Gao, S.; Philipp, Z.; Chen, W.; Zhang, L. Gene-to-Metabolite network for biosynthesis of lignans in MeJA-elicited Isatis indigotica hairy root cultures. *Front. Plant Sci.* **2015**, *6*, 952. [CrossRef]
30. Chen, J.; Dong, X.; Li, Q.; Zhou, X.; Gao, S.; Chen, R.; Sun, L.; Zhang, L.; Chen, W. Biosynthesis of the active compounds of Isatis indigotica based on transcriptome sequencing and metabolites profiling. *BMC Genom.* **2013**, *14*, 857. [CrossRef] [PubMed]
31. Zhang, K.; Luo, Z.; Guo, Y.; Mo, C.; Tu, D.; Ma, X.; Bai, L. Methyl jasmonate-induced accumulation of metabolites and transcriptional responses involved in triterpene biosynthesis in Siraitia grosvenorii fruit at different growing stages. *Acta Soc. Bot. Pol.* **2016**, *85*, 3503. [CrossRef]
32. Cao, X.W.; Jie, Y.; Lei, J.L.; Li, J.; Zhang, H.Y. De novo Transcriptome Sequencing of MeJA-Induced Taraxacum koksaghyz Rodin to Identify Genes Related to Rubber Formation. *Sci. Rep.* **2017**, *7*, 15697. [CrossRef] [PubMed]
33. Sun, G.L.; Yang, Y.F.; Xie, F.L.; Wen, J.F.; Wilson, J.Q. Deep Sequencing Reveals Transcriptome Re-Programming of Taxus x media Cells to the Elicitation with Methyl Jasmonate. *PLoS ONE* **2013**, *4*, e62865.
34. Zhao, Y.; Liu, T.; Luo, J.; Zhang, Q.; Xu, S.; Han, C.; Xu, J.; Chen, M.; Chen, Y.; Kong, L. Integration of a Decrescent Transcriptome and Metabolomics Dataset of Peucedanum praeruptorum to Investigate the CYP450 and MDR Genes Involved in Coumarins Biosynthesis and Transport. *Front. Plant Sci.* **2015**, *6*, 996. [CrossRef]
35. Wang, L.K.; Feng, Z.X.; Wang, X.; Wang, X.W. DEGseq: An R package for identifying differentially expressed genes from RNA-seq data. *Bioinformatics* **2010**, *26*, 136–138. [CrossRef]
36. Hammerschmidt, R. Induced disease resistance: How do induced plants stop pathogens? *Physiol. Mol. Plant Pathol.* **1999**, *55*, 77–84. [CrossRef]
37. Huang, X.; Stettmaier, K.; Michel, C.; Hutzler, P.; Mueller, M.J. Nitric oxide is induced by wounding and influences jasmonic acid signaling in Arabidopsis thaliana. *Planta* **2004**, *218*, 938–946. [CrossRef]
38. Rohmer, M. Mevalonate-Independent methylerythritol phosphate pathway for isoprenoid biosynthesis. Elucidation and distribution. *Pure Appl. Chem.* **2003**, *75*, 375–388. [CrossRef]
39. Li, S.-T.; Zhang, P.; Fu, C.-H.; Zhao, C.-F.; Dong, Y.-S.; Guo, A.-Y.; Yu, L.-J. Transcriptional profile of Taxus chinensis cells in response to methyl jasmonate. *BMC Genom.* **2012**, *13*, 295–305. [CrossRef]
40. Rohdich, F.; Hecht, S.; Gärtner, K.; Adam, P.; Krieger, C.; Amslinger, S.; Arigoni, D.; Bacher, A.; Eisenreich, W. Studies on the nonmevalonate terpene biosynthetic pathway: Metabolic role of IspH (LytB) protein. *Proc. Natl. Acad. Sci. USA* **2002**, *99*, 1158–1163. [CrossRef] [PubMed]
41. Percudani, R.; Peracchi, A. The B6 database: A tool for the description and classification of vitamin B6-dependent enzymatic activities and of the corresponding protein families. *BMC Bioinform.* **2009**, *10*, 273. [CrossRef]
42. Mooney, S.; Hellmann, H. Vitamin B6: Killing two birds with one stone? *ScienceDirect. Phytochem.* **2010**, *71*, 495–501. [CrossRef]
43. Raschke, M.; Boycheva, S.; Crèvecoeur, M.; Nunes-Nesi, A.; Witt, S.; Fernie, A.R.; Amrhein, N.; Fitzpatrick, T.B. Enhanced levels of vitamin B6 increase aerial organ size and positively affect stress tolerance in Arabidopsis. *Plant J.* **2011**, *66*, 414–432. [CrossRef] [PubMed]
44. Mandl, J.; Szarka, A.; Bánhegyi, G. Vitamin C: Update on physiology and pharmacology. *Br. J. Pharmacol.* **2009**, *157*, 1097–1110. [CrossRef]
45. Pastori, G.M.; Kiddle, G.; Antoniw, J.; Bernard, S.; Jovanovic, S.V.; Verrier, P.J.; Noctor, G.; Foyer, C.H. Leaf Vitamin C Contents Modulate Plant Defense Transcripts and Regulate Genes That Control Development through Hormone Signaling. *Plant Cell* **2003**, *15*, 939–951. [CrossRef] [PubMed]
46. Uchendu, E.E.; Leonard, S.W.; Traber, M.G.; Reed, B.M. Vitamins C and E improve regrowth and reduce lipid peroxidation of blackberry shoot tips following cryopreservation. *Plant Cell Rep.* **2010**, *29*, 25–35. [CrossRef]
47. Mahajan, S.; Pandey, G.K.; Tuteja, N. Calcium- and salt-stress signaling in plants: Shedding light on SOS pathway. *Arch. Biochem. Biophys.* **2008**, *471*, 146–158. [CrossRef]
48. Lushchak, V.I. Adaptive response to oxidative stress: Bacteria, fungi, plants and animals. *Comp. Biochem. Physiol. Part C Toxicol. Pharmacol.* **2011**, *153*, 175–190. [CrossRef]
49. Meyer, M.; Schreck, R.; Baeuerle, P.A. H_2O_2 and antioxidants have opposite effects on activation of NF-kappa B and AP-1 in intact cells: AP-1 as secondary antioxidant-responsive factor. *EMBO J.* **1993**, *12*, 2005–2015. [CrossRef]

50. Harfouche, A.; Rugini, E.; Mencarelli, F.; Botondi, R.; Muleo, R. Salicylic acid induces H_2O_2 production and endochitinase gene expression but not ethylene biosynthesis in Castanea sativa in vitro model system. *J. Plant Physiol.* **2008**, *165*, 734–744. [CrossRef]
51. Han, R.-B.; Yuan, Y.-J. Oxidative Burst in Suspension Culture of Taxus cuspidate Induced by a Laminar Shear Stress in Short-Term. *Biotechnol. Prog.* **2004**, *20*, 507–513. [CrossRef]
52. Luo, Y.-L.; Song, S.-Q.; Lan, Q.-Y. Possible Involvement of Enzymatic and Non-enzymatic Antioxidant System in Acquisition of Desiccation Tolerance of Maize Embryos. *Acta Bot. Yunnanica* **2009**, *31*, 253–259. [CrossRef]
53. Wang, B.; Zhang, T.X.; Du, H.W.; Zhao, Q.; Meng, X.C. Effect of peg on secondary metabolites in suspension cells of scutellaria baicalensis georgi. *Acta Med. Mediterr.* **2020**, *36*, 2307–2312.
54. Wang, B.; Zhang, T.; Li, Y.; Zhao, Q.; Meng, X. Effect of Sodium Hydrosulfite on Secondary Metabolite Flavonoids in Suspension Cells of Scutellaria baicalensis Georgi. *Lat. Am. J. Pharm.* **2020**, *39*, 1708–1714.
55. Islam, M.M.; Hoque, A.; Okuma, E.; Banu, M.N.A.; Shimoishi, Y.; Nakamura, Y.; Murata, Y. Exogenous proline and glycinebetaine increase antioxidant enzyme activities and confer tolerance to cadmium stress in cultured tobacco cells. *J. Plant Physiol.* **2009**, *166*, 1587–1597. [CrossRef] [PubMed]
56. Li, M.Y.; Yang, F.; Han, P.L.; Zhou, W.L.; Wang, J.H.; Yan, X.F.; Lin, J.X. Research progress on the mechanism of root exudates in response to abiotic stresses. *Chin. J. Appl. Environ. Biol.* **2021**, *4*, 1–13. [CrossRef]
57. Liu, X.M.; Xu, Q.L.; Li, Q.Q.; Zhang, H.; Xiao, J.X. Physiological responses of the two blueberry cultivars to inoculation with an arbuscular mycorrhizal fungus under low-temperature stress. *J. Plant Nutr.* **2017**, *40*, 2562–2570. [CrossRef]
58. Al-Gubory, K.H.; Fowler, P.A.; Garrel, C. The roles of cellular reactive oxygen species, oxidative stress and antioxidants in pregnancy outcomes. *Int. J. Biochem. Cell Biol.* **2010**, *42*, 1634–1650. [CrossRef]
59. Zhao, J.; Davis, L.C.; Verpoorte, R. Elicitor signal transduction leading to production of plant secondary metabolites. *Biotechnol. Adv.* **2005**, *23*, 283–333. [CrossRef]
60. Nagpal, P.; Ellis, C.M.; Weber, H.; Ploense, S.E.; Barkawi, L.S.; Guilfoyle, T.J.; Hagen, G.; Alonso, J.M.; Cohen, J.D.; Farmer, E.E.; et al. Auxin response factors ARF6 and ARF8 promote jasmonic acid production and flower maturation. *Development* **2005**, *132*, 4107–4118. [CrossRef]
61. Robert-Seilaniantz, A.; MacLean, D.; Jikumaru, Y.; Hill, L.; Yamaguchi, S.; Kamiya, Y.; Jones, J.D. The microRNA miR393 re-directs secondary metabolite biosynthesis away from camalexin and towards glucosinolates. *Plant J.* **2011**, *67*, 218–231. [CrossRef]
62. Yahia, A.; Kevers, C.; Gaspar, T.; Chénieux, J.-C.; Rideau, M.; Crèche, J. Cytokinins and ethylene stimulate indole alkaloid accumulation in cell suspension cultures of Catharanthus roseus by two distinct mechanisms. *Plant Sci.* **1998**, *133*, 9–15. [CrossRef]
63. Liang, Z.; Ma, Y.; Xu, T.; Cui, B.; Liu, Y.; Guo, Z.; Yang, D. Effects of Abscisic Acid, Gibberellin, Ethylene and Their Interactions on Production of Phenolic Acids in Salvia miltiorrhiza Bunge Hairy Roots. *PLoS ONE* **2013**, *8*, e72806. [CrossRef]
64. Min, L.; Li, Y.; Hu, Q.; Zhu, L.; Gao, W.; Wu, Y.; Ding, Y.; Liu, S.; Yang, X.; Zhang, X. Sugar and Auxin Signaling Pathways Respond to High-Temperature Stress during Anther Development as Revealed by Transcript Profiling Analysis in Cotton. *Plant Physiol.* **2014**, *164*, 1293–1308. [CrossRef]
65. Stephenson, P.G.; Fankhauser, C.; Terry, M.J. PIF3 is a repressor of chloroplast development. *Proc. Natl. Acad. Sci. USA* **2009**, *106*, 7654–7659. [CrossRef] [PubMed]
66. Yang, D.-L.; Yao, J.; Mei, C.-S.; Tong, X.-H.; Zeng, L.-J.; Li, Q.; Xiao, L.-T.; Sun, T.-P.; Li, J.; Deng, X.-W.; et al. Plant hormone jasmonate prioritizes defense over growth by interfering with gibberellin signaling cascade. *Proc. Natl. Acad. Sci. USA* **2012**, *109*, E1192–E1200. [CrossRef]
67. Rui, Z.; Wu, X.; Ma, B.; Shang, C.; Wang, X.; Wei, G.; Huang, L. Effects of Different Concentration of ABA on Terpenoid in Cell of Radix Folium seu Flos Tripterygii Wilfordii. *World Chin. Med.* **2018**, *13*, 264–270.
68. Gao, J.; Zhang, Y.-F.; Zhou, J.-W.; Wu, X.-Y.; Gao, W.; Huang, L.-Q. Bioinformatics and tissue distribution analysis of Tripterygium wilfordii CYP450. *China J. Chin. Mater. Med.* **2019**, *44*, 3594–3600.
69. Çoban, Ö.; Baydar, N.G. Brassinosteroid effects on some physical and biochemical properties and secondary metabolite accumulation in peppermint (*Mentha piperita* L.) under salt stress. *Ind. Crop. Prod.* **2016**, *86*, 251–258. [CrossRef]
70. Urtasun, N.; García, S.C.; Iusem, N.D.; Moretti, M.B. Predominantly Cytoplasmic Localization in Yeast of ASR1, a Non-Receptor Transcription Factor from Plants. *Open Biochem. J.* **2010**, *4*, 68–71. [CrossRef]
71. Ge, Q.; Zhang, Y.; Hua, W.-P.; Wu, Y.-C.; Jin, X.-X.; Song, S.-H.; Wang, Z.-Z. Combination of transcriptomic and metabolomic analyses reveals a JAZ repressor in the jasmonate signaling pathway of Salvia miltiorrhiza. *Sci. Rep.* **2015**, *5*, 14048. [CrossRef] [PubMed]
72. van der Fits, L.; Memelink, J. ORCA3, a Jasmonate-Responsive Transcriptional Regulator of Plant Primary and Secondary Metabolism. *Science* **2000**, *289*, 295–297. [CrossRef]
73. Leslie, V.D.F.; Memelink, J. The jasmonate-inducible AP2/ERF-domain transcription factor ORCA3 activates gene expression via interaction with a jasmonate-responsive promoter element. *Plant J.* **2010**, *25*, 43–53.
74. Turner, J.G.; Devoto, E.A. The Jasmonate Signal Pathway. *Plant Cell* **2002**, *14*, S153–S164. [CrossRef]
75. Dixon, R.A.; Achnine, L.; Kota, P.; Liu, C.-J.; Reddy, M.S.S.; Wang, L. The phenylpropanoid pathway and plant defence-a genomics perspective. *Mol. Plant Pathol.* **2002**, *3*, 371–390. [CrossRef] [PubMed]
76. Chini, A.; Fonseca, S.; Fernandez, G.; Adie, B.; Chico, J.M.; Lorenzo, O.; Garciacasado, G.; Lopezvidriero, I.; Lozano, F.M.; Ponce, M.R.; et al. The JAZ family of repressors is the missing link in jasmonate signalling. *Nature* **2007**, *448*, 666–671. [CrossRef]

77. Tian, W.-W.; Huang, W.-F.; Zhao, Y. Cloning and characterization of *HbJAZ1* from the laticifer cells in rubber tree (*Hevea brasiliensis* Muell. Arg.). *Trees Struct. Funct.* **2010**, *24*, 771–779. [CrossRef]
78. Vanholme, B.; Grunewald, W.; Bateman, A.; Kohchi, T.; Gheysen, G. The tify family previously known as ZIM. *Trends Plant Sci.* **2007**, *12*, 239–244. [CrossRef] [PubMed]
79. Kazan, K.; Manners, J.M. Jasmonate Signaling: Toward an Integrated View. *Plant Physiol.* **2008**, *146*, 1459–1468. [CrossRef]
80. Gao, W.J. *Mining and Functional Analysis of Genes Involved in the Biosynthetic of Terpenoids of Chrysanthemum Indicum Var. Aromaticum Induced by Methyl Jasmonate*; Northeast Forestry University: Haerbin, China, 2019.
81. Men, L.; Yan, S.; Liu, G. De novo characterization of *Larix gmelinii* (Rupr.) Rupr. transcriptome and analysis of its gene expression induced by jasmonates. *BMC Genom.* **2013**, *14*, 548. [CrossRef] [PubMed]
82. Wang, H.; Wang, H.; Shao, H.; Tang, X. Recent Advances in Utilizing Transcription Factors to Improve Plant Abiotic Stress Tolerance by Transgenic Technology. *Front. Plant Sci.* **2016**, *7*, 67. [CrossRef]
83. Li, J.; Chen, C.; Wang, Z.-Z. The complete chloroplast genome of the Dendrobium strongylanthum (Orchidaceae: Epidendroideae). *Mitochondrial DNA Part A* **2015**, *27*, 3048–3049. [CrossRef]
84. Zhang, S.; Ma, P.; Yang, D.; Li, W.; Liang, Z.; Liu, Y.; Liu, F. Cloning and Characterization of a Putative R2R3 MYB Transcriptional Repressor of the Rosmarinic Acid Biosynthetic Pathway from Salvia miltiorrhiza. *PLoS ONE* **2013**, *8*, e73259. [CrossRef]
85. Song, J.; Wang, Z. RNAi-Mediated suppression of the phenylalanine ammonia-lyase gene in Salvia miltiorrhiza causes abnormal phenotypes and a reduction in rosmarinic acid biosynthesis. *J. Plant Res.* **2011**, *124*, 183–192. [CrossRef]
86. Hickman, R.; Van Verk, M.C.; Van Dijken, A.J.H.; Mendes, M.P.; Vroegop-Vos, I.A.; Caarls, L.; Steenbergen, M.; Van der Nagel, I.; Wesselink, G.J.; Jironkin, A.; et al. Architecture and Dynamics of the Jasmonic Acid Gene Regulatory Network. *Plant Cell* **2017**, *29*, 2086–2105. [CrossRef] [PubMed]
87. Xu, Y.; Zhu, C.; Xu, C.; Sun, J.; Grierson, D.; Zhang, B.; Chen, K. Integration of Metabolite Profiling and Transcriptome Analysis Reveals Genes Related to Volatile Terpenoid Metabolism in Finger Citron (C. medica var. sarcodactylis). *Molecules* **2019**, *24*, 2564. [CrossRef]
88. Outchkourov, N.S.; Carollo, C.A.; Gomez-Roldan, V.; De Vos, R.C.H.; Bosch, D.; Hall, R.D.; Beekwilder, J. Control of anthocyanin and non-flavonoid compounds by anthocyanin-regulating MYB and bHLH transcription factors in Nicotiana benthamiana leaves. *Front. Plant Sci.* **2014**, *5*. [CrossRef]
89. Dubos, C.; Stracke, R.; Grotewold, E.; Weisshaar, B.; Martin, C.; Lepiniec, L. MYB transcription factors in Arabidopsis. *Trends Plant Sci.* **2010**, *15*, 573–581. [CrossRef]
90. Czemmel, S.; Heppel, S.C.; Bogs, J. R2R3 MYB transcription factors: Key regulators of the flavonoid biosynthetic pathway in grapevine. *Protoplasma* **2012**, *249*, 109–118. [CrossRef] [PubMed]
91. Escaray, F.J.; Passeri, V.; Perea-García, A.; Antonelli, C.J.; Damiani, F.; Ruiz, O.A.; Paolocci, F. The R2R3-MYB TT2b and the bHLH TT8 genes are the major regulators of proanthocyanidin biosynthesis in the leaves of Lotus species. *Planta* **2017**, *246*, 243–261. [CrossRef]
92. Jin, H.; Cominelli, E.; Bailey, P.; Parr, A.; Mehrtens, F.; Jones, J.; Tonelli, C.; Weisshaar, B.; Martin, C. Transcriptional repression by AtMYB4 controls production of UV-protecting sunscreens in Arabidopsis. *EMBO J.* **2000**, *19*, 6150–6161. [CrossRef]
93. Wang, F.; Kong, W.; Wong, G.; Fu, L.; Peng, R.; Li, Z.; Yao, Q. AtMYB12 regulates flavonoids accumulation and abiotic stress tolerance in transgenic Arabidopsis thaliana. *Mol. Genet. Genom.* **2016**, *291*, 1545–1559. [CrossRef]
94. Gonzalez, A.; Zhao, M.; Leavitt, J.M.; Lloyd, A.M. Regulation of the anthocyanin biosynthetic pathway by the TTG1/bHLH/Myb transcriptional complex in Arabidopsis seedlings. *Plant J.* **2010**, *53*, 814–827. [CrossRef]
95. Lepiniec, L.; Debeaujon, I.; Routaboul, J.-M.; Baudry, A.; Pourcel, L.; Nesi, N.; Caboche, M. Genetics and biochemistry of seed flavonoids. *Annu. Rev. Plant Biol.* **2006**, *57*, 405–430. [CrossRef]
96. Liu, C.; Jun, J.H.; Dixon, R.A. MYB5 and MYB14 Play Pivotal Roles in Seed Coat Polymer Biosynthesis in Medicago truncatula. *Plant Physiol.* **2014**, *165*, 1424–1439. [CrossRef]
97. Schluttenhofer, C.; Yuan, L. Regulation of Specialized Metabolism by WRKY Transcription Factors. *Plant Physiol.* **2015**, *167*, 295–306. [CrossRef]
98. Sun, M.; Shi, M.; Wang, Y.; Huang, Q.; Yuan, T.; Wang, Q.; Wang, C.; Zhou, W.; Kai, G. The biosynthesis of phenolic acids is positively regulated by the JA-responsive transcription factor ERF115 inSalvia miltiorrhiza. *J. Exp. Bot.* **2019**, *70*, 243–254. [CrossRef]
99. Ma, R.; Xiao, Y.; Lv, Z.; Tan, H.; Chen, R.; Li, Q.; Chen, J.; Wang, Y.; Yin, J.; Zhang, L.; et al. AP2/ERF Transcription Factor, Ii049, Positively Regulates Lignan Biosynthesis in Isatis indigotica through Activating Salicylic Acid Signaling and Lignan/Lignin Pathway Genes. *Front. Plant Sci.* **2017**, *8*, 1361. [CrossRef]
100. Dai, Y.L.; Qin, Q.L.; Kong, D.L.; Zha, L.S.; Jin, X.J. Isolation and characterization of a novel cDNA encoding methyl jasmonate-responsive transcription factor TcAP2 from Taxus cuspidata. *Biotechnol. Lett.* **2009**, *31*, 1801–1809. [CrossRef]
101. Feng, S.; Martinez, C.; Gusmaroli, G.; Wang, Y.; Zhou, J.; Wang, F.; Chen, L.; Yu, L.; Iglesias-Pedraz, J.M.; Kircher, S.; et al. Coordinated regulation of Arabidopsis thaliana development by light and gibberellins. *Nature* **2008**, *451*, 475–479. [CrossRef]
102. Yoshii, M.; Yamazaki, M.; Rakwal, R.; Kishi-Kaboshi, M.; Miyao, A.; Hirochika, H. The NAC transcription factor RIM1 of rice is a new regulator of jasmonate signaling. *Plant J.* **2010**, *61*, 804–815. [CrossRef]

genes

MDPI

Article

In Silico and Transcription Analysis of Trehalose-6-phosphate Phosphatase Gene Family of Wheat: Trehalose Synthesis Genes Contribute to Salinity, Drought Stress and Leaf Senescence

Md Ashraful Islam [1,†], Md Mustafizur Rahman [2,†], Md Mizanor Rahman [2], Xiujuan Jin [1], Lili Sun [1], Kai Zhao [1], Shuguang Wang [1], Ashim Sikdar [3], Hafeez Noor [1], Jong-Seong Jeon [2], Wenjun Zhang [1] and Daizhen Sun [1,*]

1 State Key Laboratory of Sustainable Dryland Agriculture, College of Agronomy, Shanxi Agricultural University, Taigu 030801, China; a.islam160@nwafu.edu.cn (M.A.I.); 15582408175@163.com (X.J.); 18735430724@163.com (L.S.); zk51712@163.com (K.Z.); wsg6162@126.com (S.W.); hafeeznoorbaloch@gmail.com (H.N.); zhangwenjun9876@163.com (W.Z.)

2 Graduate School of Biotechnology and Crop Biotech Institute, Kyung Hee University, Yongin 17104, Korea; mr10bau2@khu.ac.kr (M.M.R.); mizanor@khu.ac.kr (M.M.R.); jjeon@khu.ac.kr (J.-S.J.)

3 Department of Agroforestry and Environmental Science, Sylhet Agricultural University, Sylhet 3100, Bangladesh; ashim.aes@sau.ac.bd

* Correspondence: sdz64@126.com

† These authors contributed equally to this work.

Citation: Islam, M.A.; Rahman, M.M.; Rahman, M.M.; Jin, X.; Sun, L.; Zhao, K.; Wang, S.; Sikdar, A.; Noor, H.; Jeon, J.-S.; et al. In Silico and Transcription Analysis of Trehalose-6-phosphate Phosphatase Gene Family of Wheat: Trehalose Synthesis Genes Contribute to Salinity, Drought Stress and Leaf Senescence. *Genes* **2021**, *12*, 1652. https://doi.org/10.3390/genes12111652

Academic Editor: Patrizia Galeffi

Received: 22 September 2021
Accepted: 19 October 2021
Published: 20 October 2021

Publisher's Note: MDPI stays neutral with regard to jurisdictional claims in published maps and institutional affiliations.

Abstract: Trehalose-6-phosphate phosphatase (*TPP*) genes take part in trehalose metabolism and also in stress tolerance, which has been well documented in many species but poorly understood in wheat. The present research has identified a family of 31 *TPP* genes in *Triticum aestivum* L. through homology searches and classified them into five clades by phylogenetic tree analysis, providing evidence of an evolutionary status with *Hordeum vulgare, Brachypodium distachyon* and *Oryza sativa.* The exon-intron distribution revealed a discrete evolutionary history and projected possible gene duplication occurrences. Furthermore, different computational approaches were used to analyze the physical and chemical properties, conserved domains and motifs, subcellular and chromosomal localization, and three-dimensional (3-D) protein structures. *Cis*-regulatory elements (CREs) analysis predicted that *TaTPP* promoters consist of CREs related to plant growth and development, hormones, and stress. Transcriptional analysis revealed that the transcription levels of *TaTPPs* were variable in different developmental stages and organs. In addition, qRT-PCR analysis showed that different *TaTPPs* were induced under salt and drought stresses and during leaf senescence. Therefore, the findings of the present study give fundamental genomic information and possible biological functions of the *TaTPP* gene family in wheat and will provide the path for a better understanding of *TaTPPs* involvement in wheat developmental processes, stress tolerance, and leaf senescence.

Keywords: in silico; *Cis*-regulatory elements; gene transcription; trehalose-6-phosphate phosphatase; wheat

1. Introduction

Cereals are indeed the single most significant part of the diet for the majority of the global population, with about 60% to 80% of carbohydrates coming straightly from them in developing and under-developing nations, respectively [1]. According to the FAO's most current predictions, global grain production in 2021 will increase by 1.7% over 2020, achieving 2817 million tons [2]. Wheat (*Triticum aestivum* L.) is the world's largest extensively grown cereal crop and is among the most often eaten cereals by the world population [3]. The major abiotic stresses that decrease wheat productivity throughout the growing period include water shortages, high temperatures, and salinity [4]. Among them, salinity is a major barrier to crop production, especially in wheat, resulting in a yield loss of 65% in moderately saline soils, by influencing nearly every stage of plant growth and development, including germination, vegetative growth, and reproductive growth [5,6]. This abiotic stress condition results in a decrease in yield related traits that directly affect the yield of cereal crops. Thus, one of the most significant tasks for plant breeders right now is to uncover the genes associated with abiotic stress responses and to cultivate genetically engineered varieties with improved stress tolerance [7,8].

Plants generate various organic molecules, such as soluble sugar and free amino acids, in response to stress exposure. Trehalose is one of these non-reducing disaccharides composed of two molecules of α-glucose that may accumulate in the cell up to 12% of its dry mass to maintain its integrity and is associated with plant abiotic stress tolerance, including high and low temperature, drought, and osmotic stress tolerance [9–12]. Many species, including yeast, fungus, invertebrates, plants, bacteria, insects, green weed, and cyanobacteria synthesize this sugar substance [12–15]. Except for vertebrates, the synthesis of trehalose in plants and other organisms involves two phases with two catalytic enzymes, trehalose-6-phosphate synthase (TPS) and trehalose-6-phosphate phosphatase (TPP). TPS produces trehalose-6-phosphate (T6P), a phosphorylated intermediate, from Uridine diphosphate-glucose (UDPG) and Glucose-6-phosphate (G6P) in the first phase, and the TPP dephosphorylates T6P to produce trehalose in the second phase (Figure 1). Trehalose is then hydrolyzed by an enzyme called trehalase (TRE) to synthesize two molecules of glucose, which suggests that TPS, TPP, and TRE are the three enzymes involved in the trehalose biosynthesis pathway [16]. The *TPS* and *TPP* families encode multiple genes, but *TRE* is denoted by a single copy of the gene [17–19].

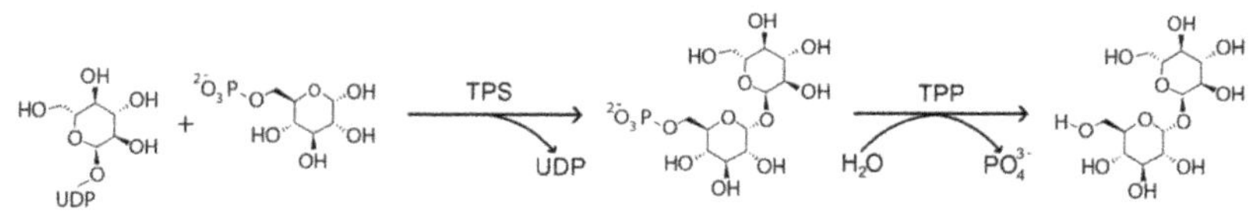

Figure 1. Trehalose biosynthesis pathway in plants. Uridine diphosphate glucose (UDP), Trehalose-6-phosphate synthase (TPS) and Trehalose-6-phosphate phosphatase (TPP).

In addition to providing a route for the production of trehalose, *TPS* and *TPP* have been shown to serve as signaling molecules in higher plants by modulating a variety of plant metabolic and developmental processes. T6P is a signaling metabolite in plants that links growth and development to carbon metabolism and serves as a signal of sucrose status at various phases of the plant's development [20–22]. *TPS* genes were discovered to be involved in the germination of seeds, stress signaling, vegetative phase separation, shoot branching, and flowering time regulation in *Arabidopsis* and rice, with *TPS1* being the most studied [23–27]. Instead, TPP was found to inhibit SnRK1 (Sn1-related protein kinase) activity, a well-known transcriptional regulatory pathway under stress and energy metabolism [28]. The Ramosal (RA1) transcription factor activates the transcription of *TPP* to regulate flower branching, which suggests that trehalose may have a role in specific developmental processes [29]. Tobacco plant overexpressing *Escherichia coli TPS* gene *ostA* improved photosynthesis efficiency by enhancing RUBISCO concentration, although *ostB*, a *TPP* gene, exhibited the opposite impact, further suggesting the significance of trehalose in plant photosynthesis [30].

Various studies have reported trehalose enzymes to enhance abiotic and biotic stress tolerance, such as in *Arabidopsis* [31,32]. For example, *ZxTPP* (*Zygophyllum xanthoxylum*) or *ostA* and *ostB* containing tobacco transgenic plants were significantly tolerant to drought [33,34]. Likewise, *ostA* and *ostB* transformed rice plants showed increased trehalose levels and enhanced performance against cold, salt, and drought stresses [35]. Exogenous trehalose triggered a signal transduction pathway including calcium and reactive oxygen species (ROS) and *OsTPP1* or *OsTPP3* transgenic rice and maize plants induced stress-related genes that conferred drought tolerance [36–38]. After drought stress, vulnerable maize seedlings had lower *ZmTPP1* expression, whereas resistant seedlings had higher expression [39]. *TPP* promoters' *Cis*-regulatory elements (CREs) stimulate trehalose metabolism and improve stress response. In *Arabidopsis*, *ABF1*, *ABF2*, and *ABF4* are ABA-responsive elements that directly influence *AtTPPI* expression to increase drought tolerance by changing stomatal apertures [40]. The transcription factor that responds to ABA in the presence of ABA, ABF2 binds directly to the *AtTPPE* promoter, triggering its expression for root elongation and stomatal movement via producing ROS [41]. DREB1A, which binds to the DRE/CR motif in the *AtTPPF* promoter, is thought to upregulate *AtTPPF* transcription in drought-stressed plants [32]. T6P role as a signal for increased carbon availability might have implications for leaf senescence control, as the accumulation of sugars has been demonstrated during leaf senescence in *Arabidopsis*, wheat, tobacco, and maize. The phenotype of mature *otsB*-overexpressing *Arabidopsis* plants included delayed senescence and decreased anthocyanin accumulation, suggesting that the role of *TPP* may perform a crucial role during leaf senescence in plants [42–45]. To date, *TaTPP-6AL1* and its functional marker have been shown to improve crop yield in wheat [46]. However, the gene structure and regulatory mechanism of wheat *TPPs* are not well studied.

The present study intends to investigate wheat *TPPs* in silico by identification of *TaTPPs*, gene duplication analysis, phylogenetic relationship with other species, subcellular localization prediction, motif and domain analyses, proteins 3-D structure modeling, investigation of CREs, and gene transcription analysis that have all been performed to better understand *TaTPPs* functions in wheat.

2. Materials and Methods

2.1. Identification of Putative TPPs in the Wheat Genome

To find putative *TPPs* in wheat, we utilized *TPPs* from *Arabidopsis* and rice. Ensembl Plants database was used to collect TPP protein sequences from *Arabidopsis* and rice and a BLASTp search was conducted against the most recent wheat assembly from the IWGSC (RefSeq v1.0) (http://plants.ensembl.org/index.html, 10^{-5} cut-off e-value and bit-score > 100, accessed on 12 March 2021). After eliminating duplicated sequences, SMART (http://smart.embl-heidelberg.de/, accessed on 12 March 2021) or InterPro (https://www.ebi.ac.uk/interpro, accessed on 12 March 2021) and NCBI CDD (https://www.ncbi.nlm.nih.gov/Structure/cdd/wrpsb.cgi, accessed on 12 March 2021) were used to examine the remaining sequences for the presence of transmembrane domains. TPP-related domain-containing protein sequences were collected and designated consecutively according to their chromosomal locations after the sequences without transmembrane domains were deleted. The ProtParam software (https://web.expasy.org/protparam/, accessed on 13 March 2021) was used to calculate the length, molecular weight, isoelectric point (pI), and grand average of hydropathicity (GRAVY) of TPP proteins.

2.2. Chromosome Localization, Gene Duplication and Synteny Analysis

TPPs genomic locations were acquired from the Ensembl Plants BioMart (http://plants.ensembl.org/biomart/martview, accessed on 14 March 2021) for chromosomal distribution. The *TPPs* were given a 'Ta' prefix and were numbered in ascending order according to their ascending chromosomal location. The *TaTPPs* on the wheat chromosomes were represented using TBtools. A NCBI BlastP search (https://blast.ncbi.nlm.nih.gov/Blast.cgi?PROGRAM=blastp&PAGE_TYPE=BlastSearch&BLAST_SPEC=&LINK_LOC=blasttab&LAST_PAGE=blastn, query conditions: percent identity between 75 and 100 and query coverage between 80 and 100, accessed on 14 March 2021) based on the proportion of query cover to the identity of the *TaTPPs* against each other was performed to check for gene duplication [47]. Based on a BLAST search and a phylogenetic tree, duplicate gene pairs were identified. TBtools was used to determine the non-synonymous substitution rate (Ka), synonymous substitution rate (Ks), and Ka/Ks ratio [48]. The synteny relationships of wheat *TPP* genes with different plant species were analyzed using TBtools.

2.3. Phylogenetic Analysis, Exon-Intron Distribution and 3-D Structure Modeling

ClustalW in MEGA X was used to align full-length protein sequences from various species [49]. Following the alignment, MEGA X was used to create a phylogenetic tree with the Maximum Likelihood method [50] and 1000 bootstrap values [51]. To examine the exon-intron distribution of *TaTPPs*, the TBtool was used to align the CDSs and genomic sequences. SWISS-MODEL Workspace web tools (https://swissmodel.expasy.org/interactive#sequence, accessed on 16 March 2021), GASS and SOPMA secondary structural method (https://npsa-prabi.ibcp.fr/cgi-bin/npsa_automat.pl?page=npsa%20_sopma.html, accessed on 16 March 2021) and MolProbity server (http://molprobity.biochem.duke.edu/, accessed on 16 March 2021) were used to conduct 3-D structure analyses of TaTPP proteins [52–57].

2.4. Subcellular Localization Prediction and Protein Domain Analysis

PredSL (http://aias.biol.uoa.gr/PredSL/index.html, accessed on 17 March 2021) was used to predict subcellular localizations. The TPP domain (trehalose-phosphatase (Trehalose PPase); PF02358) was retrieved from the Pfam database and the structures were created with TBtools [58,59]. We utilized MEME suite 5.1.1 to examine TaTPP motifs and The site distribution was set to any number of repetitions, the maximum number of motifs to locate was set to 9, the minimum width was set to 6, the maximum width was set to 50, and the maximum number of motifs to locate was set to 9 [60].

2.5. Analysis of Publicly Accessible Expression Data and Cis-Regulatory Elements (CREs)

We used the NCBI database (https://www.ncbi.nlm.nih.gov/, accessed on 19 March 2021) to obtain 2 kb upstream from start codon promoter sequences of 11 *TaTPPs*, which we subsequently submitted to PlantCARE to find the CREs [61]. Netbeans IDE 8.0 (https://netbeans.org., accessed on 25 March 2021) was used to organize data [62] and subsequently TBtools Heatmap was used for data visualization. The Genevestigator RNAseq public anatomy was used to examine gene expression [63] and the MeV tool was then used to visualize expression [64].

2.6. Plant Materials and Treatments

T. aestivum L. cultivar Jinmai39 was used to investigate the transcription of *TaTPPs* in the presence of salt, drought, and ABA treatments. The seedlings were grown in a growth chamber at 22 °C with 16 h/8 h of light/ darkness and a light intensity of 9000 lux. Wheat plants were treated with either double-distilled water (control) or a 20% PEG-6000 or a 250 mM NaCl solution at the 2–3 leaf stage for drought and salt stress, respectively. For abscisic acid (ABA) treatment, plants at the same stage are sprayed with 100 mM abscisic acid (ABA) or 0.1% (v/v) ethanol (control). To analyze the expression of *TaTPPs* during leaf senescence, the delayed senescence wheat cultivar Yannong19 was grown in field conditions and collected samples from flag leaf at 0, 7, 10, 16, 19, 22, 24, and 25 days after anthesis. All the leaves after collection are immediately frozen into liquid nitrogen and stored at −80 °C for further RNA extraction.

2.7. RNA Extraction, Quantitative Real-Time Reverse Transcription PCR Analysis and Protein Interaction Network

The Quick RNA isolation Kit (Huayueyang Biotechnology, Beijing, China) was used to extract RNA according to the manufacturer's instructions and DNase I treatment was used to remove DNA contamination. The RevertAid First Strand cDNA Synthesis Kit (Thermo Scientific, Waltham, MA, USA) was used to synthesize cDNA from a 3-μg aliquot of total RNA from each sample. To measure the expression of *TaTPPs* qRT-PCR analysis was performed with specific primers (Table S1), as described previously [65]. The ABI PRISM 7500 system (Applied Biosystems, Foster City, CA, USA) was used to generate threshold values (CT) and the transcription level of *TaTPPs* was measured using the comparative $2^{-\Delta\Delta CT}$ technique that was standardized with the *Elongation factor 1α* (*TaEF-1α*) (GenBank accession no. Q03033) [66,67] (Table S1). All of the studies were carried out three times. The TaTPP protein interaction network was examined using the STRING online server (https://string-db.org/, accessed on 27 April 2021).

3. Results

3.1. Identification and Annotation of Wheat TPPs

We identified a total of 31 TPP protein sequences in the wheat genome (Tables 1 and S2). This number is relatively large when compared to TPPs previously identified in *Arabidopsis*, rice, and maize (Table S3). Wheat has a greater ploidy level and a larger genome size as it originated from the natural hybridization of three closely related genomes (A, B, and D), which may justify this result [68]. These protein sequences were encoded by 31 genes, three of which were chosen as representatives because they showed splice variants with full domains. A detailed description of *TaTPPs* is summarized in Table 1. The ORF of *TaTPPs* ranged from 750 to 1755 bp, with protein lengths ranging from 249 to 584 amino acids (Table 1). The molecular weight of the genes ranged from 28.67 KDa to 65.02 KDa. (Table 1). Fifteen genes were found to be basic (>7) and 16 genes were found to be acidic (<7) based on the predicted pI value (Table 1).

In addition, the Aliphatic Index and Instability Index were computed. The Aliphatic Index measures how much space is taken up by aliphatic side chains in Alanine, Isoleucine, Leucine, and Valine amino acids [69]. The Aliphatic Index ranges observed were 72.28 to 86.42, and the Instability Index ranges were 32.59 to 55.74 (Table 1). The high Aliphatic Index of a protein sequence suggests that it can function at a broad range of temperatures, whereas the Instability Index shows whether the protein is stable or unstable [70]. All the TaTPPs had negative GRAVY values ranging from −0.700 to −0.142 (Table 1). A protein with a negative GRAVY value is non-polar and hydrophilic in nature [69].

3.2. Subcellular Localization Prediction and Chromosomal Distribution of TaTPPs

PredSL (http://aias.biol.uoa.gr/PredSL/index.html accessed on 22 September 2021) was used to predict subcellular localization. Subcellular localization of the TaTPPs was predicted mostly in the chloroplast, whereas, TaTPP1-A, TaTPP7-D, TaTPP10-B appeared to be localized in the mitochondrion (Table 1). Moreover, TaTPP5-B, TaTPP7-A were predicted as secreted proteins (Table 1). However, TaTPP5-A, TaTPP10-A, TaTPP10-D were predicted with unknown localization (Table 1). A schematic diagram was created to explain the chromosomal location of *TaTPPs*. The *TaTPPs* are present on 17 wheat chromosomes (Figure 2 and Table 1). On the chromosomes of the A subgenome, the highest number of *TaTPP* genes (11 genes) were mapped. B and D subgenomes had 10 *TaTPP* genes in each subgenome. The maximum 14 genes of *TaTPPs* were located on chromosome 2 (Figure 2). Chromosome 6A, 6B and 6D, had 2 genes on each chromosome and 1A, 1B, 1D, 3A, 3D, 5A, 5B, and 5D had only a single gene. On the other hand, no *TaTPPs* were found on chromosomes 3B, 4A, 4B, or 4D (Figure 2 and Table 1), suggesting that *TPP* family genes were unevenly distributed throughout the three subgenomes of wheat.

Table 1. Detailed annotations of the *TaTPPs* in wheat.

Gene Name	Gene ID	Splice Variant	PC	ORF	Chromosome Location					Introns	Exons	Length (aa)	M.W. (KDa)	PI	Instability Index	Aliphatic Index	GRAVY	SL Prediction
					Chr	Strand	Start	End	Chr Length									
TaTPP1-A	TraesCS1A02G210400.1	1	VII	1146	1A	reverse	372639121	372643307	594102056	9	10	381	42.56	5.61	46.35	83.91	−0.314	chloroplast
TaTPP1-B	TraesCS1B02G224300.2	2	VII	1317	1B	reverse	402147526	402150391	689851870	12	13	438	49.46	6.09	47.92	86.32	−0.238	mitochondrion
TaTPP1-D	TraesCS1D02G213700.1	1	VII	1146	1D	reverse	298692275	298696295	495453186	9	10	380	42.54	5.53	45.29	84.13	−0.315	chloroplast
TaTPP2-A	TraesCS2A02G161000.1	1	II	1074	2A	forward	111921839	111925164	780798557	8	9	357	39.43	6.11	34.73	79.3	−0.254	chloroplast
TaTPP2-B	TraesCS2B02G187000.1	1	II	1077	2B	forward	161722263	161725704	801256715	8	9	358	39.62	6.04	35.64	82.6	−0.216	chloroplast
TaTPP2-D	TraesCS2D02G168100.1	1	II	1077	2D	forward	111588533	111591881	651852609	8	9	358	39.66	5.77	35.52	80.42	−0.249	chloroplast
TaTPP3-A	TraesCS2A02G161100.1	1	II	1077	2A	forward	112744717	112747766	780798557	8	9	358	39.63	7.14	35.67	79.61	−0.282	chloroplast
TaTPP3-B	TraesCS2B02G187100.1	1	II	1077	2B	forward	162007594	162010969	801256715	8	9	358	39.61	6.57	33.2	79.61	−0.264	chloroplast
TaTPP3-D	TraesCS2D02G168200.1	1	II	1077	2D	forward	112099169	112102442	651852609	8	9	358	39.50	6.84	34.77	79.08	−0.283	chloroplast
TaTPP4-A	TraesCS2A02G161200.1	1	II	1077	2A	reverse	113309228	113312068	780798557	8	9	358	39.63	6.77	35.05	79.86	−0.281	chloroplast
TaTPP4-B	TraesCS2B02G187200.1	1	II	1077	2B	reverse	162445597	162448929	801256715	8	9	358	39.65	7.12	32.59	79.86	−0.257	chloroplast
TaTPP4-D	TraesCS2D02G168300.1	2	II	1179	2D	reverse	112177996	112181078	651852609	7	8	392	43.58	6.44	39.59	82.65	−0.190	chloroplast
TaTPP5-A	TraesCS2A02G167100.1	1	V	750	2A	reverse	119307539	119314162	780798557	8	9	249	28.67	8.28	43.34	75.9	−0.700	other
TaTPP5-B	TraesCS2B02G193300.1	1	V	1680	2B	reverse	168831609	168853380	801256715	10	11	559	62.28	8.61	45.14	85.64	−0.230	secreted
TaTPP6-A	TraesCS2A02G412100.1	1	VI	1113	2A	forward	669749666	669753186	780798557	9	10	370	41.23	5.70	49.54	83.46	−0.307	chloroplast
TaTPP6-B	TraesCS2B02G430700.1	1	VI	1116	2B	forward	619679522	619682850	801256715	9	10	371	41.19	5.88	51.27	82.21	−0.297	chloroplast
TaTPP6-D	TraesCS2D02G409300.1	1	VI	1113	2D	forward	524105415	524108680	651852609	9	10	370	41.11	5.58	55.74	84.27	−0.232	chloroplast
TaTPP7-A	TraesCS3A02G085700.1	1	V	1662	3A	forward	55223622	55255982	750843639	11	12	553	61.55	8.06	36.82	83.67	−0.291	secreted
TaTPP7-D	TraesCS3D02G085800.1	1	V	1755	3D	forward	43259299	43283531	615552423	13	14	584	65.02	8.86	32.8	77.4	−0.400	mitochondrion
TaTPP8-A	TraesCS5A02G190000.1	1	I	1122	5A	reverse	394181080	394183400	709773743	5	6	373	40.85	8.95	41.99	79.81	−0.144	chloroplast
TaTPP8-B	TraesCS5B02G193100.1	3	I	1248	5B	reverse	348448002	348450302	713149757	4	5	373	40.93	8.97	41.35	79.81	−0.145	chloroplast
TaTPP8-D	TraesCS5D02G200800.1	1	I	1122	5D	reverse	303758772	303761166	566080677	5	6	373	40.86	8.96	42.04	79.81	−0.142	chloroplast
TaTPP9-A	TraesCS6A02G248400.1	1	VI	1119	6A	forward	461143866	461147635	618079260	8	9	372	41.11	5.68	48.98	85.65	−0.218	chloroplast
TaTPP9-B	TraesCS6B02G276300.1	1	VI	1119	6B	reverse	500209451	500213722	720988478	8	9	372	41.28	5.89	49.51	85.89	−0.232	chloroplast
TaTPP9-D	TraesCS6D02G230500.1	1	VI	1119	6D	forward	323712099	323716021	473592718	8	9	372	41.05	5.56	49.53	86.42	−0.192	chloroplast
TaTPP10-A	TraesCS6A02G301800.1	1	I	1251	6A	reverse	535151913	535154867	618079260	8	9	416	45.37	9.26	44.89	72.28	−0.305	other
TaTPP10-B	TraesCS6B02G330900.1	3	I	1224	6B	reverse	581079293	581082545	720988478	7	8	407	44.29	8.64	42.6	77.69	−0.147	mitochondrion
TaTPP10-D	TraesCS6D02G281100.1	1	I	1110	6D	reverse	388537685	388540648	473592718	8	9	369	40.25	8.79	37.81	76.99	−0.164	other
TaTPP11-A	TraesCS7A02G180800.1	1	I	1086	7A	reverse	135006112	135008690	736706236	9	10	361	39.54	8.41	50.69	81.39	−0.219	chloroplast
TaTPP11-B	TraesCS7B02G085800.1	1	I	1095	7B	reverse	97972425	97975038	750620385	9	10	364	40.16	8.11	50.16	79.09	−0.274	chloroplast
TaTPP11-D	TraesCS7D02G182600.1	1	I	1092	7D	reverse	136013159	136015620	638686055	9	10	363	39.91	8.60	51.06	79.56	−0.268	chloroplast

PC, Phylogenetic clade; ORF, Open Reading Frame; No, Number; bp, Base pair; Chr, Chromosome; aa, Amino Acid; M.W., Molecular Weight; Pi, Iso electric point; GRAVY, Grand average of hydropathy, SL, Subcellular Localization.

We further investigated the duplication events in the *TaTPP* gene family in the context of wheat being hexapolyploid and having big genomes. Genes are usually considered duplicated when the query cover and identity value of gene sequences are more than 80% [71]. It has also been reported that genes are considered duplicated when protein sequence similarity and identity are more than 70% and 75%, respectively [72]. By analyzing the sequences, we found 27 pairs of *TaTPPs* with a sequence identity ranges from 82.14% to 95.25% and 100% query cover within all gene pairs (Tables S4 and S5) and identified in the same phylogenetic tree clade (Figure 3). We further computed the non-synonymous (Ka) and synonymous (Ks) substitutions, as well as the Ka/Ks ratios, for the 27 *TaTPP* gene pairs to determine the selection pressure on the duplicated *TaTPPs* (Table S5). These gene pairs Ka/Ks ratios were smaller than one, indicating that they developed under functional restriction with negative or purifying selection. The divergence period ranged from 2.93 to 13.3 million years ago (MYA), showing that these gene pairs were duplicated recently (Table S5).

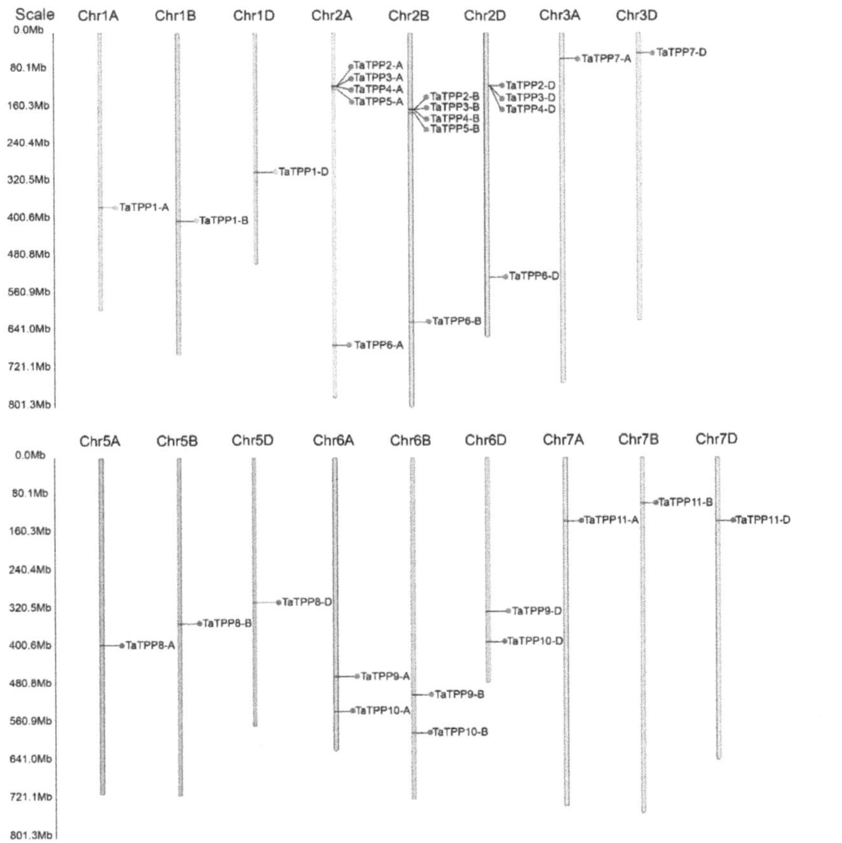

Figure 2. Graphical presentations of *TaTPPs* chromosomal distribution of on wheat chromosomes. The name of the gene on the right side and the location of the *TaTPPs* is indicated by the colored circular circle on the chromosomes. The three subgenomes chromosomal numbers are shown at the top of each bar.

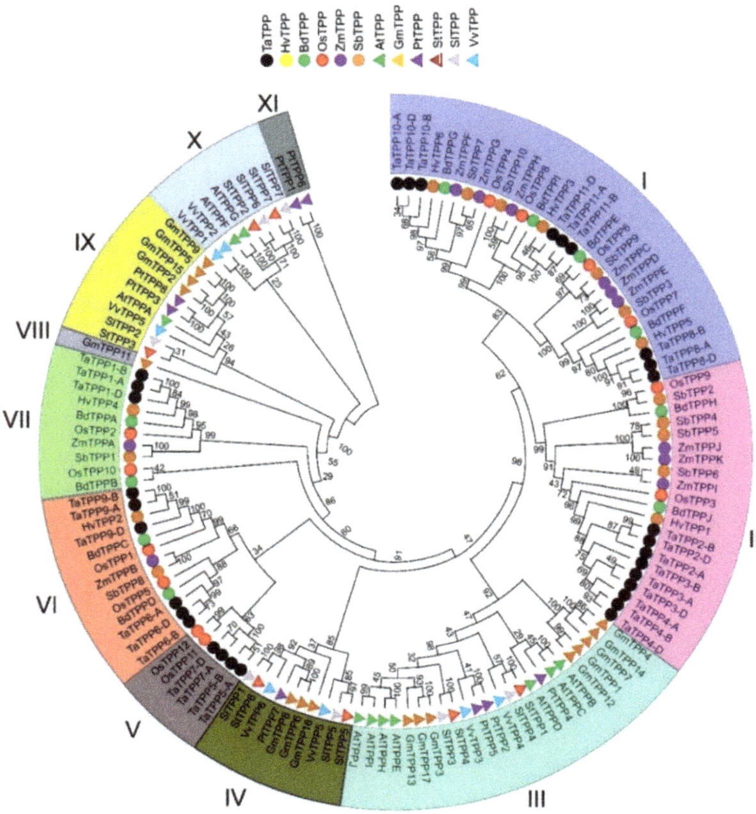

Figure 3. Phylogenetic analysis of TaTPP proteins. The tree was generated using MEGA X by the maximum likelihood method with 1000 bootstrap values. All the species and protein ID used for constructing tree were presented in Table S6.

3.3. Phylogenetic and Conserved Domain Analyses of TaTPP Proteins

A phylogenetic tree containing full length TPP protein sequences from twelve plant species was constructed by the maximum likelihood method to better understand the evolutionary relations among the TaTPP proteins with other species (Figure 3, Table S6), including five species from monocot: *Hordeum vulgare, Brachypodium distachyon, Oryza sativa, Zea maize, Sorghum bicolor*; and 6 species from dicot: *Arabidopsis thaliana, Glycine max, Populus trichocarpa, Solanum tuberosum, Solanum lycopersicum, Vitis vinifera*. The results indicated that TPP proteins were divided into eleven clades, where clade I was the largest with 30 members. Clades II to XI (total 10 clades in order) included 21, 24, 10, 6, 13, 10, 1, 10, 8, and 2 members, respectively (Figure 3).

Plants classified as dicots and monocots were divided into distinct clades. Proteins from monocot plants were grouped into clade I, clade II, clade V, clade VI, and clade VII, whereas proteins from dicot plants were grouped into clade III, clade IV, clade VIII, clade IX, clade X, and clade XI. The highest number of TaTPP proteins were grouped into clade I and clade II, which had nine proteins in each clade. In addition, clades V, VI, VII contained four, six, and three TaTPP proteins, respectively (Figure 3). Most of the wheat TPP proteins were closely related to *H. vulgare*, *B. distachyon*, and *O. sativa*, suggesting their conserved function with those plant species and offering information that can be used to conduct a more in-depth functional analysis. All the *TaTPPs* were assembled into 11 groups, as sequences from A, B, and D subgenome of 11 groups clustered together in the phylogenetic tree (Figure 3) and protein sequence identity was more the 88% between A, B, and D subgenome of each group (Table S4). Thus, we considered the protein sequences from A, B, and D subgenome of each group are copies of separate *TaTPP* genes and named them according to the ascending order of the chromosomal location (Table 1).

Further, the Pfam database was utilized to find the important component domains of TaTPP proteins [59]. All the TaTPP proteins contain a specific Trehalose PPase domain (PF02358). In addition, a stress antifungal domain was found in TaTPP-5A, TaTPP7-A and TaTPP7-D (Figure 4a). We used MEME suite 5.1.1 to evaluate motif sequences for 31 TaTPPs and found six significant motifs (motifs 1–6) (Figure 4b). All the motifs were found to be conserved in all TaTPP proteins except for TaTPP5-A, which lacks lacks motif 3 (Figure 4b).

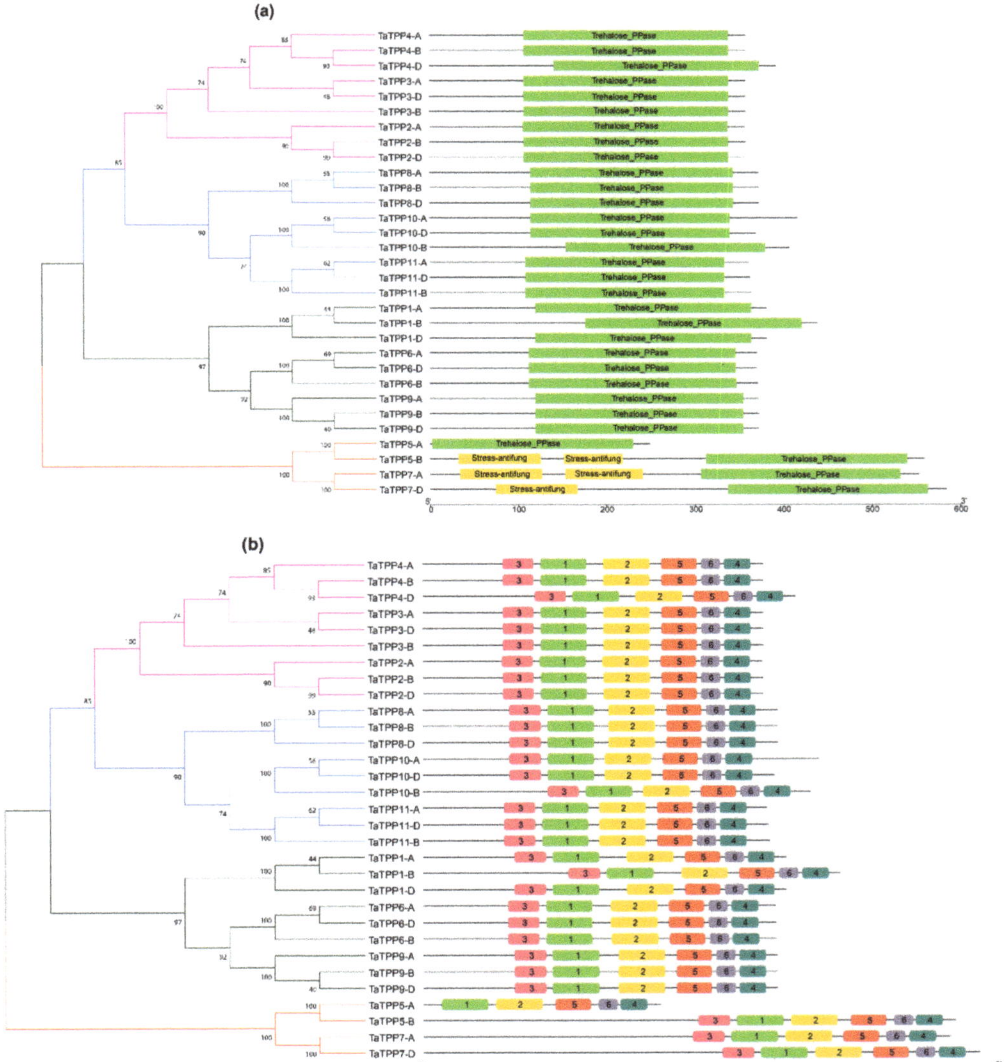

Figure 4. Conserved domain and motif of TaTPPs. (**a**) The conserved domain of TaTPP members was identified from Pfam and SMART databases and presented using TBtools. (**b**) The conserved motifs of TaTPP members. Six motifs were identified using MEME program and presented with different colored boxes.

3.4. Gene Structure and Evolution Analyses of *TaTPPs*

The exon-intron structures of *TaTPPs* were studied to better understand their structural features (Figure 5). The *TaTPP* gene family had a lot of variation in terms of gene structure, according to gene structure analyses as introns ranged from 4 to 13. Most of the *TaTPPs* contain eight or nine introns. A maximum of 13 introns was found in *TaTPP7-D* and a minimum of four introns were observed in *TaTPP8-B* (Figure 5). Moreover, different *TaTPPs* showed different intron phase patterns. *TaTPP1-A, TaTPP1-D, TaTPP5-A, TaTPP6, TaTPP8, TaTPP9* showed phase 0 and *TaTPP2, TaTPP3, TaTPP4, TaTPP5-B, TaTPP7-A, TaTPP10, 11* showed phase 2 patterns, whereas *TaTPP1-B* and *TaTPP7-D* exhibited all phases (Phase 0,1,2) (Figure 5).

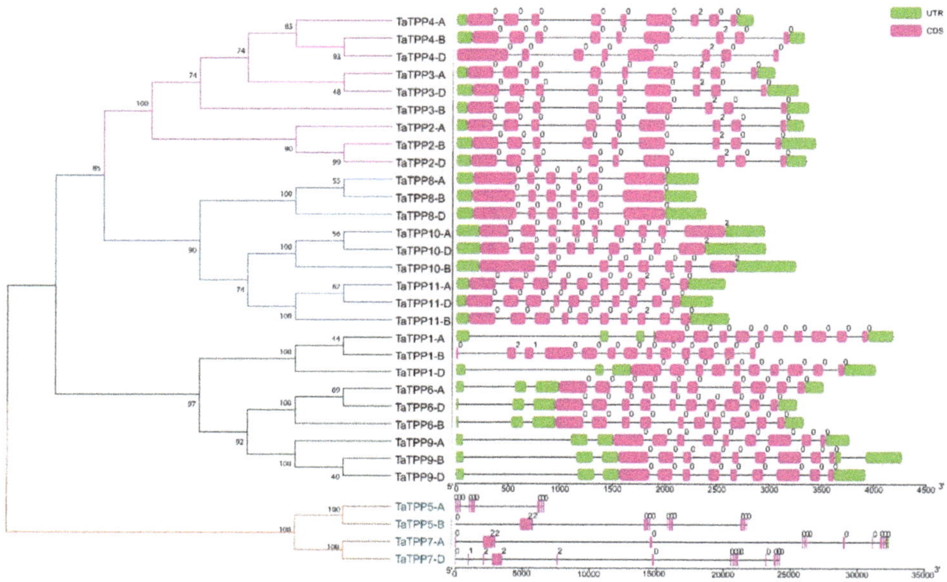

Figure 5. Structural organizations of *TaTPPs*. The introns are shown by black lines, whereas the exons are represented by pink boxes and untranslated regions (UTRs) are represented with green boxes. Intron phase, 0: phase 0, 1: phase 1 and 2: phase 2 denotes that a codon is not disrupted by introns, a codon between the first and second bases is disrupted by an intron and a codon between the second and third bases is disrupted by an intron, respectively.

Further, the Multiple Collinearity Scan toolkit was used to investigate the synteny networks between *TaTPPs* and other wheat relatives and model plants. The results showed that 27, 26, 13, 33, 26, and 22 orthologous gene pairs were identified between *TaTPPs* and other *TPPs* in *B. distachyon, O. sativa, A. thaliana, H. vulgare, Z. mays,* and *S. bicolor*, respectively (Figure 6 and Table S7). A collinear relation was observed for 19, 18, 9, 22, 17 and 19 *TaTPPs* with other *TPPs* in *B. distachyon, O. sativa, A. thaliana, H. vulgare, Z. mays,* and *S. bicolor*, respectively. *TaTPP6, TaTPP9, TaTPP10* and *TaTPP11* were shown to have more than one pair of orthologs. Thus, these *TaTPPs* might have a crucial role in the evolution of *TPPs*. These findings imply that *TaTPPs* in wheat may have evolved from other plant species orthologous genes.

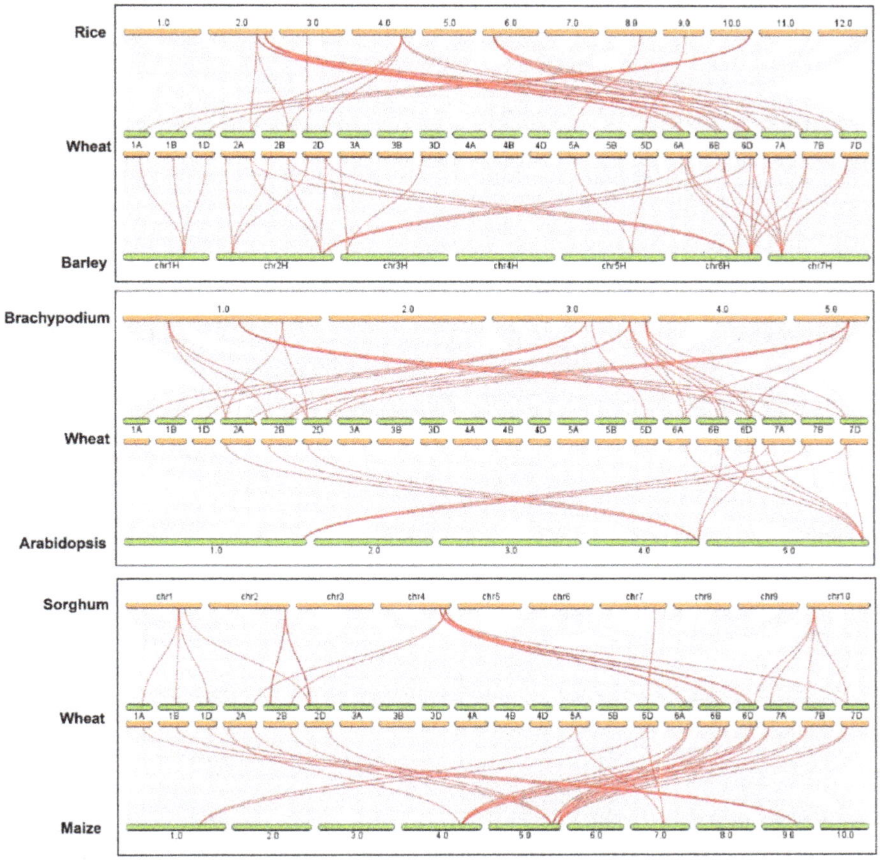

Figure 6. Syntenic relationship between *TaTPPs* with rice, *Arabidopsis*, *Brachypodium*, sorghum, and maize. The collinear blocks within wheat and other plant genomes are shown by gray lines in the background, while the syntenic *TaTPP* gene pairs are highlighted by red lines.

3.5. 3-D Protein Structure Analysis

The 3-D structure reveals a few key residues linked to biological processes or intended outcomes [73]. Thus, we used SWISS-MODEL to identify the 3-D model of TaTPP proteins (Figure S1a). For all TaTPP proteins, the 3-D structures were analyzed using template "5gvx.1.A." and predicted 3-D structures covering the N-terminus and C-terminus regions of 31 TaTPP proteins (Figure S1a). Within 4 A°, three conserved residues that worked as ligands were identified. The interaction of those ligands with chain A and the magnesium ion (Mg^{2+}) indicates that TaTPP proteins have distinct catalytic activities, which have also been reported for AtTPP, ZmTPP, ScTPP, CaTPP, and EcTPP that have a catalytic function and they are all similar to one other by 80% [39,74]. Further, we used SOPMA to calculate the secondary structure elements of protein sequences (Table S8). TaTPP proteins were found to contain a range of 35.70% to 47.99% α helix, 13.41% to 18.18% extended strand, 6.62% to 9.93% β turn and 8.38% to 41.21% random coil (Table S8). All TaTPPs except TaTPP5, TaTPP7, TaTPP9-A, and TaTPP10-B had a coiled coil-like structure in the C-terminus and one Mg^{2+} ligand each was observed in all the TaTPPs (Figure S1a).

To validate TaTPP protein structures, we employed SWISS-MODEL analysis and the MolProbity server (Figure S1b and Table S9). The produced Ramachandran plot has an average favored region of 94.07%, an average allowed region of 99.08%, and an average outer region of 0.91% (Table S9). The average sequence identity was 34.95%, with a similarity of 37%, covering 68% of the query sequences obtained by the X-ray Method in 2.6 A° (Table S9). The ligand interaction between chain A and Mg^{2+} was confirmed with the Protein–Ligand Interaction Pipeline (PLIP), and the residue site was noticed to be highly conserved. We investigated these conserved residues further in all TaTPP protein sequence alignments and found that they include aspartic acid (D/Asp), which is conserved in motif 3 and motif 6. (Figure S2, Table S9). For improved visual clarity, the side chains of the catalytic triads were expanded with the TaTPP1-A residues (Figure S3).

3.6. Analysis of Cis-Regulatory Elements

To examine the responses of *TaTPPs* members to various stimuli, the 2 kb promoter sequences upstream of the start codon of these genes were submitted to the PlantCARE service to predict their *Cis*-regulatory elements (CREs). A total of 90 CREs with a frequency of 1985 were identified in all *TaTPP* promoters (Figure 7, Table S10). Among them, 72 CREs were related to phytohormones, stress, growth, and development (Figure 7, Table S10). All of the identified CREs were divided into five groups according to their known functions (Tables S10 and S11). Group I contained four core *Cis*-elements, including AT~TATA-box, CAAT-box, TATA, TATA-box. TATA-box (which comprises TATA and AT TATA-box) is a critical promoter element found in approximately 30% of transcription start sites and the CAAT-box is a kind of promoter that may influence the choice of transcription start location [75]. TATA-box and CAAT-box are generally present 25–30 bp and ~75 bp upstream of the transcription start site, respectively, and both of them are found in a wide range across all the promoters.

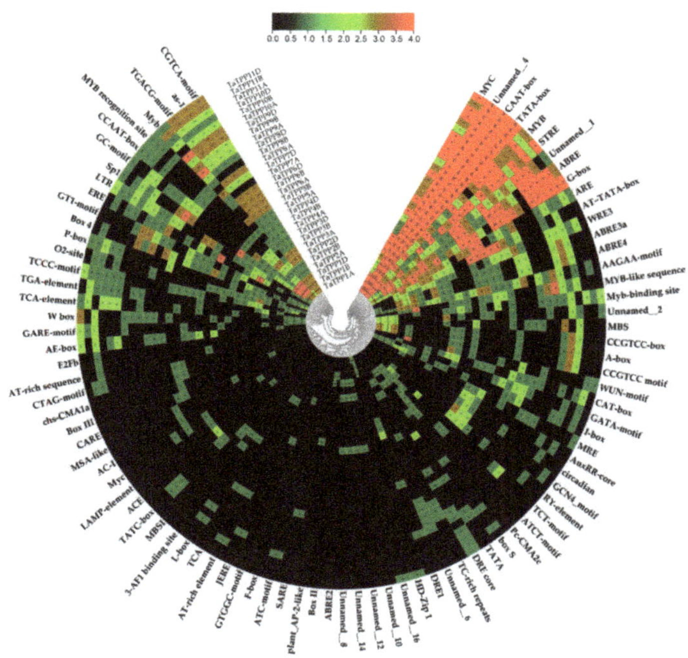

Figure 7. Putative *Cis*-acting regulatory elements (CREs) of *TaTPPs*. The CREs were identified with the 2 kb upstream sequences of the start codon using the PlantCARE online server and presented using TBtools. Red color indicates the CREs with high frequency, while black color indicates CREs with zero frequency.

Group II contained 44 stress-related CREs, among them 20 were light-responsive *Cis*-elements such as 3-AF1 binding site, 3-AF1 binding site, ABRE4, ACE, ATCT-motif, Box 4, AE-box, Box II, LAMP-element. The stress-responsive CREs consist of one anaerobic-responsive element (ARE), one low-temperature-responsive element (LRT), one drought-responsive element (MBS), two wound-responsive elements (WUN-motif, box S), one cold- and dehydration-responsive (DRE core) and 17 defense- and stress-responsive elements (as-1, TC-rich repeats, W box, CCAAT-box, MYB, MYB recognition site, Myb, Myb-binding site, MYB-like sequence, MYC, Myc, STRE, WRE3, Unnamed_1, Unnamed_8, GC-motif, AT-rich sequence). There were 12 CREs in group III, involved in cell development including seed specific expression (AAGAA-motif, RY-element), cellular development and cell cycle regulation (AC-I, MSA-like), meristem expression (CAT-box, CCGTCC-box, CCGTCC-box), circadian control (circadian), differentiation of palisade mesophyll cells (HD-Zip I), cell cycle regulation (MSA-like), and endosperm expression (GCN4_motif).

Additionally, the hormone-responsive CREs in group IV included 16 CREs such as abscisic-acid-responsive element (ABRE, ABRE2), auxin-responsive elements (AuxRR-core and TGA-element), salicylic-acid-responsive element (TCA-element, TCA, SARE), methyl-jasmonate-responsive elements (TGACG and CGTCA motifs), ethylene-responsive element (ERE) and gibberellin-responsive elements (GARE-motif, P-box, and TATC-box). There were also 14 CREs in group V with unknown functions. CTAG-motif and A-box might act as a CRE, Unnamed_2 might act as an antisense transcript, BOX III might function as a protein binding site and Unnamed_16 was found to be involved in sugar transporter family genes. Most *TaTPPs* possessed one or more CREs associated with hormone and stress-related activities, suggesting that *TaTPPs* may be engaged in a variety of physiological processes as a result of diverse environmental adaptations.

3.7. Transcriptional Patterns of TaTPPs in Different Organs and Developmental Stages of Wheat

To investigate the transcription level of *TaTPP* genes in different wheat organs and development stages, mRNA transcripts data was collected from Genevestigator and visualized with a heatmap in Figures 8 and S4. The transcript data were divided into six groups. Group I included callus, Group II included primary cells (cell culture, spike cell, spikelet cell, floret cell, stamen cell, anther cell, meiocyte, microspore), Group III included seedlings (seedling, coleoptile, root, radicle, radicle tip), Group IV included inflorescence (inflorescence, spike, rachis, spikelet, floret, stamen, anther, pistil, ovary, lemma, awn, glume, caryopsis, embryo, endosperm, aleurone layer, starchy endosperm, endosperm transfer layer, pericarp, outer pericarp), Group V included shoot (shoot, culm (stem), internode, peduncle, leaf, blade (lamina), sheath, flag leaf, blade (lamina), ligule, sheath, crown, shoot apex, shoot apical meristem, axillary bud) and Group VI included rhizome (rhizome, roots, nodal root, unspecified root type, root tip, root, apical meristem). Our results showed that *TaTPP1*, *TaTPP8*, *TaTPP9-A*, *TaTPP9-D* and *TaTPP10-B* had the highest transcriptions in most of the organs compared to other *TaTPPs* (Figure 8). In addition, high expression was observed for *TaTPP2-D*, *TaTPP5-B*, *TaTPP6-A* and *TaTPP6-B* only in the rhizome group. In contrast, other *TaTPPs* had no expression in most of the organs (Figure 8).

Further, we observed the mRNA transcripts level of *TaTPPs* during different developmental stages of wheat, such as germination, seedling growth, tillering, stem elongation, booting, inflorescence emergence, anthesis, milk development, dough development, and ripening (Figure S4). A number of *TaTPPs* were expressed differently at various stages of wheat development. For example, *TaTPP8* and *TaTPP4-A* were found to be expressed in all stages, whereas *TaTPP1* and *TaTPP9* were induced in all except the ripening stage. *TaTPP4-D* was expressed in all except stem elongation and *TaTPP3-D* was expressed in all except tillering and ripening stages. *TaTPP5* and *TaTPP7* showed very low expression in all wheat developmental stages and other *TaTPPs* were either slightly or highly expressed in one or more developmental stages (Figure S4). These findings suggest that various *TaTPPs* may have a role in the development of various tissues at different development stages.

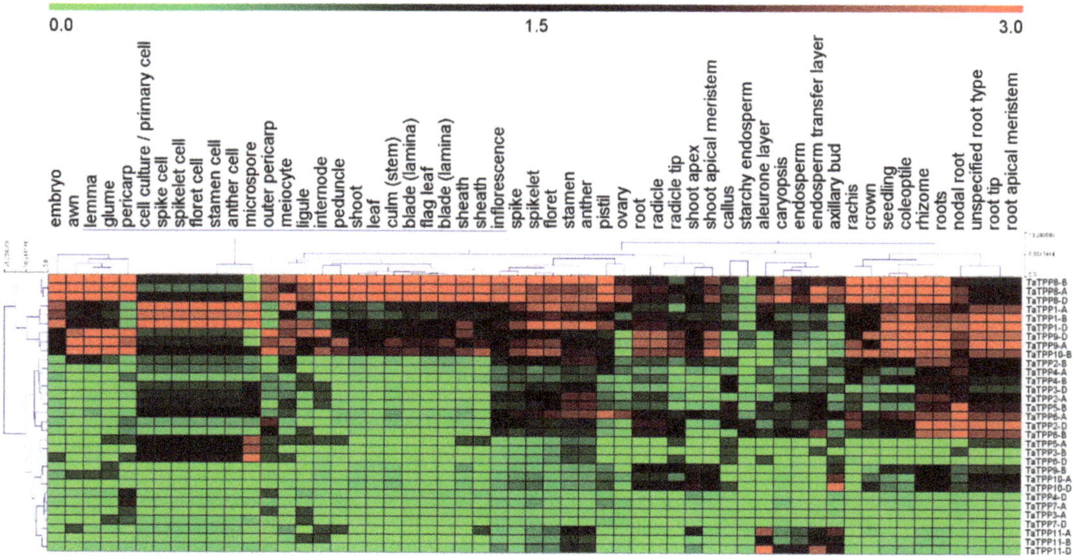

Figure 8. Transcription profiles of *TaTPPs* in different wheat organs. mRNA transcription data of *TaTPPs* in different wheat organs were retrieved from Genevestigator and presented using MeV software.

3.8. Transcriptions of TaTPPs Were Induced in Response to ABA, Abiotic Stresses and Leaf Senescence

Wheat seedlings treated with ABA or abiotic stress (drought and salinity) were used to analyze the transcriptional pattern of the *TaTPPs* in wheat. Under ABA treatment, *TaTPP1* and *TaTPP4* exhibited upregulated transcriptions at most of the time points and significant upregulation was observed from 3 to 12 hpt (hours post treatment). Moreover, three *TaTPPs* (*TaTPP7*, *TaTPP8* and *TaTPP9*) were upregulated immediately after ABA treatment and transcriptions decreased with an increase in ABA treatment time points. Transcriptions were significantly downregulated for *TaTPP2*, *TaTPP3* and *TaTPP11* at most of the time points after ABA treatments compared to control (0 hpt) (Figure 9a). The transcriptional patterns of *TaTPP* members were examined following drought stress in wheat to provide insight into the underlying functional roles of wheat *TPPs* in response to drought stress. During the drought stress treatment, only *TaTPP1* and *TaTPP4* showed significant upregulations at a later time post treatment. A slight upregulation or significant downregulation was observed for all other *TaTPPs* after drought stress in wheat (Figure 9b). The transcriptional levels of *TaTPP* members were examined to elucidate the mechanism of gene responses to leaf senescence in wheat. Most of the *TaTPP* members were slightly or highly induced during leaf senescence, *TaTPP1* showed obvious upregulated transcriptions at 19 and 22 days after anthesis compared to the control (0 days after anthesis) (Figure 9c).

Further, we analyzed the transcriptions of *TaTPP* members under salt stress by qRT-PCR to observe the involvement of *TaTPPs* in wheat salt tolerance (Figure 10). Significant upregulation of the transcripts was observed for *TaTPP1*, *TaTPP2*, *TaTPP4* and *TaTPP9* at an early stage of salt treatment and downregulations were observed at the later stage of salt treatment. Moreover, *TaTPP7* showed a significant upregulation only at 12 h post treatment (hpt) compared to the control (0 hpt). In contrast, either no changes or significant downregulations were observed for other *TaTPP* members compared to the control (Figure 10). However, no expression was observed for *TaTPP5* and *TaTPP6* by qRT-PCR in all aspects. Overall, these findings suggest that *TaTPPs* act as an important regulator of wheat abiotic stress and leaf senescence responses.

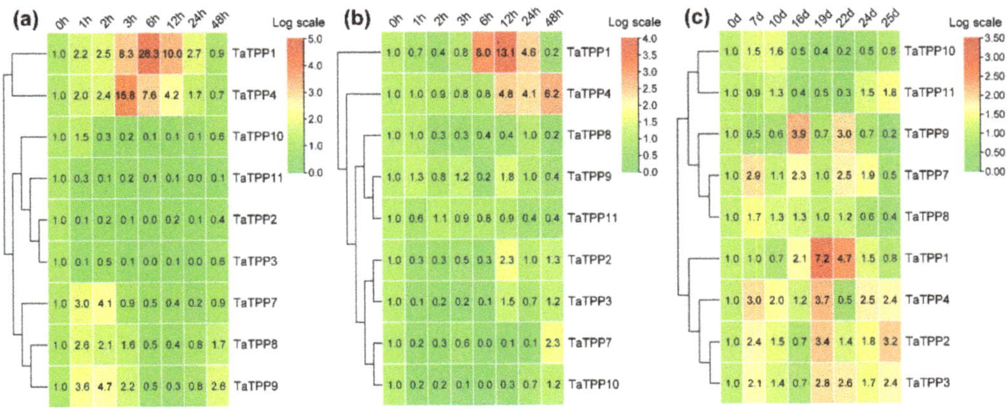

Figure 9. Relative transcript profiles of *TaTPPs* in response to (**a**) abscisic acid (ABA), (**b**) drought stress and (**c**) leaf senescence. The relative transcripts of all genes were analyzed using qRT-PCR. The relative transcript levels of *TaTPPs* were measured using the comparative threshold ($2^{-\Delta\Delta CT}$) method. Data normalized with the transcripts of wheat elongation factor, *TaEF-1α*. The 0 h post treatment (**a**,**b**) or 0 days after anthesis (**c**) was used as a control and standardized with 1. Red and green colors denote strong and weak transcription of *TaTPPs*, respectively. The heat map was generated with TBtools and tree was constructed with the average linkage clustering method.

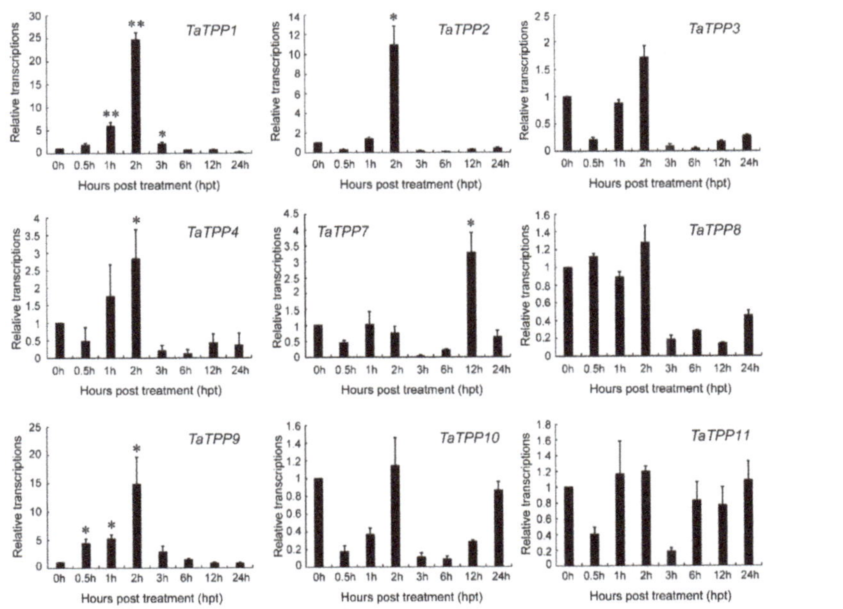

Figure 10. Relative transcript profiles of *TaTPPs* in response to salt stress. The relative transcripts of all genes were analyzed using qRT-PCR. The relative transcript levels of *TaTPPs* were measured using the comparative threshold ($2^{-\Delta\Delta CT}$) method. Data normalized with the transcripts of wheat elongation factor, *TaEF-1α*. The 0 h post treatment was used as a control and standardized with 1. Values represent the mean ± SD from three independent biological samples. Asterisks ($p < 0.05$) or double asterisks ($p < 0.01$) designate significant differences from 0 hpt by the Student's *t*-test.

3.9. Protein–Protein Interaction Analysis of TaTPPs

The STRING database was used to build a network to study protein–protein interactions between TaTPPs and other wheat proteins (Figure S5 and Table S12). From prediction results, it was found that TaTPPs can interact with five other wheat proteins. Traes_1AL_7531AC097.1, Traes_1BL_2AE952A77.1 and Traes_1DL_50B29C62B.2 have encoded an enzyme called TRE, which is hydrolyzed Trehalose to synthesize two molecules of glucose. Moreover, Traes_6DL_33F8A5EF4.1 has encoded TPS enzyme which produces T6P, a phosphorylated intermediate, from UDPG and G6P and Traes_4AS_4B8E78B13.1 was an unknown protein. Thus, our results suggesting that TaTPPs might interact with other enzymes that are involved trehalose biosynthesis pathway to accelerate the trehalose biosynthesis process.

4. Discussion

The *TPP* gene family has been characterized as catalytic enzymes that mainly function in trehalose biosynthesis [18,76,77]. Despite their catalytic function, a portion of *TPP* genes has been identified to be involved in growth and development, response in abiotic and biotic stress and senescence [27–29,31–33,35,42,78,79]. Although wheat is one of the most economically important cereal crops, systemic studies on *TPP* homologs in wheat have not been reported yet.

In the present study, we analyzed wheat *TPPs* with other species and identified 31 *TPPs* in wheat based on the Chinese Spring genome sequence (Table 1). The highest number of *TPPs* were found in wheat and these genes were distributed over 17 chromosomes (Figure 2). In comparison to previously described *TPPs* in *Arabidopsis*, rice, maize, and purple false brome, the wheat *TPP* gene family has been significantly extended with relatively more *TPPs* [38,74,80–82]. The major driving forces for extending the gene family in various plant species are gene duplication mechanisms, which include segmental, tandem, and whole-genome duplications [83,84]. All the *TaTPPs* are distributed unevenly on the wheat chromosome and the number ranges from 1 to 5 on each chromosome (Figure 2). Gene duplication analysis revealed that 27 pairs of *TaTPPs* duplicated within the wheat genome (Tables S4 and S5). The gap between genes on the chromosomal map of common wheat was higher than 200 kb (Figure 2), indicating that these genes were not formed via tandem duplication [85]. In addition, Ka/Ka ratio was less than one for all pairs of duplicated genes, suggesting that *TaTPPs* were subjected to a rigorous purifying selection (Table S5) and a comparable segmental duplication event was also observed for *TPPs* in rice [74]. Natural whole-genome duplicating processes might have led to the expansion of the *TaTPP* gene family. Thus, these findings suggest that whole-genome and segmental duplications might be vital in the expansion and evolution of *TaTPPs*.

Phylogenetic analysis of 31 TaTPP proteins and 11 other plant species showed that these proteins clustered into 11 groups, where TPPs from monocots and dicots species were grouped into separate clades (Figure 3). TaTPP proteins were grouped into clade I, clade II, clade V, clade VI, and clade VII and closely related to *Brachypodium*, rice, and barley TPPs, suggesting that TaTPP proteins might originate from a common ancestor. TaTPP5 and TaTPP7 have moved far away from the cluster of all other TPPs in the radiation tree (Figure S6) that was similar to OsTPP11 and OsTPP12 as previously reported [74]. The *TaTPP* gene structure study demonstrated that the majority of *TaTPPs* had highly conserved gene structures. The size of an intron has a significant impact on the size of a gene. The number of introns in *TaTPPs* ranged from 4 to 13 and most of the *TaTPPs* had 8 or 9 introns (Table 1). The difference in total intron length between the largest gene *TaTPP7-A* (32 kb) and the shortest gene *TaTPP-8B* (2.3 kb), resulted in a significant variation in gene size. Further, multiple alignments of TaTPP protein sequences revealed that the Trehalose_PPase domain and conserved motif are conserved within the *TaTPPs* (Figure S2). Among the identified six motifs, all the motifs were highly conserved in all *TaTPPs* except *TaTPP5-A*, which lacks motif 3. All the *TaTPPs* had a complete Trehalose_PPase domain, suggesting the various proteins' functional equivalence and evolutionary relationships. In addition, *TaTPP5-B* and *TaTPP7* had a stress-antifungal domain which has been reported to be involved in disulphide bridges and response to salt stress [86,87]. Subcellular localization prediction showed that most of the TaTPPs are localized in the chloroplast, whereas some of them are found in the mitochondrion or secreted protein (Table 1). In *Arabidopsis* or rice, different localizations were also detected. For instance, AtTPPD and AtTPPE were localized in the chloroplast whereas AtTPPA, AtTPPB, AtTPPC, AtTPPF, and AtTPPH were found in the cytosol and AtTPPG, AtTPPI, and AtTPPJ showed localization in the nucleus [88]. This variation in the localization of TaTPPs might be due to a lack of conserved N-terminus (Figure S2). According to the 3-D structure analysis, all TaTPPs were highly conserved and showed Mg^{2+} ligand-binding sites in SWISSMODEL (Figure S1a), which are shown to have a role in catalysis by activating or inhibiting a variety of enzymes [89,90]. To investigate the *TPP* gene synteny relationship in wheat and other plant species, we identified 27, 26, 13, 33, 26 and 22 orthologous gene pairs between *TaTPPs* and other *TPPs* in *B. distachyon*, *O. sativa*, *A. thaliana*, *H. vulgare*, *Z. mays*, and *S. bicolor*, respectively (Figure 6 and Table S7). These findings imply that *TaTPPs* in wheat might have evolved from other plant species orthologous genes.

A non-coding DNA sequence found in the promoter region of a gene is known as a CREs. Different CREs distribution in promoter regions may indicate variations in gene regulation and function [91]. To identify the CREs, we used 2kb promoter regions of all *TaTPPs* and classified into five groups according to their known functions (Figure 7, Table S10). Stress related CREs were identified in high frequency compared to cellular development and hormone related CREs, suggesting the involvement of *TaTPPs* in response to stress. ABRE, At~ABRE, ABRE3a, and ABRE4 are ABA-responsive CREs that play important roles in seed dormancy, stomatal closure, leaf senescence, and plant biotic and abiotic stress responses. Multiple ABREs or their combinations have been reported to act as CEs (Coupling Elements) in the formation of ABA-responsive complex (ABRC) [92–97]. The ABRE CREs were predicted in all *TaTPPs* promotor with high frequency and ABRE3a and ABRE4 CREs were found in most of the *TaTPPs* except *TaTPP5*, *TaTPP7*, *TaTPP9* and *TaTPP11*. Following that, we also discovered TGA-element, and AuxRR-core (auxin-responsive element), TCA-element (salicylic acid responsiveness), CGTCA-motif (MeJA responsiveness) and *p*-box and TATC-box (gibberellin responsive element), among other hormone-related CREs [98], that might potentially induce possible signal transduction pathways for wheat *TPPs* during stress response. Furthermore, other CREs linked to a variety of development and stress were also predicted in *TaTPP* promoters with high frequency, including MBS (drought inducibility), MYC (drought-responsive CRE) MYB and STRE (stress response element), as-1(Defense response), Unnamed_1 (ABRE-like CRE, responsible for biotic and abiotic stress responses), ARE (anaerobic induction CRE), Unnamed_4 (might responsible for tissue specific expression) and AAGAA-motif (involved in seed specific expression) [91]. These findings suggest that *TPP* gene family members in wheat may be controlled by a variety of developmental events, hormones, and stress; however, additional experimental investigations will be required to validate this.

Higher transcriptional levels of *TaTPPs* were observed in different wheat organs and developmental stages. *TaTPP1*, *TaTPP8* and *TaTPP9* were expressed in most organs and developmental stages but predominantly expressed in the roots, suggesting that they could be important for root physiology. (Figures 8 and S4). Previous evidence showed that *AtTPPE* modulates ABA-mediated root growth and a rice *TPP*, *OsTPP7*, enhanced the anaerobic germination [41,99]. Wang et. al. [27] reported that seed germination was regulated by *OsTPP1* via crosstalk with the ABA catabolism pathway. In addition, high expression was also observed for *TaTPP1*, *TaTPP8* and *TaTPP9* in all developmental stages except dough development and ripening, suggesting that these genes might have a significant association with wheat developmental processes. Plants have developed sensory and response systems that enable them to adjust physiologically to environmental stress conditions such as drought, excessive salt, and low temperature stress. Previous studies in rice and *Arabidopsis* demonstrated the involvement of *TPP* genes in various environmental stresses and ABA signaling [32,37,38,40,41,100]. Under ABA treatment, *TaTPP1* and *TaTPP4* exhibited upregulated transcriptions at most of the time points and significant upregulations were observed from 3 to 12 hpt. Moreover, three *TaTPPs* (*TaTPP7*, *TaTPP8* and *TaTPP9*) were upregulated immediately after ABA treatment and transcriptions were decreased with the duration after ABA treatment and a significant downregulated transcription level was observed for *TaTPP2*, *TaTPP3* and *TaTPP11* at most of the time points after ABA treatment (Figure 9a). During the drought stress treatment, only *TaTPP1* and *TaTPP4* showed significant upregulations at later time post treatment. A slight upregulation or significant downregulation was observed for all other *TaTPPs* after drought stress in wheat (Figure 9b). An obvious significant upregulation was observed for *TaTPP1*, *TaTPP2*, *TaTPP4* and *TaTPP9* at an early stage of salt stress and downregulated transcriptions were observed at the latter stage of salt treatment. A similar expression pattern was also observed for rice *TPP* and *BdTPPC* genes that were upregulated in the first hour under abiotic stress [28,38]. Moreover, *TaTPP7* showed a significant upregulation only at 12 dpt. In contrast, either no changes or significant downregulation was observed for other *TaTPP* members compared to the control (Figure 10). The phenotype of mature *otsB*-overexpressing *Arabidopsis* plants included delayed senescence and decreased anthocyanin accumulation, suggesting that the role of *TPP* may perform a crucial role during leaf senescence in plants [42–45]. Our results showed that most of the *TaTPP* members were slightly or highly induced during leaf senescence, especially *TaTPP1* showed an obvious upregulated transcription (Figure 9c). Overall, these findings suggest that *TaTPPs* might act as an important regulator of wheat abiotic stress and leaf senescence responses and could be good candidate genes for wheat improvement under environmental stimuli. Moreover, protein network prediction revealed that TaTPP proteins possible interact with TaTPS or TaTRE protein which involved in trehalose biosynthesis pathway to accelerate the trehalose biosynthesis process (Figure S5).

Furthermore, we suggested a feasible working model based on *TaTPPs* transcription profiling to illustrate the roles of *TaTPPs* in a range of biological processes in wheat (Figure 11). TPS produces T6P, a phosphorylated intermediate, from UDPG and G6P and then TPP dephosphorylates T6P to produce trehalose in the second phase. Trehalose is then hydrolyzed by an enzyme called TRE to synthesize two molecules of glucose [16]. The expression of *TaTPPs* was induced by both endogenous and exogenous stimuli in this model. These signals were detected by multiple *Cis*-regulatory elements, which then regulated the transcription and functions of *TaTPPs* involved in numerous plant developmental stages and stress situations, affecting plant growth and tolerance mechanisms (Figure 11).

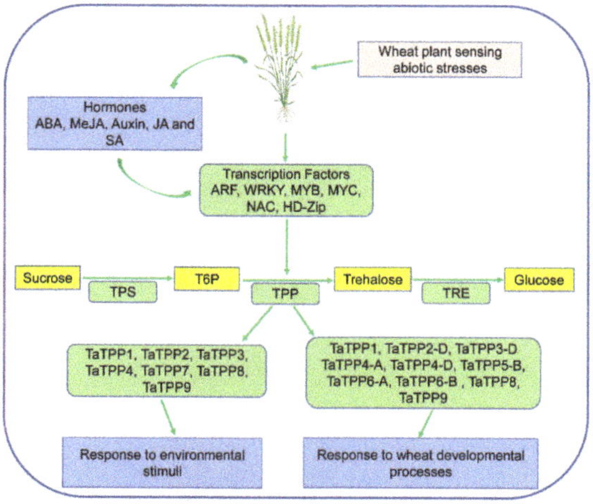

Figure 11. A proposed model for *TaTPP* genes functions in various wheat developmental processes and diverse stress conditions. ABA: Abscisic acid; MeJA: Methyl jasmonate; JA: Jasmonic acid; SA: Salicylic acid; ARF: Auxin response factors; MYB: Myeloblastosis; NAC: No apical meristem; TPS: trehalose-6-phosphate synthase; T6P: trehalose-6-phosphate; TPP: trehalose-6-phosphate phosphatase; TRE: trehalase. Yellow boxes indicate carbohydrates and light green boxes indicate proteins.

5. Conclusions

In conclusion, a relatively comprehensive analysis of the *TaTPP* gene family was performed in this study, which may help to explain the biological activities of TaTPP proteins in developmental processes, stress responses and leaf senescence of wheat. However, our knowledge of their precise biological role is still lacking. Thus, in order to give important insights to help wheat breeders for developing resistant crops cultivars to unfavorable stress conditions, an extensive functional validation study of *TaTPPs* is necessary.

Supplementary Materials: The following are available online at https://www.mdpi.com/article/10.3390/genes12111652/s1 Figure S1. Schematic illustration of 3-D protein structures and the Ramachandran plot for TaTPP proteins, Figure S2. Schematic representation of protein alignment of all TaTPPs, Figure S3. Schematic representation of TaTPP1-A side chains catalytic triads, Figure S4. Heat map of *TaTPPs* transcriptions in wheat development stages, Figure S5. Protein–protein interaction analysis of TaTPP proteins, Figure S6. Radiation tree of TaTPPs along with TPPs from other species, Table S1. List of primers used for *TaTPPs* qRT-PCR analysis, Table S2. TPP protein sequences identified from wheat genome, Table S3. Number of TPP proteins in different plant species, Table S4. Sequence identity and query cover of TaTPP proteins, Table S5. Pairwise identities and divergence between *TaTPP* genes and details about the duplication of those genes, Table S6. Phylogenetic tree member with their gene ID, Table S7. The synteny relationships of wheat *TPP* genes with different plant species, Table S8. Details of the calculated secondary structure elements TaTPPs by SOPMA, Table S9. Validation of TaTPP protein structures, Table S10. Frequency of all identified *Cis*-Regulatory Elements (CREs) in different *TaTPPs* promoters, Table S11. *Cis*-Regulatory Elements (CREs) with sequences and functions, Table S12: The protein-protein interaction network between TaTPPs and other proteins in wheat.

Author Contributions: The present study was conceptualized by D.S. and M.A.I.; bioinformatics analysis and visualization were conducted by M.M.R. (Md Mustafzur Rahman), M.A.I. and M.M.R. (Md Mizanor Rahman); experiments were investigated by M.A.I., X.J., L.S., K.Z., S.W., and H.N.; writing—original draft was prepared by M.A.I. and M.M.R. (Md Mustafzur Rahman); writing—reviewed and edited by J.-S.J., M.M.R. (Md Mizanor Rahman), D.S., A.S. and W.Z.; funding was acquired by D.S. All authors have read and agreed to the published version of the manuscript.

Funding: This research was sponsored by the State Key Laboratory of Sustainable Dryland Agriculture, Shanxi Agricultural University (No. 202002-2).

Institutional Review Board Statement: Not applicable.

Informed Consent Statement: Not applicable.

Data Availability Statement: Not applicable.

Conflicts of Interest: The authors declare no conflict of interest.

References

1. USDA. World Agricultural Production. 2019. Available online: https://www.fas.usda.gov/data/world-agricultural-production (accessed on 20 January 2020).
2. FAO. World Food Situation. 2021. Available online: http://www.fao.org/worldfoodsituation/csdb/en/ (accessed on 10 July 2021).
3. Yin, J.-L.; Fang, Z.-W.; Sun, C.; Zhang, P.; Zhang, X.; Lu, C.; Wang, S.-P.; Ma, D.-F.; Zhu, Y.-X. Rapid identification of a stripe rust resistant gene in a space-induced wheat mutant using specific locus amplified fragment (SLAF) sequencing. *Sci. Rep.* **2018**, *8*, 1–9. [CrossRef]
4. Tester, M.; Bacic, A. Abiotic Stress Tolerance in Grasses. From Model Plants to Crop Plants. *Plant Physiol.* **2005**, *137*, 791–793. [CrossRef] [PubMed]
5. Shafi, M.; Zhang, G.; Bakht, J.; Khan, M.A.; Islam, U.; Khan, M.D.; Raziuddin, G. Effect of cadmium and salinity stresses on root morphology of wheat. *Pak. J. Bot.* **2010**, *42*, 2747–2754.
6. Foolad, M.R. Recent Advances in Genetics of Salt Tolerance in Tomato. *Plant Cell Tissue Organ Cult.* **2004**, *76*, 101–119. [CrossRef]
7. Genc, Y.; Taylor, J.; Lyons, G.; Li, Y.; Cheong, J.; Appelbee, M.; Oldach, K.; Sutton, T. Bread Wheat with High Salinity and Sodicity Tolerance. *Front. Plant Sci.* **2019**, *10*, 1280. [CrossRef]
8. Rahneshan, Z.; Nasibi, F.; Moghadam, A.A. Effects of salinity stress on some growth, physiological, biochemical parameters and nutrients in two pistachio (*Pistacia vera* L.) rootstocks. *J. Plant Interact.* **2018**, *13*, 73–82. [CrossRef]

9. Wingler, A. The function of trehalose biosynthesis in plants. *Phytochemistry* **2002**, *60*, 437–440. [CrossRef]
10. Mostafa, M.R.; Mervat, S.S.; Safaa, R.E.-L.; Ebtihal, M.A.E.; Magdi, T.A. Exogenous α-tocopherol has a beneficial effect on *Glycine max*(L.) plants irrigated with diluted sea water. *J. Hortic. Sci. Biotechnol.* **2015**, *90*, 195–202. [CrossRef]
11. Patist, A.; Zoerb, H. Preservation mechanisms of trehalose in food and biosystems. *Colloids Surf. B Biointerfaces* **2005**, *40*, 107–113. [CrossRef]
12. Iturriaga, G.; Suárez, R.; Nova-Franco, B. Trehalose Metabolism: From Osmoprotection to Signaling. *Int. J. Mol. Sci.* **2009**, *10*, 3793–3810. [CrossRef]
13. Lunn, J.E.; Delorge, I.; Figueroa, C.M.; Van Dijck, P.; Stitt, M. Trehalose metabolism in plants. *Plant J.* **2014**, *79*, 544–567. [CrossRef]
14. Klähn, S.; Hagemann, M. Compatible solute biosynthesis in cyanobacteria. *Environ. Microbiol.* **2010**, *13*, 551–562. [CrossRef]
15. Chang, S.-W.; Chang, W.-H.; Lee, M.-R.; Yang, T.-J.; Yu, N.-Y.; Chen, C.-S.; Shaw, J.-F. Simultaneous Production of Trehalose, Bioethanol, and High-Protein Product from Rice by an Enzymatic Process. *J. Agric. Food Chem.* **2010**, *58*, 2908–2914. [CrossRef]
16. Elbein, A.D.; Pan, Y.T.; Pastuszak, I.; Carroll, D. New insights on trehalose: A multifunctional molecule. *Glycobiology* **2003**, *13*, 17R–27R. [CrossRef]
17. Lunn, J.E. Gene families and evolution of trehalose metabolism in plants. *Funct. Plant Biol.* **2007**, *34*, 550–563. [CrossRef]
18. Avonce, N.; Mendoza-Vargas, A.; Morett, E.; Iturriaga, G. Insights on the evolution of trehalose biosynthesis. *BMC Evol. Biol.* **2006**, *6*, 109. [CrossRef] [PubMed]
19. Ponnu, J.; Wahl, V.; Schmid, M. Trehalose-6-Phosphate: Connecting Plant Metabolism and Development. *Front. Plant Sci.* **2011**, *2*, 70. [CrossRef] [PubMed]
20. Lunn, J.E.; Feil, R.; Hendriks, J.H.M.; Gibon, Y.; Morcuende, R.; Osuna, D.; Scheible, W.-R.; Carillo, P.; Hajirezaei, M.-R.; Stitt, M. Sugar-induced increases in trehalose 6-phosphate are correlated with redox activation of ADPglucose pyrophosphorylase and higher rates of starch synthesis in *Arabidopsis thaliana*. *Biochem. J.* **2006**, *397*, 139–148. [CrossRef] [PubMed]
21. Meitzel, T.; Radchuk, R.; McAdam, E.L.; Thormählen, I.; Feil, R.; Munz, E.; Hilo, A.; Geigenberger, P.; Ross, J.J.; Lunn, J.E.; et al. Trehalose-6-phosphate promotes seed filling by activating auxin biosynthesis. *New Phytol.* **2021**, *229*, 1553–1565. [CrossRef] [PubMed]
22. Yadav, U.P.; Ivakov, A.; Feil, R.; Duan, G.Y.; Walther, D.; Giavalisco, P.; Piques, M.; Carillo, P.; Hubberten, H.-M.; Stitt, M.; et al. The sucrose–trehalose 6-phosphate (Tre6P) nexus: Specificity and mechanisms of sucrose signalling by Tre6P. *J. Exp. Bot.* **2014**, *65*, 1051–1068. [CrossRef]
23. Acevedo Hernández, G.J.; León, P.; Herrera-Estrella, L.R. Sugar and ABA responsiveness of a minimal RBCS light-responsive unit is mediated by direct binding of ABI4. *Plant J.* **2005**, *43*, 506–519. [CrossRef]
24. Gómez, L.D.; Gilday, A.; Feil, R.; Lunn, J.E.; Graham, I.A. AtTPS1-mediated trehalose 6-phosphate synthesis is essential for embryogenic and vegetative growth and responsiveness to ABA in germinating seeds and stomatal guard cells. *Plant J.* **2010**, *64*, 1–13. [CrossRef] [PubMed]
25. Fichtner, F.; Olas, J.J.; Feil, R.; Watanabe, M.; Krause, U.; Hoefgen, R.; Stitt, M.; Lunn, J.E. Functional Features of TREHALOSE-6-PHOSPHATE SYNTHASE1, an Essential Enzyme in *Arabidopsis* [OPEN]. *Plant Cell* **2020**, *32*, 1949–1972. [CrossRef]
26. Fichtner, F.; Barbier, F.F.; Annunziata, M.G.; Feil, R.; Olas, J.J.; Mueller-Roeber, B.; Stitt, M.; Beveridge, C.A.; Lunn, J.E. Regulation of shoot branching in *Arabidopsis* by trehalose-6-phosphate. *New Phytol.* **2021**, *229*, 2135–2151. [CrossRef] [PubMed]
27. Wang, G.; Li, X.; Ye, N.; Huang, M.; Feng, L.; Li, H.; Zhang, J. OsTPP1 regulates seed germination through the crosstalk with abscisic acid in rice. *New Phytol.* **2021**, *230*, 1925–1939. [CrossRef]
28. Wang, S.; Ouyang, K.; Wang, K. Genome-Wide Identification, Evolution, and Expression Analysis of *TPS* and *TPP* Gene Families in *Brachypodium distachyon*. *Plants* **2019**, *8*, 362. [CrossRef] [PubMed]
29. Satoh-Nagasawa, N.; Nagasawa, N.; Malcomber, S.; Sakai, H.; Jackson, D. A trehalose metabolic enzyme controls inflorescence architecture in maize. *Nature* **2006**, *441*, 227–230. [CrossRef]
30. Pellny, T.K.; Ghannoum, O.; Conroy, J.P.; Schluepmann, H.; Smeekens, S.; Andralojc, J.; Krause, K.P.; Goddijn, O.; Paul, M. Genetic modification of photosynthesis with *E. coli* genes for trehalose synthesis. *Plant Biotechnol. J.* **2004**, *2*, 71–82. [CrossRef]
31. Miranda, J.A.; Avonce, N.; Suárez, R.; Thevelein, J.M.; Van Dijck, P.; Iturriaga, G. A bifunctional TPS–TPP enzyme from yeast confers tolerance to multiple and extreme abiotic-stress conditions in transgenic *Arabidopsis*. *Planta* **2007**, *226*, 1411–1421. [CrossRef]
32. Lin, Q.; Yang, J.; Wang, Q.; Zhu, H.; Chen, Z.; Dao, Y.; Wang, K. Overexpression of the trehalose-6-phosphate phosphatase family gene *AtTPPF* improves the drought tolerance of *Arabidopsis thaliana*. *BMC Plant Biol.* **2019**, *19*, 381. [CrossRef]
33. Pilon-Smits, E.A.; Terry, N.; Sears, T.; Kim, H.; Zayed, A.; Hwang, S.; van Dun, K.; Voogd, E.; Verwoerd, T.C.; Krutwagen, R.W.; et al. Trehalose-producing transgenic tobacco plants show improved growth performance under drought stress. *J. Plant Physiol.* **1998**, *152*, 525–532. [CrossRef]
34. Wu, B.; Su, X. Identification of drought response genes in *Zygophyllum xanthoxylum* by suppression subtractive hybridization. *J. Plant Biol.* **2016**, *59*, 377–385. [CrossRef]
35. Garg, A.K.; Kim, J.-K.; Owens, T.G.; Ranwala, A.P.; Choi, Y.D.; Kochian, L.; Wu, R.J. Trehalose accumulation in rice plants confers high tolerance levels to different abiotic stresses. *Proc. Natl. Acad. Sci. USA* **2002**, *99*, 15898–15903. [CrossRef]
36. Shi, Y.; Sun, H.; Wang, X.; Jin, W.; Chen, Q.; Yuan, Z.; Yu, H. Physiological and transcriptomic analyses reveal the molecular networks of responses induced by exogenous trehalose in plant. *PLoS ONE* **2019**, *14*, e0217204. [CrossRef]

37. Jiang, D.; Chen, W.; Gao, J.; Yang, F.; Zhuang, C. Overexpression of the trehalose-6-phosphate phosphatase *OsTPP3* increases drought tolerance in rice. *Plant Biotechnol. Rep.* **2019**, *13*, 285–292. [CrossRef]
38. Ge, L.; Chao, D.-Y.; Shi, M.; Zhu, M.-Z.; Gao, J.-P.; Lin, H.-X. Overexpression of the trehalose-6-phosphate phosphatase gene *OsTPP1* confers stress tolerance in rice and results in the activation of stress responsive genes. *Planta* **2008**, *228*, 191–201. [CrossRef] [PubMed]
39. Acosta-Pérez, P.; Camacho-Zamora, B.D.; Espinoza-Sánchez, E.A.; Gutiérrez-Soto, G.; Zavala-García, F.; Abraham-Juárez, M.J.; Sinagawa-García, S.R. Characterization of Trehalose-6-phosphate Synthase and Trehalose-6-phosphate Phosphatase Genes and Analysis of its Differential Expression in Maize (*Zea mays*) Seedlings under Drought Stress. *Plants* **2020**, *9*, 315. [CrossRef] [PubMed]
40. Lin, Q.; Wang, S.; Dao, Y.; Wang, J.; Wang, K.; Wang, S. *Arabidopsis thaliana* trehalose-6-phosphate phosphatase gene *TPPI* enhances drought tolerance by regulating stomatal apertures. *J. Exp. Bot.* **2020**, *71*, 4285–4297. [CrossRef]
41. Wang, W.; Chen, Q.; Xu, S.; Liu, W.C.; Zhu, X.; Song, C.P. Trehalose-6-phosphate phosphatase E modulates ABA-controlled root growth and stomatal movement in *Arabidopsis*. *J. Integr. Plant Biol.* **2020**, *62*, 1518–1534. [CrossRef]
42. Wingler, A.; Delatte, T.L.; O'Hara, L.; Primavesi, L.; Jhurreea, D.; Paul, M.; Schluepmann, H. Trehalose 6-Phosphate Is Required for the Onset of Leaf Senescence Associated with High Carbon Availability. *Plant Physiol.* **2012**, *158*, 1241–1251. [CrossRef] [PubMed]
43. Noodén, L.D.; Guiamét, J.J.; John, I. Senescence mechanisms. *Physiol. Plant.* **1997**, *101*, 746–753. [CrossRef]
44. Pourtau, N.; Jennings, R.; Pelzer, E.; Pallas, J.; Wingler, A. Effect of sugar-induced senescence on gene expression and implications for the regulation of senescence in *Arabidopsis*. *Planta* **2006**, *224*, 556–568. [CrossRef] [PubMed]
45. Wingler, A.; Purdy, S.; MacLean, J.A.; Pourtau, N. The role of sugars in integrating environmental signals during the regulation of leaf senescence. *J. Exp. Bot.* **2006**, *57*, 391–399. [CrossRef]
46. Zhang, P.; He, Z.; Tian, X.; Gao, F.; Xu, D.; Liu, J.; Wen, W.; Fu, L.; Li, G.; Sui, X.; et al. Cloning of *TaTPP-6AL1* associated with grain weight in bread wheat and development of functional marker. *Mol. Breed.* **2017**, *37*, 78. [CrossRef]
47. Johnson, M.; Zaretskaya, I.; Raytselis, Y.; Merezhuk, Y.; McGinnis, S.; Madden, T.L. NCBI BLAST: A better web interface. *Nucleic Acids Res.* **2008**, *36*, W5–W9. [CrossRef] [PubMed]
48. Chen, C.; Chen, H.; Zhang, Y.; Thomas, H.R.; Frank, M.H.; He, Y.; Xia, R. TBtools: An Integrative Toolkit Developed for Interactive Analyses of Big Biological Data. *Mol. Plant* **2020**, *13*, 1194–1202. [CrossRef] [PubMed]
49. Kumar, S.; Stecher, G.; Li, M.; Knyaz, C.; Tamura, K. MEGA X: Molecular Evolutionary Genetics Analysis across Computing Platforms. *Mol. Biol. Evol.* **2018**, *35*, 1547–1549. [CrossRef]
50. Guindon, S.; Gascuel, O. A Simple, Fast, and Accurate Algorithm to Estimate Large Phylogenies by Maximum Likelihood. *Syst. Biol.* **2003**, *52*, 696–704. [CrossRef]
51. Felsenstein, J. Confidence limits on phylogenies: An approach using the bootstrap. *Evolution* **1985**, *39*, 783–791. [CrossRef]
52. Combet, C.; Blanchet, C.; Geourjon, C.; Deléage, G. NPS@: Network Protein Sequence Analysis. *Trends Biochem. Sci.* **2000**, *25*, 147–150. [CrossRef]
53. Izidoro, S.; De Melo-Minardi, R.C.; Pappa, G.L. GASS: Identifying enzyme active sites with genetic algorithms. *Bioinformatics* **2015**, *31*, 864–870. [CrossRef]
54. Lovell, S.C.; Davis, I.W.; Arendall III, W.B.; De Bakker, P.I.; Word, J.M.; Prisant, M.G.; Richardson, J.S.; Richardson, D.C. Structure validation by Cα geometry: ϕ, ψ and Cβ deviation. *Proteins* **2003**, *50*, 437–450. [CrossRef]
55. Moraes, J.P.A.; Pappa, G.L.; Pires, D.E.V.; Izidoro, S.C. GASS-WEB: A web server for identifying enzyme active sites based on genetic algorithms. *Nucleic Acids Res.* **2017**, *45*, W315–W319. [CrossRef] [PubMed]
56. Waterhouse, A.; Bertoni, M.; Bienert, S.; Studer, G.; Tauriello, G.; Gumienny, R.; Heer, F.T.; De Beer, T.A.P.; Rempfer, C.; Bordoli, L.; et al. SWISS-MODEL: Homology modelling of protein structures and complexes. *Nucleic Acids Res.* **2018**, *46*, W296–W303. [CrossRef]
57. Williams, C.J.; Headd, J.J.; Moriarty, N.W.; Prisant, M.G.; Videau, L.L.; Deis, L.N.; Verma, V.; Keedy, D.A.; Hintze, B.J.; Chen, V.B.; et al. MolProbity: More and better reference data for improved all-atom structure validation. *Protein Sci.* **2017**, *27*, 293–315. [CrossRef] [PubMed]
58. Ren, J.; Wen, L.; Gao, X.; Jin, C.; Xue, Y.; Yao, X. DOG 1.0: Illustrator of protein domain structures. *Cell Res.* **2009**, *19*, 271–273. [CrossRef]
59. Finn, R.D.; Bateman, A.; Clements, J.; Coggill, P.; Eberhardt, R.Y.; Eddy, S.R.; Heger, A.; Hetherington, K.; Holm, L.; Mistry, J.; et al. Pfam: The protein families database. *Nucleic Acids Res.* **2014**, *42*, D222–D230. [CrossRef] [PubMed]
60. Bailey, T.L.; Boden, M.; Buske, F.A.; Frith, M.; Grant, C.E.; Clementi, L.; Ren, J.; Li, W.W.; Noble, W.S. MEME SUITE: Tools for motif discovery and searching. *Nucleic Acids Res.* **2009**, *37*, w202–w208. [CrossRef]
61. Lescot, M.; Déhais, P.; Thijs, G.; Marchal, K.; Moreau, Y.; Van de Peer, Y.; Rouzé, P.; Rombauts, S. PlantCARE, a database of plant *cis*-acting regulatory elements and a portal to tools for in silico analysis of promoter sequences. *Nucleic Acids Res.* **2002**, *30*, 325–327. [CrossRef]
62. NetBeans, IDE 8.0. Oracle Co., Redwood City, CA, 2015. Available online: https://netbeans.org (accessed on 25 March 2021).
63. Hruz, T.; Laule, O.; Szabo, G.; Wessendorp, F.; Bleuler, S.; Oertle, L.; Widmayer, P.; Gruissem, W.; Zimmermann, P. Genevestigator V3: A Reference Expression Database for the Meta-Analysis of Transcriptomes. *Adv. Bioinform.* **2008**, *2008*, 1–5. [CrossRef]
64. Howe, E.; Holton, K.; Nair, S.; Schlauch, D.; Sinha, R.; Quackenbush, J. MeV: Multi Experiment Viewer. In *Biomedical Informatics for Cancer Research*; Springer: Berlin/Heidelberg, Germany, 2010; pp. 267–277.

65. Duan, Y.-H.; Guo, J.; Ding, K.; Wang, S.-J.; Zhang, H.; Dai, X.-W.; Chen, Y.-Y.; Govers, F.; Huang, L.-L.; Kang, Z.-S. Characterization of a wheat *HSP70* gene and its expression in response to stripe rust infection and abiotic stresses. *Mol. Biol. Rep.* **2010**, *38*, 301–307. [CrossRef]
66. Livak, K.J.; Schmittgen, T.D. Analysis of Relative Gene Expression Data Using Real-Time Quantitative PCR and the $2^{-\Delta\Delta CT}$ Method. *Methods* **2001**, *25*, 402–408. [CrossRef]
67. Paolacci, A.R.; Tanzarella, O.A.; Porceddu, E.; Ciaffi, M. Identification and validation of reference genes for quantitative RT-PCR normalization in wheat. *BMC Mol. Biol.* **2009**, *10*, 11. [CrossRef] [PubMed]
68. Fleury, D.; Jefferies, S.; Kuchel, H.; Langridge, P. Genetic and genomic tools to improve drought tolerance in wheat. *J. Exp. Bot.* **2010**, *61*, 3211–3222. [CrossRef]
69. Bhattacharya, M.; Hota, A.; Kar, A.; Chini, D.S.; Malick, R.C.; Patra, B.C.; Das, B.K. In silico structural and functional modelling of Antifreeze protein (AFP) sequences of Ocean pout (Zoarces americanus, Bloch & Schneider 1801). *J. Genet. Eng. Biotechnol.* **2018**, *16*, 721–730. [CrossRef] [PubMed]
70. Yaqoob, U.; Kaul, T.; Nawchoo, I.A. *In-Silico* Analysis, Structural Modelling and Phylogenetic Analysis of Acetohydroxyacid Synthase Gene of *Oryza sativa*. *Med. Aromat. Plants* **2016**, *5*, 2167-0412. [CrossRef]
71. Kong, X.; Lv, W.; Jiang, S.; Zhang, D.; Cai, G.; Pan, J.; Li, D. Genome-wide identification and expression analysis of calcium-dependent protein kinase in maize. *BMC Genom.* **2013**, *14*, 433. [CrossRef]
72. Gu, Z.; Cavalcanti, A.; Chen, F.-C.; Bouman, P.; Li, W.-H. Extent of Gene Duplication in the Genomes of *Drosophila*, Nematode, and Yeast. *Mol. Biol. Evol.* **2002**, *19*, 256–262. [CrossRef]
73. Bordoli, L.; Schwede, T. Automated Protein Structure Modeling with SWISS-MODEL Workspace and the Protein Model Portal. In *Homology Modeling*; Springer: Berlin/Heidelberg, Germany, 2011; Volume 857, pp. 107–136. [CrossRef]
74. Rahman, M.M.; Rahman, M.M.; Eom, J.S.; Jeon, J.S. Genome-wide identification, expression profiling and promoter analysis of trehalose-6-phosphate phosphatase gene family in rice. *J. Plant Biol.* **2021**, *64*, 55–71. [CrossRef]
75. Grace, M.L.; Chandrasekharan, M.; Hall, T.C.; Crowe, A.J. Sequence and Spacing of TATA Box Elements Are Critical for Accurate Initiation from the β-Phaseolin Promoter. *J. Biol. Chem.* **2004**, *279*, 8102–8110. [CrossRef]
76. Vogel, G.; Aeschbacher, R.A.; Müller, J.; Boller, T.; Wiemken, A. Trehalose-6-phosphate phosphatases from *Arabidopsis thaliana*: Identification by functional complementation of the yeast *tps2* mutant. *Plant J.* **1998**, *13*, 673–683. [CrossRef]
77. Svanström, Å.; van Leeuwen, M.R.; Dijksterhuis, J.; Melin, P. Trehalose synthesis in *Aspergillus niger*: Characterization of six homologous genes, all with conserved orthologs in related species. *BMC Microbiol.* **2014**, *14*, 90. [CrossRef]
78. Wang, Y.-J.; Hao, Y.-J.; Zhang, Z.-G.; Chen, T.; Zhang, J.-S.; Chen, S.-Y. Isolation of trehalose-6-phosphate phosphatase gene from tobacco and its functional analysis in yeast cells. *J. Plant Physiol.* **2005**, *162*, 215–223. [CrossRef]
79. Streeter, J.; Gomez, M. Three enzymes for trehalose synthesis in *Bradyrhizobium* cultured bacteria and in bacteroids from soybean nodules. *Appl. Environ. Microbiol.* **2006**, *72*, 4250–4255. [CrossRef] [PubMed]
80. Henry, C.; Bledsoe, S.W.; Siekman, A.; Kollman, A.; Waters, B.; Feil, R.; Stitt, M.; Lagrimini, L.M. The trehalose pathway in maize: Conservation and gene regulation in response to the diurnal cycle and extended darkness. *J. Exp. Bot.* **2014**, *65*, 5959–5973. [CrossRef] [PubMed]
81. Schluepmann, H.; van Dijken, A.; Aghdasi, M.; Wobbes, B.; Paul, M.; Smeekens, S. Trehalose Mediated Growth Inhibition of *Arabidopsis* Seedlings Is Due to Trehalose-6-Phosphate Accumulation. *Plant Physiol.* **2004**, *135*, 879–890. [CrossRef] [PubMed]
82. Leyman, B.; Van Dijck, P.; Thevelein, J. An unexpected plethora of trehalose biosynthesis genes in *Arabidopsis thaliana*. *Trends Plant Sci.* **2001**, *6*, 510–513. [CrossRef]
83. Lawton-Rauh, A. Evolutionary dynamics of duplicated genes in plants. *Mol. Phylogenetics Evol.* **2003**, *29*, 396–409. [CrossRef]
84. Moore, R.C.; Purugganan, M.D. The early stages of duplicate gene evolution. *Proc. Natl. Acad. Sci. USA* **2003**, *100*, 15682–15687. [CrossRef]
85. Holub, E.B. The arms race is ancient history in *Arabidopsis*, the wildflower. *Nat. Rev. Genet.* **2001**, *2*, 516–527. [CrossRef]
86. Miyakawa, T.; Miyazono, K.-I.; Sawano, Y.; Hatano, K.-I.; Tanokura, M. Crystal structure of ginkbilobin-2 with homology to the extracellular domain of plant cysteine-rich receptor-like kinases. *Proteins Struct. Funct. Bioinform.* **2009**, *77*, 247–251. [CrossRef] [PubMed]
87. Zhang, L.; Tian, L.-H.; Zhao, J.-F.; Song, Y.; Zhang, C.-J.; Guo, Y. Identification of an Apoplastic Protein Involved in the Initial Phase of Salt Stress Response in Rice Root by Two-Dimensional Electrophoresis. *Plant Physiol.* **2009**, *149*, 916–928. [CrossRef]
88. Krasensky-Wrzaczek, J.; Broyart, C.; Rabanal, F.; Jonak, C. The Redox-Sensitive Chloroplast Trehalose-6-Phosphate Phosphatase AtTPPD Regulates Salt Stress Tolerance. *Antioxid. Redox Signal.* **2014**, *21*, 1289–1304. [CrossRef]
89. Cowan, J.A. Metal Activation of Enzymes in Nucleic Acid Biochemistry. *Chem. Rev.* **1998**, *98*, 1067–1088. [CrossRef] [PubMed]
90. Bertini, G.; Gray, H.B.; Gray, H.; Valentine, J.S.; Stiefel, E.I.; Stiefel, E. *Biological Inorganic Chemistry: Structure and Reactivity*; University Science Books: Sausalito, CA, USA, 2007.
91. Hernandez-Garcia, C.M.; Finer, J.J. Identification and validation of promoters and *cis*-acting regulatory elements. *Plant Sci.* **2014**, *217*, 109–119. [CrossRef]
92. Choi, H.-I.; Hong, J.-H.; Ha, J.-O.; Kang, J.-Y.; Kim, S.Y. ABFs, a Family of ABA-responsive Element Binding Factors. *J. Biol. Chem.* **2000**, *275*, 1723–1730. [CrossRef]
93. Hobo, T.; Asada, M.; Kowyama, Y.; Hattori, T. ACGT-containing abscisic acid response element (ABRE) and coupling element 3 (CE3) are functionally equivalent. *Plant J.* **1999**, *19*, 679–689. [CrossRef]

94. Guiltinan, M.J.; Marcotte, W.R.; Quatrano, R.S. A Plant Leucine Zipper Protein That Recognizes an Abscisic Acid Response Element. *Science* **1990**, *250*, 267–271. [CrossRef]
95. Marcotte, W.R., Jr.; Russell, S.H.; Quatrano, R.S. Abscisic acid-responsive sequences from the em gene of wheat. *Plant Cell* **1989**, *1*, 969–976.
96. Narusaka, Y.; Nakashima, K.; Shinwari, Z.K.; Sakuma, Y.; Furihata, T.; Abe, H.; Narusaka, M.; Shinozaki, K.; Yamaguchi-Shinozaki, K. Interaction between two *cis*-acting elements, ABRE and DRE, in ABA-dependent expression of *Arabidopsis rd29A* gene in response to dehydration and high-salinity stresses. *Plant J.* **2003**, *34*, 137–148. [CrossRef]
97. Shen, Q.; Zhang, P.; Ho, T. Modular nature of abscisic acid (ABA) response complexes: Composite promoter units that are necessary and sufficient for ABA induction of gene expression in barley. *Plant Cell* **1996**, *8*, 1107–1119. [PubMed]
98. Banerjee, J.; Sahoo, D.K.; Dey, N.; Houtz, R.L.; Maiti, I.B. An Intergenic Region Shared by *At4g35985* and *At4g35987* in *Arabidopsis thaliana* Is a Tissue Specific and Stress Inducible Bidirectional Promoter Analyzed in Transgenic Arabidopsis and Tobacco Plants. *PLoS ONE* **2013**, *8*, e79622. [CrossRef] [PubMed]
99. Kretzschmar, T.; Pelayo, M.A.F.; Trijatmiko, K.; Gabunada, L.F.M.; Alam, R.; Jimenez, R.; Mendioro, M.S.; Slamet-Loedin, I.; Sreenivasulu, N.; Bailey-Serres, J.; et al. A trehalose-6-phosphate phosphatase enhances anaerobic germination tolerance in rice. *Nat. Plants* **2015**, *1*, 15124. [CrossRef]
100. Shima, S.; Matsui, H.; Tahara, S.; Imai, R. Biochemical characterization of rice trehalose-6-phosphate phosphatases supports distinctive functions of these plant enzymes. *FEBS J.* **2007**, *274*, 1192–1201. [CrossRef] [PubMed]

genes

MDPI

Article

Transcriptome Profiling of Maize (*Zea mays* L.) Leaves Reveals Key Cold-Responsive Genes, Transcription Factors, and Metabolic Pathways Regulating Cold Stress Tolerance at the Seedling Stage

Joram Kiriga Waititu [1], Quan Cai [2], Ying Sun [1], Yinglu Sun [3], Congcong Li [1], Chunyi Zhang [1], Jun Liu [3] and Huan Wang [1,4,*]

1 Biotechnology Research Institute, Chinese Academy of Agricultural Sciences, Beijing 100081, China; joram.kiriga@gmail.com (J.K.W.); sunying0624@126.com (Y.S.); licongcong01@caas.cn (C.L.); zhangchunyi@caas.cn (C.Z.)
2 Maize Research Institute, Heilongjiang Academy of Agricultural Sciences, Harbin 150086, China; cq6539@163.com
3 National Key Facility for Crop Resources and Genetic Improvement, Institute of Crop Science, Chinese Academy of Agricultural Sciences, Beijing 100081, China; sunny_5211995@sina.com (Y.S.); liujun@caas.cn (J.L.)
4 National Agricultural Science and Technology Center, Chengdu 610213, China
* Correspondence: wanghuan@caas.cn

Citation: Waititu, J.K.; Cai, Q.; Sun, Y.; Sun, Y.; Li, C.; Zhang, C.; Liu, J.; Wang, H. Transcriptome Profiling of Maize (*Zea mays* L.) Leaves Reveals Key Cold-Responsive Genes, Transcription Factors, and Metabolic Pathways Regulating Cold Stress Tolerance at the Seedling Stage. *Genes* **2021**, *12*, 1638. https://doi.org/10.3390/genes12101638

Academic Editor: Patrizia Galeffi

Received: 13 August 2021
Accepted: 11 October 2021
Published: 18 October 2021

Abstract: Cold tolerance is a complex trait that requires a critical perspective to understand its underpinning mechanism. To unravel the molecular framework underlying maize (*Zea mays* L.) cold stress tolerance, we conducted a comparative transcriptome profiling of 24 cold-tolerant and 22 cold-sensitive inbred lines affected by cold stress at the seedling stage. Using the RNA-seq method, we identified 2237 differentially expressed genes (DEGs), namely 1656 and 581 annotated and unannotated DEGs, respectively. Further analysis of the 1656 annotated DEGs mined out two critical sets of cold-responsive DEGs, namely 779 and 877 DEGs, which were significantly enhanced in the tolerant and sensitive lines, respectively. Functional analysis of the 1656 DEGs highlighted the enrichment of signaling, carotenoid, lipid metabolism, transcription factors (TFs), peroxisome, and amino acid metabolism. A total of 147 TFs belonging to 32 families, including MYB, ERF, NAC, WRKY, bHLH, MIKC MADS, and C_2H_2, were strongly altered by cold stress. Moreover, the tolerant lines' 779 enhanced DEGs were predominantly associated with carotenoid, ABC transporter, glutathione, lipid metabolism, and amino acid metabolism. In comparison, the cold-sensitive lines' 877 enhanced DEGs were significantly enriched for MAPK signaling, peroxisome, ribosome, and carbon metabolism pathways. The biggest proportion of the unannotated DEGs was implicated in the roles of long non-coding RNAs (lncRNAs). Taken together, this study provides valuable insights that offer a deeper understanding of the molecular mechanisms underlying maize response to cold stress at the seedling stage, thus opening up possibilities for a breeding program of maize tolerance to cold stress.

Keywords: cold; stress; differentially expressed genes; transcriptome; transcription factors

1. Introduction

Maize (*Zea mays* L.) is the world's most commonly grown cereal crop, with an estimated global annual production of about 1186.86 million metric tons in 2020/2021 [1]. The high dependence on maize for human, animal, and industrial consumption makes it one of the most critical food crops. However, maize growth and yield are highly dependent on sufficient environmental factors [2]. Thus, the current and expected scarcity of water sources and arable land due to the increasing world population and the recurrent extreme weather caused by global warming is projected to increase the incidence of abiotic stresses, such as drought, cold, and freezing during the planting, flowering, and grain-filling stages, in many corn-growing areas [3]. These abiotic stresses typically serve as crucial impediments to maize production and geographical distribution [4] and restrict agricultural yields worldwide.

Since maize has a tropical origin, cold stress is a significant risk factor among the several abiotic stresses in the development of maize. A previous report has shown that cold stress adversely affects maize growth from germination to harvest, resulting in significant yield losses due to low and slow germination and poor grain filling [5]. Corn production losses can surpass 20% in the most prolonged cold temperatures [6]. Therefore, the development of high-yielding cultivars tolerant to cold stress may help in augmenting maize production in vulnerable regions and act as an essential maize-breeding target.

The optimum maize growth temperatures range from 21 to 27 °C, while sub-optimal temperatures of about 10–20 °C decrease biomass production, thereby leading to growth retardation [7]. Cold stress induces multiple abnormalities in physiological, molecular, and biochemical processes, which harm plant growth and yield. Cell membranes may become disorganized, proteins may be denatured, oxidative defense and osmotic stress may be altered, photosynthesis possibly restricted, and metabolism may become dysfunctional, all of which subsequently disrupt growth and development, decrease fertility, and cause premature senescence and even plant death [8–11]. All of these cold stress-associated changes occur through accurate gene expression regulation and are therefore genetically regulated. Thus, screening cold stress-related candidate genes may help identify essential regulators and pathways as potential targets for breeding resistant varieties adaptable to environments with fluctuating temperatures.

As a result of their sessile nature, plants have developed complex cold acclimation mechanisms, which entail the interaction of multiple biochemical pathways in an organ-, genetic-, and environmental-specific manner [12]. They sense cold stress through changes in membrane fluidity and the accumulation of calcium signatures, leading to downstream activation of cold signaling pathways [13]. The enhanced cytosolic Ca^{2+} levels then induce C-repeat binding factors (CBFs), which act as core regulators for expressing cold-response genes [14,15]. The stress receptors, in conjunction with the cell membrane transporters, facilitate the perception of stress signals and their transmission to target genes. Multiple protein kinases, including CaMs, CMLs, CBLs, CDPKs, and MAPKs, phosphorylate other kinases and/or various TFs, resulting in activation of the cold-responsive genes [16]. Moreover, the transcriptional factor (TF) families bHLH, CAMTA, MADS, WRKY, NAC, TRAF, C3H, and AP2/ERF are critical in cold response mechanisms, while phytohormones, such as abscisic acid (ABA), regulate specific pathways that lead to cold tolerance [17,18]. Cellular redox homeostasis is protected by synthesizing defense enzymes and other antioxidant systems, while soluble sugars serve as a stabilizer of cellular components and plasma membrane [19]. Secondary metabolites, such as lignin, anthocyanin, terpenoids, chaperones, and late embryogenesis abundant (LEA), provide cold tolerance by protecting cellular components from cold-induced cellular damage [20,21]. Transporters, such as the ATP-binding cassette (ABC) transporter, play integral roles in plant growth and development, homeostasis of phytohormones, and resistance to abiotic stress [22]. Tremendous progress has been made in elucidating the mechanisms underlying cold tolerance in plants. However, the complex molecular mechanisms of cold tolerance in maize seedlings are still elusive and require comprehensive research.

Moreover, considering the genetic diversity of maize inbred lines, it will be interesting to identify the stress-responsive genes that have a consistent function in a variety of inbred lines in spite of their genetic background. Differential transcriptome analysis using RNA-sequencing (RNA-seq) approaches has recently emerged as robust, reliable, and responsive to broader levels of gene expression [23]. This effective technology makes it easier to rapidly classify stress-responsive genes and decode metabolic pathways associated with biotic and abiotic stresses [24]. Currently, little is known about the transcriptomic responses of maize seedlings to cold stress. In this study, we used RNA-seq analysis to decipher the expression profiles of the differentially expressed genes (DEGs) responsible for the contrasting cold response of 46 maize inbred lines (24 tolerant and 22 sensitive) at the seedling stage. The common cold-responsive genes were then characterized by their patterns of expression and evaluated for their functional significance. The current study provides valuable clues for the in-depth characterization of molecular responses of maize seedlings to cold stress, which could lead to effective strategies for breeding and developing cold-tolerant maize varieties.

2. Materials and Methods

2.1. Plant Materials and Treatments

The maize inbred lines were derived from the hybrid 19NL, which was provided by Heilongjiang Academy of Agricultural Sciences. The maize hybrid 19NL is a highly suitable cultivar for the spring season in northeastern China. Field experiments were conducted at open field stations during the spring maize growing season (March–June 2019) in Heilongjiang (Harbin, China). In April 2019, approximately 6000 inbred lines of 19NL were planted in rows, with 40 cm between the rows and each row containing 20 seedlings. However, in the late spring of May 2019, due to climate change, the Heilongjiang region of Harbin was affected by a cold spell of below 10 °C for more than three days. We observed that the cold spell impacted the seedlings differently, with survival rates varying from one inbred line to another. This indicated that their response to varying degrees of cold stress was different, regardless of being from the same germplasm. When seedlings had six fully expanded leaves (40 days old), all inbred lines were sampled, and 46 inbred lines with contrasting cold tolerance were selected and classified into cold-tolerant (24 lines) and cold-sensitive (22 lines) lines based on the survival rates of the seedlings, as well as a visual observation of the phenotypic changes of the leaves. The top fully expanding leaves of the 24 cold-resistant and 22 cold-sensitive inbred lines were harvested, frozen into liquid nitrogen, and later stored at −80 °C for further use.

To validate the expression of our cold-responsive DEGs, we planted B73 and CIMBL116 maize inbred lines, which have previously been reported as being cold-sensitive and -tolerant lines, respectively [25,26]. The seeds from these two maize inbred lines were provided by the Chinese Academy of Agricultural Sciences' crop science institute). Ten seeds from each inbred line (B73 and CIMBL116) were surface sterilized with 75% (v/v) ethanol for three minutes before being rinsed three times with distilled water. Seeds were then placed between two layers of damp paper at 25 °C and left to germinate in the dark for 3 days. Uniformly germinated seeds with 2–3 cm coleoptiles were selected and sown in pots filled with peat, vermiculite, and perlite (10:1:1 by vol.). The seedlings were then grown in a growth chamber with a controlled temperature of 25/20 °C (day/night), 450 L mol m^{-2} s^{-1} light density, and a 12/12 h (light/dark) photoperiod until the third leaves were fully developed. The seedlings from the two inbred lines were divided into two groups, with the first receiving a cold stress treatment of 4 °C for 2 h followed by 25 °C for 2 days. The other group, which served as a control, was kept in the growth chamber under the same conditions as described above. The control and cold treatment samples were harvested at the same time after 2 days and immediately frozen in liquid nitrogen before being stored at −80 °C for total RNA isolation.

2.2. RNA Extraction, Library Construction, and Illumina Sequencing

TRIzol reagent (Invitrogen, CA, USA) was used to isolate total RNA from 24 tolerant and 22 sensitive leaf samples as per the manufacturer's standard. The samples were treated with RNase-free DNaseI (Takara, Kusatsu, Japan) to remove the genomic DNA. The NanoDrop 1000 (NanoDrop Technologies, Wilmington, DE, USA) and Agilent 2100 Bioanalyzer (Agilent Technologies, Santa Clara, CA, USA) were used to assess RNA concentration and integrity, respectively. The cDNA libraries were constructed and sequenced using the Illumina HiSeq™ 2500 platform to generate 150 bp paired-end reads. Moreover, the above procedure was also employed to extract RNA from B73 and CIMBL116 at control and treatment levels for the purpose of qRT-PCR.

2.3. Reads Processing, Mapping, and Gene Expression Quantification

We used FastQC (V0.11.3) to evaluate the quality of raw reads, while Trimmomatic (V0.32) was utilized to eliminate low-quality and adapter-containing reads [27]. The Phred quality scores, including Q20 (99% base call accuracy), Q30 (99.9% base call accuracy), and the GC content of the clean data, were calculated. Consequently, high-quality clean data were used in all the subsequent analyses. The maize reference genome (B73_v4) was downloaded from the maize database (http://www.maizegdb.org/genome/genome_assembly/Zm-B73-REFERENCE-GRAMENE-4.0, accessed on 15 November 2019). All the clean reads obtained from the 46 samples were aligned to the maize B73_v4 reference genome using HISAT2 (V2.0.5) [28] with default parameters. The aligned reads were assembled into transcripts, and the transcripts from all samples were merged using Cufflinks [29]. The assembled transcripts were compared to the reference annotation by Cuffcompare. The HTSeq tool [30] was used to count the number of fragments mapped to each gene, and the transcripts per million (TPM) for each unigene as the expression level was calculated. Principal component analysis (PCA) of gene expression levels was performed to calculate the distance between samples using the clustering method. Differential expression analysis was performed using the DESeq2 R package to identify DEGs in general for the whole transcriptomes of the tolerant and sensitive samples. To estimate expression level, the DESeq2 program was used to normalize the number of counts of each sample gene using the base means. The difference was calculated, and the statistical significance was determined using the negative binomial distribution test [29,31]. Genes with p-value ≤ 0.05 and an absolute value of log2 fold change ≥ 1 or ≤ -1 between tolerant and sensitive samples were considered differentially expressed.

2.4. Functional Annotation of the DEGs

For functional annotation, the 2237 transcripts that qualified to be our DEGs were annotated against the maize genome (AGPv4, B73 RefGen_v4) (http://ensembl.gramene.org/Zea_mays/Info/Index, accessed on 17 November 2019). In total, 1656 (74%) DEGs were annotated, and 581 DEGs were unannotated. To elucidate the function of the 581 unannotated genes, we applied several previously published procedures to identify high confidence lncRNAs [32]. Briefly, (i) unannotated DEG lengths were confirmed to be longer than 200 nucleotides for further analysis; (ii) DEGs that encode open reading frames (ORFs) of 120 or fewer amino acids were retained as lncRNA candidates; (iii) DEGs with similarity to known proteins based on BlastX against the SWISS-PROT database were filtered out; (iv) all the 581 unannotated DEGs were further evaluated using Coding Potential Calculator (CPC) (http://cpc.gao-lab.org/, accessed on 20 March 2020) [33], which assesses the coding probability of transcripts; (v) a total of 337 high confidence drought-responsive lncRNAs were obtained by comparing the output of the two procedures.

2.5. Gene Ontology (GO) Enrichment and KEGG Pathway Enrichment Analyses

The GO enrichment analysis of DEGs was conducted by agriGOv2 (http://systemsbiology.cau.edu.cn/agriGOv2/, accessed on 12 December 2019) [34]. Significant enriched GO terms were determined by the p-value ≤ 0.05 with the Fisher's exact test and the Bonferroni multi-test adjustment. Redundant GO terms were removed using the web tool Revigo [35]. Significantly enriched GO terms were assigned to the GO categories of biological process (BP), molecular function (MF), and cellular component (CC). The KEGG (http://www.Genome.jp/kegg/, accessed on 15 December 2019) database [36] was used to analyze the functional involvement of DEGs in various metabolic pathways. Furthermore, the statistical enrichment of DEGs in KEGG pathways was tested using the KOBAS 3.0 webserver (http://kobas.cbi.pku.edu.cn/waitkobas.php, accessed on 15 December 2019) [37], while the criteria for substantially enriched KEGG pathways was a p-value ≤ 0.05. A co-expression network was constructed using the

R package based on a weighted gene co-expression network analysis (WGCNA) to identify significant hub genes associated with cold tolerance in maize.

2.6. Validation of Cold-Responsive DEGs by Quantitative Real-Time PCR (qRT-PCR)

To validate the reliability and repeatability of the RNA-Seq data, six DEGs were randomly selected for verification by qRT-PCR. The gene-specific primers (Table S1) were designed using Primer Premier 5.0 software (Premier Biosoft International, Palo Alto, CA, USA). The qRT-PCR was conducted in triplicate using 2× ChamQ SYBR qPCR Master Mix Kit (Low ROX Premixed) (Vazyme Biotechnology Co., Nanjing, China) on an Applied Biosystems QuantStudio® 6 Flex (Thermo Lifetech, Waltham, MA, USA). The reaction system utilized was as previously described by Zhang et al. [38]. The internal reference β-actin was utilized to normalize the expression data. The relative expression levels of the six DEGs were calculated according to the $2^{-\Delta\Delta CT}$ (cycle threshold) method [39].

3. Results

3.1. Phenotypic Analysis of Maize Population under Cold Stress Conditions

Forty-six inbred lines cultivated in Heilongjiang Province, China, were selected and evaluated for cold tolerance based on their seedling survival rate and physiological response (visual observations of the leaves). From the results, 24 inbred line seedlings displayed little phenotypic changes, maintained fully expanded green leaves, had intact plant architecture, strong vigor, and high survival rates (average of 0.8) (Table S2). These inbred lines were therefore classified as cold tolerant and were labeled with an extension of −1 (Figure 1 and Figure S1A). The remaining 22 inbred line seedlings showed visible phenotypic damage, including shriveled, curled, and yellowish spots on the leaves, and low survival rates (average of 0.3) (Table S2). These seedlings were classified as cold sensitive and were labeled with an extension of −2 (Figure 1 and Figure S1B). Collectively, the susceptible lines were more severely damaged by cold stress than the tolerant lines, as evidenced by the shriveled, curled, and yellowish patches on their leaves, as well as low seedling survival rates of about 0.3 on average.

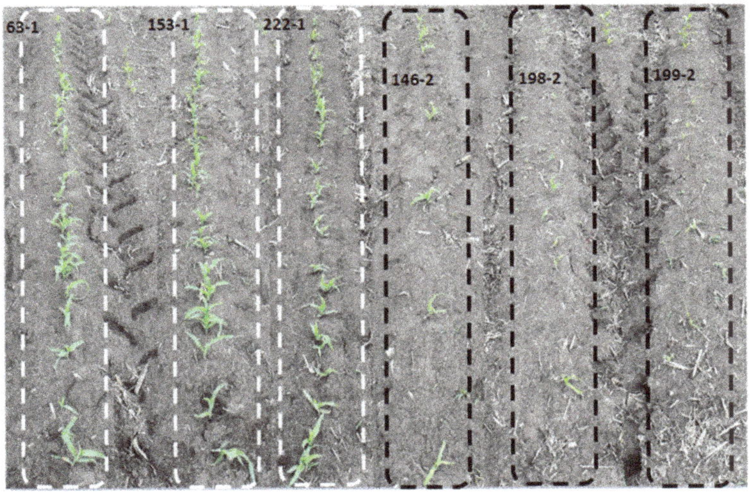

Figure 1. Maize seedling performance under cold stress. The performances were based on seedling survival rates and visual observations of the seedling leaves. The white and black dotted lines represent the cold-tolerant and cold-sensitive lines, respectively. The majority of the tolerant lines maintained a strong vigor with higher survival rates, while most of the sensitive line died, and their leaves had yellowish spots.

3.2. RNA-Seq Analysis and Alignment of Unique Reads to the Maize Reference Genome

The RNA for the RNA-seq analysis was extracted from the top fully expanding leaves of six-leaf-stage maize seedlings of the 24 tolerant and 22 sensitive maize inbred lines mentioned in Section 2.1 above. The cDNA libraries developed from the RNA described above were constructed and used for Illumina Genome Analyzer (HiSeq™ 2500) deep sequencing. The resulting raw data were deposited into the Genome Sequence Archive under accession number CRA003678 and are publicly accessible at https://bigd.big.ac.cn/gsa, accessed on 5 January 2021). After the filtration of low-quality sequences and adaptor sequences, the 24 tolerant and the 22 sensitive libraries produced 1.13 and 1.12 billion paired-end reads, respectively (Table S3). A total of 2.247 billion paired-end reads with a length of 2 × 150 base pairs (bp) and an average of 48.84 million clean reads per library were obtained (Table S3). The Q20 percentages (sequencing error rates lower than 1%) were more than 95.8%, while the Q30 base percentage, which is an indicator of the overall reproducibility and quality of the assay, was greater than 90%. Moreover, the GC content of each library was 52.6% on average (Table S3). About 1.8 billion clean reads (82%) were mapped to the maize B73_v4 reference genome using HISAT2. Multiple mapped clean reads in each library were excluded from further analysis, and only uniquely mapped clean reads were used for subsequent analysis.

3.3. Identification, Annotation, and Differential Analysis of DEGs

The transcription level was calculated via HTSeq-count as transcript counts, and a total of 53,037 transcripts were obtained and normalized with DESeq2. The expression patterns of these transcripts were investigated in both 24 tolerant and 22 sensitive samples. A definite expression pattern of a single transcript was established after a group comparison analysis (24 tolerant versus 22 sensitive samples), and its base mean, log2 fold change, *p*-value, and padj values were acquired. From this information, a gene was considered differentially expressed if the *p*-value was ≤ 0.05 and the log2 fold change value was ≥ 1 or ≤ -1 between the 24 tolerant and the 22 sensitive samples. A total of 2237 DEGs were detected in all 46 samples. A principal component analysis (PCA) plot was generated based on the gene expression levels of the 46 samples, and the PC1 and PC2 explained 22% of the total variance (Figure 2). The PCA revealed that the correlation of the 46 samples was based on their response to cold stress. Both the tolerant and sensitive samples clustered together, implying a differential response to cold stress. A heatmap of the 2237 DEGs revealed two distinctive clusters, with 1064 DEGs possessing a positive log2 fold change enriched in the tolerant samples, and 1173 DEGs with a negative log2 fold change were enriched in the sensitive samples (Figure 3). Collectively, cold stress upregulated 1064 DEGs in the tolerant lines, while it downregulated them in the sensitive lines. Similarly, cold stress upregulated 1173 DEGs in the sensitive lines, while downregulating them in the tolerant lines.

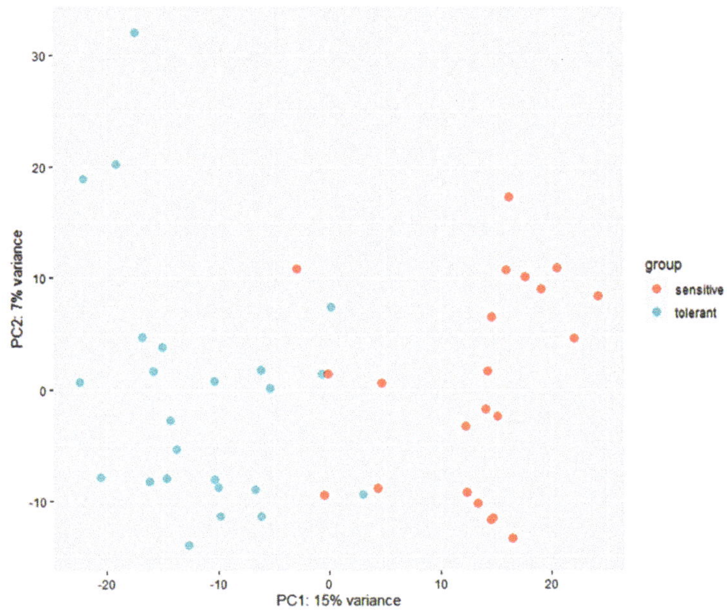

Figure 2. Principal component analysis (PCA) of pairwise genetic distance. The grouping of the 46 maize inbred line populations is indicated using blue (tolerant) and orange (sensitive). The proportion of variance captured is given as a percentage for both the first and second principal components (PC1 and PC2).

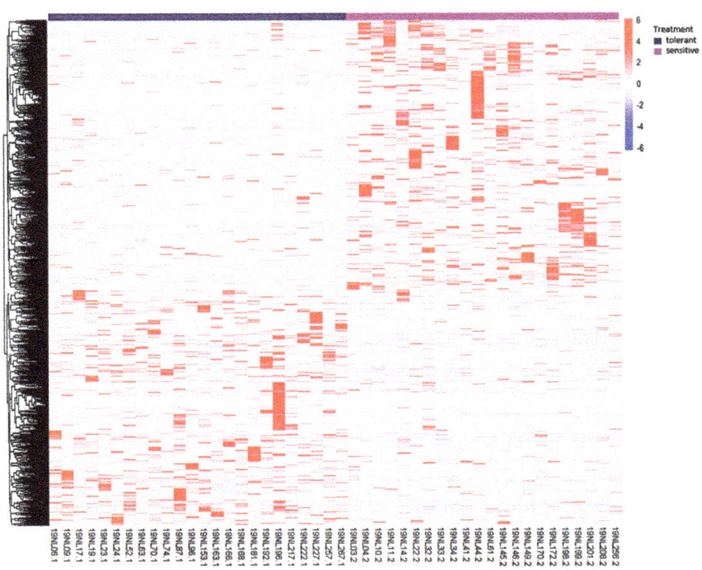

Figure 3. Heatmap showing the clustering analysis of 2237 common cold-responsive genes. The *x*-axis represents different maize samples. The purple color denotes the cold-tolerant lines while the pink color represents the cold-sensitive lines. The red and blue color scale represents high and low expressions, respectively.

The annotation of the 2237 DEGs with the maize reference genome B73 RefGen_v4 model resulted in 1656 (74.0%) annotated and 581 (26%) unannotated DEGs. The differential analysis of the 1656 annotated DEGs was carried out based on the two categories of cold-tolerant and cold-sensitive lines. Resultantly, 779 and 877 DEGs were significantly enhanced in the tolerant and sensitive lines, respectively. Further analysis of the 1656, 779, and 877 DEGs was carried out to establish the pathways involved in the cold stress response among the 46 samples.

For the 581 unannotated DEGs, their sequences ($\geq$200 bp) were uploaded to the CPC website for classification as protein-coding or non-coding RNAs. A total of 271 DEGs were classified as protein-coding, while 310 were classified as long non-coding RNAs (lncRNAs) (Figure 4). Moreover, the 581 unannotated DEGs were scanned for the open reading frame (ORF). A total of 402 DEGs with an ORF greater than 120 amino acids (aa) were discarded. The remaining 179 DEGs (ORF length $\leq$ 120 aa) were aligned to the SWISS-PROT database for the identification of homologous proteins. A total of 56 DEGs were discarded after being homologous to known proteins (E-value $\leq$ 0.001), while the remaining 123 DEGs were classified as lncRNAs (Figure 4). In total, 337 putative cold-responsive lncRNAs (Figure 4) were identified from the 581 unannotated DEGs, implying the role of lncRNAs in the cold stress response. We analyzed the top-most 20 DEGs regulated by cold stress in both tolerant and sensitive lines and half were unannotated (Table 1). However, most of the annotated DEGs remain uncharacterized, which implies more research is required to unravel the molecular mechanism of cold tolerance.

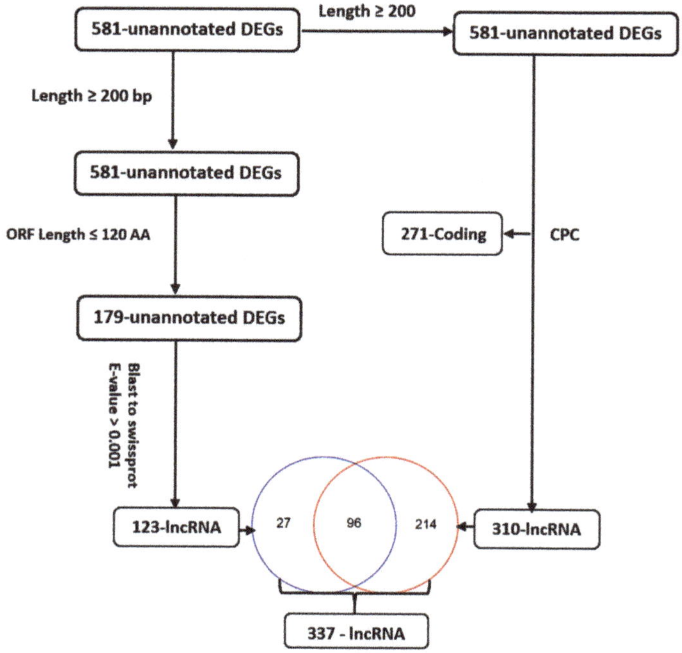

Figure 4. Analysis of 581 unannotated cold-responsive DEGs. CPC generated 271 coding and 310 long non-coding RNAs. The ORF method generated 123 long non-coding RNAs. In total, 337 putative long non-coding RNAs were generated by the two methods.

Table 1. Transcription factor gene families identified from 2237 DEGs in maize under cold stress.

TF Family	DEGs in the Tolerant Line	DEGs in the Sensitive Line	Total
B3	0	2	2
GRF	1	0	1
ERF	3	25	28
DBB	1	0	1
Dof	1	0	1
HSF	0	3	3
LBD	2	1	3
LFY	1	0	1
MYB	7	12	19
NAC	2	11	13
RAV	0	1	1
SBP	1	0	1
WOX	0	1	1
TCP	0	1	1
E2F/DP	1	0	1
GATA	1	0	1
GRAS	2	2	4
bZIP	1	2	3
C_2H_2	1	7	8
bHLH	8	5	13
Nin-like	1	0	1
NF-YC	0	1	1
ZF-HD	1	0	1
G2-like	1	1	2
CO-like	1	0	1
HD-ZIP	2	2	4
WRKY	2	19	21
TALE	1	0	1
Trihelix	1	1	2
MIKC_MADS	3	0	3
MYB_related	1	3	4
Total	47	100	147

3.4. Gene Ontology (GO) Analysis of DEGs

To identify the DEGs' significantly enriched GO terms, the functions of the 1656, 779, and 877 (46 inbred lines, 24 cold-tolerant lines, and 22 cold-sensitive lines, respectively) DEGs were analyzed using AgrigoV2 software. All DEGs were classified into three main GO categories: cellular components, molecular functions, and biological processes. The GO terms related to the response to cold (GO: 0009409), homeostatic process (GO: 0042592), response to temperature stimulus (GO: 0009266), regulation of biological quality (GO: 0065008), response to abiotic stimulus (GO: 0009628), multicellular organismal process (GO: 0032501), response to stress (GO: 0006950), G-protein-coupled receptor protein signaling pathway (GO: 0007186), transcription (GO: 0006350), and cell surface receptor-linked signaling pathway (GO: 0007166) were among the most common significantly enriched terms in the biological process (BP) category of all the three groups of DEGs named above (Figure 5). Within the molecular function (MF) category, catalytic activity (GO: 0003824), protein tyrosine kinase activity (GO: 0004713), G-protein-coupled receptor activity (GO: 0004930), water binding (GO: 0050824), and transcription regulator activity (GO: 0030528) were the most significantly enriched in all the three groups of DEGs (Figure 5). In the cellular component category (CC), integral to membrane (GO: 0016021) and intrinsic to membrane (GO: 0031224) were the common significantly enriched categories (Figure 5). The GO enrichment analysis of the tolerant and sensitive lines (Figure 6) showed that the GO term of metal ion transport (GO: 0030001) was significantly enriched in the tolerant lines (Figure 6A), while the GO terms of apoptosis (GO: 0006915), proteolysis (GO: 0006508), cell death (GO: 0008219), death (GO: 0016265), and programmed cell death (GO: 0012501) were significantly enriched in the sensitive lines (Figure 6B).

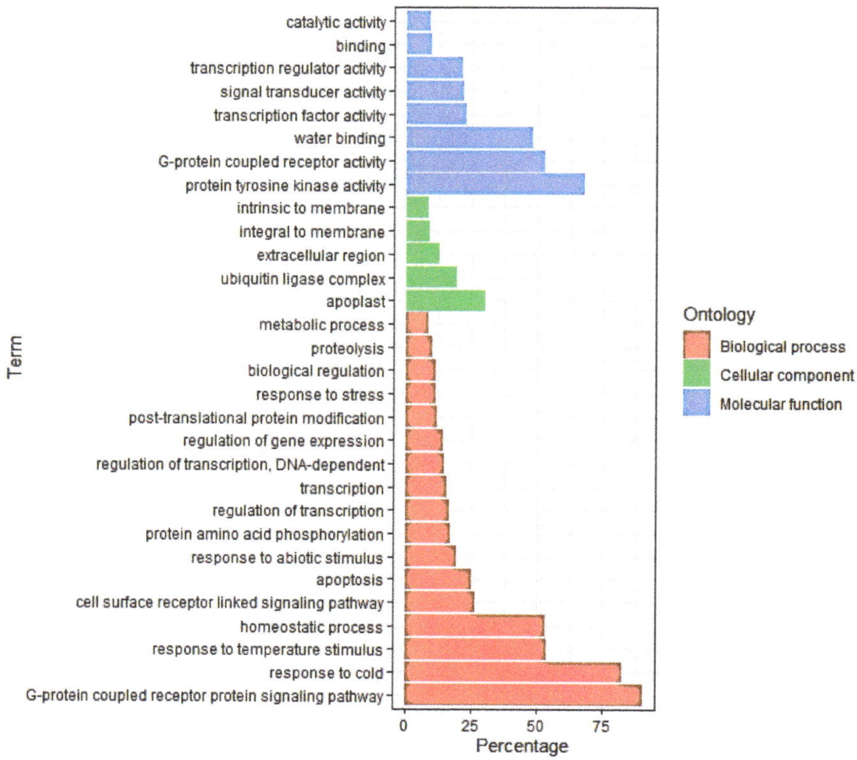

Figure 5. Gene Ontology (GO) enrichment analysis of the 1656 common cold-responsive genes. The GO terms shown here are the top-most biological process (BP), molecular function (MF), and cellular component (CC) categories.

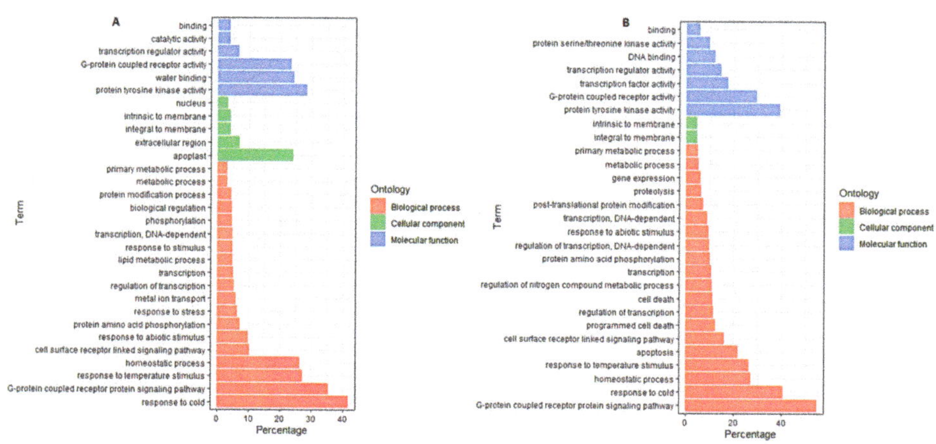

Figure 6. Gene Ontology (GO) enrichment analysis. (**A**) The 779 DEGs highly enriched in the tolerant lines. (**B**) The 877 DEGs highly enriched in the sensitive lines. The GO terms shown here are the top-most biological process (BP), molecular function (MF), and cellular component (CC) categories.

3.5. KEGG Pathway Analysis of DEGs

To further explore the biological pathways of the 1656, 779, and 877 DEGs involved in maize response to cold stress, we assessed the number of DEGs in each KEGG pathway. For the 1656 DEGs, the KEGG pathways of lipid metabolism (linoleic acid and arachidonic acid), biosynthesis of other secondary metabolites (isoquinoline alkaloid, betalain, and phenylpropanoid biosynthesis), signal transduction (MAPK signaling pathway-plant), amino acid metabolism (alanine, aspartate, glutamate, glycine, serine, threonine, and phenylalanine metabolism), membrane transport (ABC transporters), terpenoids and polyketide metabolism (carotenoid biosynthesis), and metabolism of other amino acids (glutathione metabolism) were significantly enriched (Figure 7). For the 779 DEGs obtained from the cold-tolerant lines, lipid metabolism (linoleic acid, α-linolenic, ether lipid, arachidonic acid, and glycerophospholipid metabolism), replication and repair (base excision repair), biosynthesis of other secondary metabolites (monobactam and phenylpropanoid biosynthesis), membrane transport (ABC transporters), metabolism of terpenoids and polyketides (carotenoid biosynthesis), and metabolism of other amino acids (glutathione metabolism) were the most significantly enriched pathways (Figure 8A). Within the 887 DEGs from the cold-sensitive lines, the KEGG pathways of biosynthesis of other secondary metabolites (isoquinoline alkaloid biosynthesis), signal transduction (MAPK signaling pathway-plant), transport and catabolism (peroxisome), translation (ribosome), and carbon metabolism were significantly expressed (Figure 8B).

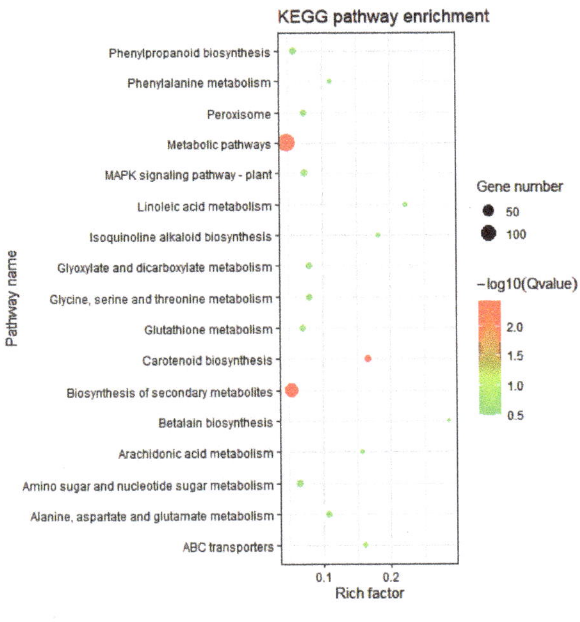

Figure 7. KEGG pathway enrichment analysis of the 1656 common cold-responsive genes. The experimental comparisons were based on the hypergeometric test, while the significance of the enrichment of the KEGG pathway was based on q value, $q < 0.05$. The color gradient represents the size of the q value; the color ranges from green to red, and the closer to green, the smaller the q value and the higher the significant degree of enrichment of the corresponding KEGG pathway. The "rich factor" represents the percentage of DEGs to total genes in a given pathway.

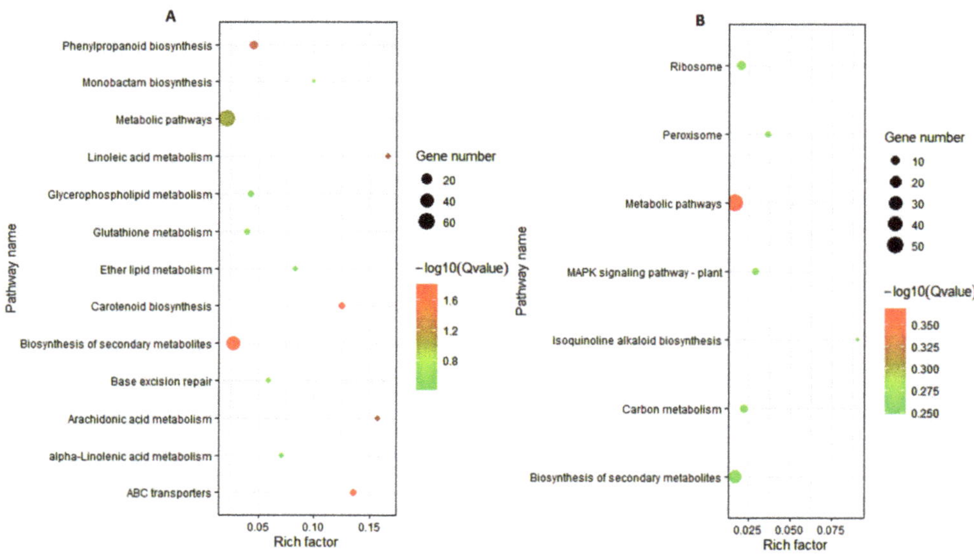

Figure 8. KEGG pathway enrichment analysis. (**A**) The 779 DEGs highly enriched in the tolerant lines. (**B**) The 877 DEGs highly enriched in the sensitive lines. The experimental comparisons were based on the hypergeometric test, while the significance of the enrichment of the KEGG pathway was based on q value, $q < 0.05$. The color gradient represents the size of the q value; the color ranges from green to red, and the closer to green, the smaller the q value and the higher the significant degree of enrichment of the corresponding KEGG pathway. The "rich factor" represents the percentage of DEGs to total genes in a given pathway.

3.6. Dynamic Expression of Signaling and Transcription Factors Genes in Response to Cold Stress

Our GO analysis highlighted a significant number of GO terms related to signaling, such as G-protein-linked receptor protein signaling pathway, cell surface receptor-linked signaling pathway, and protein amino acid phosphorylation (Figures 5 and 6). However, KEGG pathway analysis highlighted the MAPK signaling pathway-plant as one of the most significant pathways (Figure 8B). Stress sensing and signal transduction form crucial adaptive mechanisms in the tolerance of abiotic stresses. Cold stress causes a change in membrane fluidity and cytoskeleton rearrangement, thereby producing signals perceived in the cell membrane by either G-protein-coupled receptors (GPCRs) or osmotic sensors. In this study, 27 DEGs (11 enhanced and 16 suppressed) encoding GPCR were regulated by cold stress (Table S4). Activation of the sensors leads to the generation of reactive oxygen species (ROS), plant hormonal signaling, and cell wall integrity sensing (CWI) [40]. A substantial number of protein kinases, including 21 (11 enhanced and 10 suppressed) leucine-rich repeat protein kinase family proteins, 12 (6 enhanced and 6 suppressed) protein kinase superfamily proteins, 9 (2 enhanced and 7 suppressed) mitogen-activated protein kinases, 7 (2 enhanced and 5 suppressed) S-locus lectin protein kinase family proteins, and 5 wall-associated kinases (WAKs), were regulated by cold stress (Table S4). The high expression of protein kinase suggests that cold stress is primarily regulated at the protein level.

Transcription factors play a vital role in regulating gene expression in response to abiotic stress conditions, such as cold stress. These TFs are DNA-binding proteins that interact with cis-acting elements of genes to activate or inhibit gene transcription, hence regulating plant growth and development and response to the external environment. In the present study, we analyzed for putative TFs in the 1656 common cold-responsive genes based on the 3308 maize TFs and 56 families available in the PlantTFDB 4.0 [41]. Resultantly, 147 TFs, which fell into 32 families, were enriched from our 1656 DEGs (Table 2). ERF

(18.37%), WRKY (14.29%), MYB (12.93%), NAC (8.84%), bHLH (8.84%), C_2H_2 (5.44%) and GRAS (2.72%) were the most abundant TF families (Table 2; Figure 9). However, the expression levels of the TF families MYB, ERF, NAC, WRKY, bHLH, and C_2H_2 were higher in the sensitive lines than those in the tolerant lines (Figure 9). On the contrary, the TF families MIKC_MADS, CO-like, DBB, Dof, E2F/DP, GATA, GRF, LFY, SBP, Nin-like, TALE, and ZF-HD were only induced in the tolerant lines.

Table 2. List of top 20 most regulated DEGs by cold stress.

Locus ID	Gene ID	log2 Fold Change (T/S)	*p*_Value	Chr	Start	End	Annotation
XLOC_000983	Zm00001d027606	23.73473512	6.81×10^{-31}	Chr1	8,915,719	8,918,057	transmembrane protein
XLOC_056725	-	−23.7343301	9.38×10^{-29}	Chr7	170,162,017	170,164,579	-
XLOC_046285	Zm00001d017622	−2.670016539	9.74×10^{-25}	Chr5	201,997,628	201,998,691	OSJNBa0088A01
XLOC_018249	-	24.01063195	1.05×10^{-24}	Chr2	137,103,173	137,104,433	-
XLOC_058213	Zm00001d021006	−1.554984072	3.70×10^{-22}	Chr7	140,181,294	140,183,018	MTD1
XLOC_034912	-	23.0239833	5.49×10^{-21}	Chr4	176,582,912	176,584,805	-
XLOC_056365	Zm00001d021394	22.79156651	7.11×10^{-21}	Chr7	150,891,109	150,895,395	hypothetical protein
XLOC_067464	-	20.05873411	6.49×10^{-20}	Chr9	20,669,455	20,671,879	-
XLOC_049980	-	24.72008389	2.60×10^{-17}	Chr6	114,599,197	114,601,220	-
XLOC_018351	-	24.4739764	5.35×10^{-17}	Chr2	150,920,460	150,923,282	-
XLOC_055724	-	22.96632545	9.06×10^{-17}	Chr7	95,897,276	95,897,694	-
XLOC_018159	Zm00001d004620	22.6740887	1.68×10^{-16}	Chr2	122,442,335	122,447,585	uncharacterized protein
XLOC_004771	-	−23.94951263	2.29×10^{-16}	Chr1	1,487,983	1,491,882	-
XLOC_007853	Zm00001d033411	−23.88106934	2.77×10^{-16}	Chr1	262,279,726	262,300,105	hypothetical protein
XLOC_005508	Zm00001d028673	23.87942745	2.96×10^{-16}	Chr1	42,581,898	42,586,482	small nuclear protein G
XLOC_036989	-	−23.84216727	3.11×10^{-16}	Chr4	108,230,941	108,236,238	-
XLOC_014865	Zm00001d025968	23.6089741	6.37×10^{-16}	Chr1	134,994,735	134,998,645	hypothetical protein
XLOC_018011	Zm00001d004338	23.58304546	6.85×10^{-16}	Chr2	104,459,685	104,460,441	hypothetical protein
XLOC_046112	Zm00001d017287	−4.317481518	2.92×10^{-13}	Chr5	191,750,113	191,750,625	uncharacterized protein
XLOC_027574	Zm00001d043525	−1.350259753	4.09×10^{-13}	Chr3	202,741,852	202,743,157	oxidative stress 3

Comparison of DEGs between tolerant (T) and sensitive (S) inbred lines after cold stress. The *p*-value is less than 0.05.

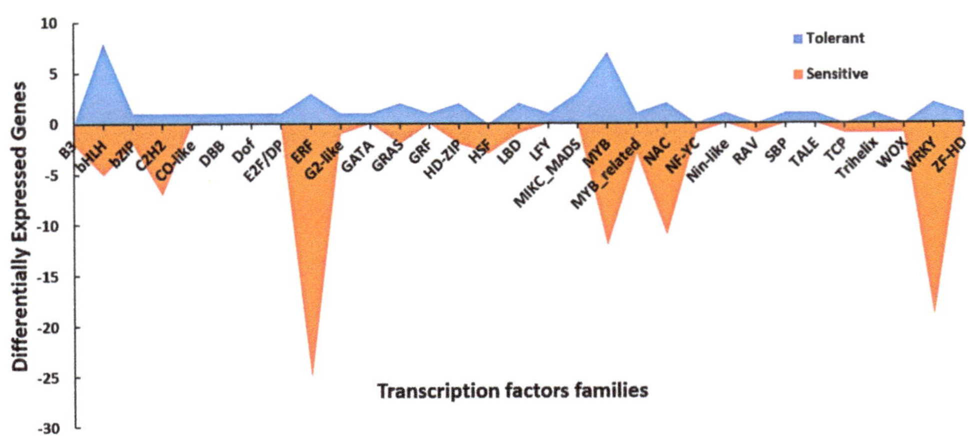

Figure 9. Area map of the various transcription factor families regulated by cold stress. The x-axis represents the TF families, while the number of genes per family is represented by the y-axis. The blue and orange colors indicate the TFs regulated in the tolerant and sensitive lines, respectively.

Additionally, GO terms, such as regulation of transcription, transcription, and transcription factor activity, which are related to the regulation of gene expression at the transcription level, were well represented in our GO analysis (Figures 5 and 6). The TF families bHLH, ERF, bZIP, and WRKY were highly regulated by cold stress in the tolerant and sensitive lines (Table S5). As observed above, DEGs encoding the TF families LFY, MIKC_MADS, GRAS, GATA, GRF, E2F/DP, TALE, and Dof were significantly induced by cold stress only in the tolerant lines (Table S5). In-depth research is required to uncover the roles of these TFs in the response of maize to cold stress at the seedling stage. Contrary, DEGs encoding the TF families AP2, B3, HSF, and RAV were all repressed by cold stress in the tolerant lines (Table S5). The tolerant lines were, to a lesser extent, affected by cold stress, which might be why fewer TFs were expressed. Otherwise, our results highlight the critical roles of TFs in the regulation of cold stress response in maize seedlings.

3.7. DEGs Related to "Response to Cold"

The GO terms of "response to cold" and "response to temperature stimulus" were well represented in our GO analysis (Figures 5 and 6). These GO terms contain crucial cold-related genes, which play significant roles in the cold regulation mechanism and may contribute to the cold tolerance of maize seedlings. The expression patterns of DEGs in these GO terms (upregulated or downregulated) may shed light on the difference between the cold response of tolerant and sensitive inbred lines. Signal transduction-related proteins, such as transmembrane proteins, serine/threonine protein kinases, leucine-rich repeat receptor-like proteins, and receptor-like kinases, were regulated by cold stress (Table S6). This highlights the critical roles of the signal transduction network in the activation of various cold-responsive genes. The involvement of TFs in regulating cold stress was highlighted by the significant expression of C_2H_2, ERF, HD-ZIP, MYB, ERF, MYB, and NAC (Table S6). Antioxidant-related enzymes, such as glutathione peroxidase, thioredoxin, and peroxidase, were regulated by cold stress, suggesting the involvement of detoxification proteins in maize response to cold stress. Carotenoid related genes, such as β-carotene isomerase and β-carotene 3-hydroxylase, further highlighted the role of ABA as a key regulator of cold stress (Table S6). Otherwise, transporter and cell surface proteoglycan-related genes were also observed in this study (Table S6).

3.8. Expression Analysis of Genes Involved in Metabolism, Transport and Functional Impacts of Co-Expressed Gene Hubs

Our KEGG analysis highlighted significant regulation of multiple metabolism pathways, including lipid, carotenoid, ABC transport, and amino acid pathways, during cold stress conditions (Figures 7 and 8). Cold stress activated the phenylpropanoid biosynthetic pathway, an essential way to accumulate various phenolic compounds during cold stress conditions. In total, 14 DEGs encoding this pathway were regulated by cold stress, and 11 of which showed high expression levels in the tolerant lines (Table S7). Moreover, there were six DEGs encoding the metabolism of alanine, aspartate, and glutamate, including β-alanine aminotransferase and glutamate decarboxylase, which are vital genes in osmotic adjustment (Table S7).

Lipid metabolism is a dynamic and complicated process involving lipid biosynthesis, transport, accumulation, turnover, and excretion, regulating the growth and tolerance of plants to different environmental stresses. In this study, 26 lipid metabolism-encoding genes were regulated by cold stress, with all of them except one being enhanced in the tolerant lines (Table S7). Among them were allene oxide synthase (AOS), an essential gene in the biosynthesis of jasmonic acid (JA), secretory phospholipase A2 (PLA2), α/β-Hydrolases, and phospholipase C (PLC), which are critical components of the signaling cascade, and 3-ketoacyl-CoA synthase (KCS) genes, which are related to the biosynthesis of cuticular wax (Table S7). Moreover, six and eight DEGs encoding ABC transporters and carotenoid biosynthesis, respectively, were regulated by cold stress. ABC transporters are associated with phytohormone homeostasis, while carotenoid genes are vital for the biosynthesis of ABA, an essential hormone in cold tolerance. Otherwise, cold stress also regulated glutathione- and peroxisome-related genes, which mediate the harmful effects of ROS. Detailed information about the metabolism genes can be found in Table S7.

WGCNA has been used to dissect the abiotic stress response in plants, thereby highlighting the power of the co-expression networks to provide deep insights into these complex processes. In this study, WGCNA identified multiple significant functional gene hubs related to the cold stress response. A total of 116 critical DEGs that were enriched by GO and KEGG into signaling, TFs, response to cold, and metabolism were defined as hub genes, highlighting the importance of their regulatory impacts on the cold stress response (Figure S2, Table S8). Thus, WGCNA elucidated the higher order relationships between genes based on their co-expression relationships and permitted a robust view of transcriptome organization in the response of cold stress. Collectively, the combination of transcriptome analysis with WGCNA represents an opportunity to achieve a higher resolution analysis that can better predict the most important functional genes that might provide a more robust bio-signature for cold tolerance in maize, thus providing more suitable biomarker candidates for future studies.

3.9. Validation of DEGs by Quantitative Real-Time PCR (qRT-PCR)

To confirm the reliability and validity of the RNA-seq results in maize seedlings, six genes were randomly selected to perform qRT-PCR. Of these, XLOC_018851, *Zm00001d002748*, and *Zm00001d027330* all had a positive log2 fold change between tolerant and sensitive lines, suggesting that their expression was enhanced by cold stress treatment in tolerant lines and declined in sensitive lines. In contrast, XLOC_058556, *Zm00001d024324*, and *Zm00001d024522* portrayed a negative log2 fold change between tolerant and sensitive lines, indicating that cold stress treatment increased their expression in sensitive lines but decreased their expression in tolerant lines. As a result, the fold change ratio used in this paper emphasizes the expression pattern in cold-tolerant lines, while the inverse of that ratio reflects the expression pattern in sensitive lines. We planted B73 and CIMBL116 maize inbred lines, which have previously been reported as cold-sensitive and -tolerant lines, respectively, to validate the expression pattern of these DEGs (Section 2.1 above). These two inbred lines were planted in a cold environment as well as without cold treatment (control), allowing us to compare the expression patterns of our six DEGs before (control) and after cold treatment. Resultantly, cold stress increased the expression of XLOC 018851, *Zm00001d027330*, and *Zm00001d002748* in the tolerant line (CIMBL116), while it decreased the expression of XLOC 058556, *Zm00001d024324*, and *Zm00001d024522* (Figure S3). A reverse expression pattern was observed in the sensitive line (B73) (Figure S3). Thus, the expression trend of our DEGs shown by our RNA-Seq results was in agreement with that shown by the qRT-PCR analyses.

Furthermore, the fold change ratio of the six DEGs between the control and the cold-treated samples was calculated and compared to the fold change obtained from RNA-Seq. As a result, the cold-sensitive line's RNA-Seq and qRT-PCR results showed an inverse expression pattern (Figure 10A) because the fold change ratio highlighted in this study emphasizes the expression pattern in cold-tolerant lines. However, if you take the inverse of that fold change ratio, which will highlight the expression pattern in cold-sensitive lines, then the RNA-Seq and qRT-PCR of the sensitive line will have a similar expression pattern to that of the tolerant line (Figure 10B). For example, cold stress enhanced the fold change of XLOC_018851, *Zm00001d002748*, and *Zm00001d027330* and declined that of XLOC_058556, *Zm00001d024324*, and *Zm00001d024522* in the tolerant lines (Figure 10B). However, cold stress inversely regulated the fold change of the above-named DEGs in the sensitive lines (Figure 10A). These results validate the authenticity of the DEGs obtained in this study, as the relative fold change in qRT-PCR matched the RNA-Seq results, implying that transcript identification and abundance estimation were remarkably precise. Furthermore, the DEGs identified in this study are universal in maize seedlings of various genetic backgrounds in response to cold stress, according to the qRT-PCR results.

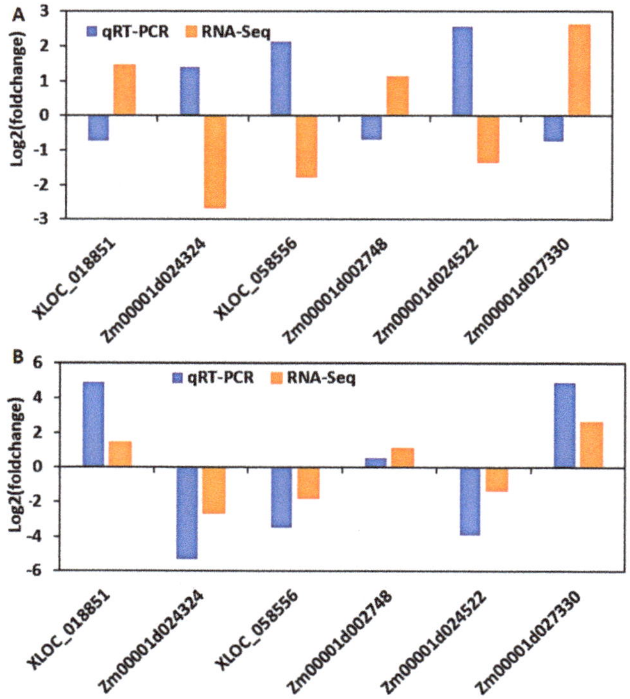

Figure 10. Validation of RNA-Seq results by qRT-PCR. Each log2 fold change calculated from qRT-PCR was compared with the log2 fold change of the RNA-Seq data. (**A**) The inverse expression patterns of the RNA-Seq and qRT-PCR results from the cold-sensitive line (B73) is because the fold change ratio highlighted in this study emphasizes the expression pattern in cold-tolerant lines. However, if you take the inverse of that fold change ratio, which will highlight the expression pattern in cold-sensitive lines, then the RNA-Seq and qRT-PCR expression trends would be identical. (**B**) The cold-tolerant line's (CIMBL116) RNA-Seq and qRT-PCR results show a similar expression trend. Orange and blue bars represent RNA-Seq and qRT-PCR data, respectively.

4. Discussion

Maize has surpassed rice and wheat as the world's most significant cereal crop. However, cold stress affects maize at any stage of development, including the germination, vegetative, and reproductive stages. Screening for cold-tolerant maize cultivars, such as rice, is difficult due to the lack of linkage between cold resistance and developmental periods [42]. Furthermore, cold tolerance is a quantitative trait influenced by the interactions of numerous genes as well as the environment. Nevertheless, RNA-seq research, which evaluates the main genes and regulatory pathways at the transcriptome level, has been widely used to investigate the molecular basis of maize response to abiotic stress [43,44]. To gain a deeper understanding of maize's cold stress tolerance mechanisms and to develop cold-tolerant maize cultivars, we conducted a comparative transcriptome analysis of 24 cold-tolerant and 22 cold-sensitive maize inbred lines to uncover the common cold-responsive DEGs and pathways. To the best of our knowledge, this is among the first studies to profile a broad set of maize inbred lines at the seedling stage with varying levels of resistance to cold stress in the field. Previous research has usually profiled two samples with differing tolerances to specific environmental stress [45]. As a result, our study provides valuable insight into the in-depth characterization of the molecular responses of maize seedlings to cold stress, because, to date, there remains a scarcity of data on the expression patterns of critical genes and pathways across a gradient of various genotypes with varied responses to specific stress conditions.

Plants perceive abiotic stress via cell wall receptors, which activate internal signaling components through several mechanisms. The cyclic nucleotide–gated channel (CNGC) and glutamate receptors (GLRs) are the primary cell membrane cold receptors reported in plants [46]. These receptors mediate membrane Ca^{2+} fluxes and produce several endogenous signals responsible for cold tolerance [46]. Multiple G-protein-coupled receptors were shown to be strongly regulated by cold stress in our study (Table S4). The regulation of these receptors may have caused Ca^{2+} fluxes across membranes and generated multiple signals important for maize's cold stress response. A G-protein subunit γ gene (*Zm00001d032072*) and a trihelix GT-2 (*Zm00001d027335*) were both upregulated in the tolerant lines (Table S4), indicating their role in cold tolerance. In a previous study, transgenic cucumber plants overexpressing *CsGG3.2* had increased CBF gene expression and were more tolerant to chilling stress [47]. COLD1 provided chilling tolerance in rice by encoding a signal regulator for guanine nucleotide-binding proteins (G-proteins) on the plasma membrane [48]. A GT2 family (*AtGT2L*) gene in *Arabidopsis* has been reported to interact with calcium/calmodulin, allowing plants to withstand cold and salt stress [49]. Two soybean GT2 genes provided abiotic stress resistance to transgenic *Arabidopsis* plants [50]. Thus, in this study, increased GPCR protein expression may have increased cytosolic calcium levels in the tolerant lines, activating a quick and diverse signaling mechanism responsible for cold acclimation.

Cold stress perceived by membrane sensors triggers an influx of Ca^{2+} into the cytoplasm, which then generates Ca^{2+} signatures that induce the activation of downstream genes, such as CBF/COR genes, in the cold signaling pathway [15]. Proteins with an EF-hand domain, such as CaMs, CMLs, CBLs, and CDPKs, act as Ca^{2+} sensors in response to cold stress [51–53]. Following the binding of Ca^{2+}, these proteins interact with other proteins, thereby regulating downstream activities of multiple genes, which provides cold tolerance. In this study, CaMs, CMLs, CBLs, CDPKs, 21 leucine-rich repeat receptor-like kinase proteins (LRR-RLKs), 8 lectin receptor-like kinases (LecRLKs), 12 protein kinases (PKs), and 5 WAKs were regulated by cold stress (Table S4). In *Camellia japonica*, protein phosphorylation by CDPKs and CIPKs improve cold tolerance [54], whereas in plants, WAKs play an important function in abiotic stress tolerance [55]. Interestingly, TMK1 (*Zm00001d033777*, *Zm00001d007313*), CBL10 (*Zm00001d023353*, *Zm00001d010459*), RKL1 (*Zm00001d048054*), NIK3 (*Zm00001d018635*), and PSKR (*Zm00001d018635*) displayed a high expression pattern in the tolerant lines. In a previous study on *Arabidopsis*, CBL10 mediated salt tolerance [56], while RKL1 regulated cold and salicylic acid stresses [57]. Recent research has highlighted the key roles of CBL10 in plant abiotic stress tolerance through the regulation of Na^{+} and Ca^{2+} homeostasis [58], whereas under cold stress, TMK1 significantly regulated plant development [59]. MAPKs, including MAPKK and MAPKKK, are key players in cold tolerance [48], where they phosphorylate other kinases and/or various TFs. In this study, 14 DEGs encoding MAPKs were regulated by cold stress (Table S4). Multiple MAPKs have been implicated in improving cold tolerance in rice and Chinese jujube, according to previous research [60,61]. In this study, RBOH (*Zm00001d009248*), MAP3 (*Zm00001d001978*), and MKK3 (*Zm00001d013510* and *Zm00001d028026*) had higher expression levels in the tolerant lines (Table S4). In a previous study, two strawberry RBOHs were reported to enhance cold stress tolerance and defense responses [62]. MKK3 was substantially expressed in tolerant lines during a comparative transcriptome investigation of two cotton cultivars with differing responses to cold [63]. The transgenic tobacco over-expressing MAP3K gene demonstrated improved tolerance to a variety of environmental stresses, including cold stress [64]. PP2C adversely controls stress-induced MAPK and SnRK2 protein kinases [4,65]. A previous study on maize has shown that stress-induced proline accumulation and tolerance to hyperosmotic stress were negatively controlled by maize PP2C [66]. This might explain why five PP2C-encoding genes were suppressed by cold stress in the tolerant lines (Table S4). Heptahelical protein 2 (HHP2) (*Zm00001d046852*), a FASCICLIN-like arabinogalactan (FLA) (*Zm00001d016059*, *Zm00001d052819*, and *Zm00001d006009*), and a membrane-associated kinase (*Zm00001d044176*) were enhanced in tolerant lines (Table S6). A previous study on *Arabidopsis* reported that the HHP2-MYB module is involved in integrating cold and abscisic acid signaling to activate the CBF–COR pathway [67]. Cold stress activated a membrane-associated kinase in rice, and FLAs were found to improve banana resistance to low temperatures by activating a cold signal pathway [68,69]. Collectively, the increased expression of Ca^{2+} signaling protein transcripts in maize tolerant lines activated a complex signaling cascade that regulated various downstream cold tolerance responsive genes. These genes will have significant implications for future research into maize cold tolerance at the seedling stage.

Transcription factors are key regulators of cold stress as they control multiple downstream stress-responsive genes [70]. In this study, 147 TFs belonging to 32 TF families, including AP2/ERF (27), MYB (19), bHLH (13), WRKY (21), C_2H_2 (8), and NAC (13), were largely regulated by cold stress (Table 2, Figure 5). Similar to our findings, in previous studies, comparative transcriptome analysis of rice, Chinese jujube, and peanut under cold stress conditions identified the above TF families to be the most regulated gene members [60,61,71,72]. These TF genes in their respective families are divided into diverse subgroups based on their specific motif structures, showing that they may perform their specific biological activities under cold stress. Moreover, individual TFs from these families have previously been reported to play a crucial function in controlling plant cold tolerance. In this study, four ERFs, namely ERF38 (*Zm00001d002748*), ERF022 (*Zm00001d048991*), DREB26 (*Zm00001d018191*), and ERF (*Zm00001d029669*), had higher expression in the tolerant lines (Table S5). They all encode for DREB elements, which enhance cold tolerance by activating CORs. ERF38, ERF022, and DREB26 effectively regulate COR genes and sugar and proline accumulation in *Arabidopsis*, resulting in abiotic stress tolerance [73,74]. Thus, these genes may have impacted maize cold tolerance by modulating COR genes and osmotic regulators. WRKYs are yet another important class of plant TFs with diverse roles in plant response to cold stress. In this study, 18 WRKY genes were regulated by cold stress (Table S5), with WRKY70 (*Zm00001d023332*) and WRKY53 (*Zm00001d023336*) genes showing higher expression in the tolerant lines. In previous studies, the expression of WRKY70 and *TcWRKY53* was induced by cold stress in wheat and *Thlaspi caerulescens*, respectively [75,76]. Moreover, peanut cold stress tolerance was regulated by WRKY70 and WRKY53 via the plant–pathogen interaction pathway [72]. Moreover, MYB4 (*Zm00001d041853*), bHLH57 (*Zm00001d027419*), PIF4 (*Zm00001d013130*), PTF1 (*Zm00001d045046*), AGL22 (*Zm00001d018142*), GRF6 (*Zm00001d000238*), ATHB4 (*Zm00001d002754*), Scl7 (*Zm00001d033834*), BTB/POZ (*Zm0000 1d023313*, *Zm00001d030864*), and C_2H_2 (*Zm00001d024883*) were all enhanced in the tolerant lines in this study (Tables S5 and S6). *SIPIF4* and *ZmPTF1* have been attributed to cold and drought stress tolerance in tomatoes and maize, respectively, via the modulation of ABA synthesis and signaling pathways [77,78]. In *Arabidopsis thaliana*, rice MYB4 was induced by cold stress, which in return transactivated the expression of COR genes, such as RD29A, COR15a, and PAL2 [79]. The over-expression of finger millet bHLH57 caused salinity and drought stress in tobacco [80], while *SlGRF6* was significantly regulated by cold stress in *Solanum Lycopersicum* [81]. C_2H_2 zinc finger proteins targeted C-repeat/DRE-binding factor genes (CBFs) to provide cold resistance in plants [82], while BTB/POZ significantly accumulated in resistant cotton cultivars during chilling stress conditions [83]. Thus, all these differentially expressed TFs identified in tolerant lines in response to cold stress could represent a useful genetic resource for breeding cold-tolerant crops. Nevertheless, many cold stress-regulating TFs are yet to be identified along with known TFs whose functions are not yet known (Table S5). Understanding the role of the above-mentioned cold-regulating TFs at the molecular level will be pivotal in improving maize performance under cold stress conditions.

Abscisic acid is an essential plant hormone that regulates cold stress via interactions between ABA-dependent and ABA-independent pathways [84]. Moreover, exogenous ABA treatment at normal temperature improves freezing tolerance [85]. In this study, carotenoid biosynthesis genes, such as ZEP (*Zm00001d025968*), NCED (*Zm00001d042076* and *Zm00001d018819*), β-carotene isomerase (*Zm00001d007549* and *Zm00001d007560*), β-carotene 3-hydroxylase (*Zm00001d048469*), and ABA 8′-hydroxylase (*Zm00001d051554* and *Zm00001d050021*) were all enhanced by cold stress in the tolerant lines except NCED (Tables S6 and S7). In *Arabidopsis*, ABA regulates cold tolerance by improving the levels of ABA 8′-hydroxylase [86]. However, Alfalfa-related ZEP is regulated in response to drought, cold, and heat [87]. Moreover, β-carotene hydroxylase regulates the biosynthesis of a carotenoid precursor of abscisic acid called zeaxanthin. A previous report highlighted that a β-carotene hydroxylase gene caused drought and oxidative stress in rice by elevating the synthesis of ABA and xanthophylls [88]. Higher ABA levels induced cold tolerance in herbaceous plants [89]. The elevated expression of ABA-related genes was correlated to cold adaptation in a comparative transcriptome investigation of tea and tobacco plants [90,91]. Therefore, elevated carotenoid biosynthesis genes triggered ABA accumulation, and the transcriptional regulation of ABA-related gene expression is one factor that contributed to cold stress tolerance in maize. Otherwise, the suppression of NCED genes in the tolerant lines reflects the complexity of cold tolerance in plants.

Cold stress triggers rapid and intermittent ROS production that can damage plant cellular components and structures, but ROS also act as signaling molecules for abiotic stress tolerance [92]. Nevertheless, plants deploy a cascade of antioxidant machinery consisting of enzymatic and non-enzymatic defense systems to diminish the deleterious effects of ROS on plant cells [93]. The antioxidant enzymes include SOD, CAT, APX, GPX, GST, and GPX, which can trap and scavenge free radicals [94]. In this study, antioxidant genes, such as SOD (*Zm00001d014632*), GST (*Zm00001d029699*, *Zm00001d043787*, and *Zm00001d018809*), GPX (*Zm00001d029089*), PRX (*Zm00001d008266*, *Zm00001d028348*, *Zm00001d031635*, *Zm00001d032406* and *Zm00001d041827*), and APX (*Zm00001d024253*), were all enhanced by cold stress in the tolerant lines (Table S7). The activities of SOD and GST were reported to reduce cold injury in cold acclimatized wheat [95], while the over-expression of GST in transgenic rice enhanced growth and development at a low temperature [96]. In cassava, chilling and oxidative stress was correlated with increased levels of SOD and APX genes [97]. PRXs play a role in phytoalexin-mediated plant defense and ROS metabolism [98]. Thioredoxin (TRX), however, functions as a redox transmitter [99]. Thus, the enhanced expression of TRX (*Zm00001d011352* and *Zm00001d007800*) genes in the tolerant lines might be crucial in cold acclimation through redox regulation. A previous study reported that soybean TRX genes (*Sb03g004670* and *Sb06g029490*) were significantly regulated in the cold acclimation of different accessions [100]. These findings confirm that ROS-mediated signaling could activate antioxidant enzymes, which might be responsible for imparting cold stress tolerance in maize seedlings.

The phenylpropanoid pathway and its branches of secondary metabolites are activated under cold stress, leading to the accumulation of various phenolic compounds for protection and mechanical support [101]. In this study, 14 phenylpropanoid genes, including CCR (*Zm00001d019669*, *Zm00001d008435*), PAL (*Zm00001d033286*), trans-cinnamate 4-monooxygenase (*Zm00001d016471* and *Zm00001d032468*), and β-glucosidase (*Zm00001d028199*), and 5 PRXs were significantly enhanced in the tolerant lines (Table S7). Elevated PAL expression stimulates the biosynthesis of phenolic compounds, such as suberin and lignin, which reinforce the cell wall and prevent cell collapse during cold stress. Similarly, enhanced expression of the CCR gene has previously been correlated with lignin biosynthesis under abiotic stress [102]. Trans-cinnamate 4-monooxygenase is positioned at the turning point of phenylalanine, lignin biosynthesis, and flavonoid metabolism, making it one of the key enzymes in the synthesis of lignin and flavonoids [103]. Moreover, PRXs in the presence of H_2O_2 catalyze the oxidative polymerization of phenols, such as lignin precursors, which improve cell wall rigidity by boosting the cross-linking of cell wall components [104]. This increased suberin or lignin biosynthesis increases the thickness of the cell wall, preventing chilling injury and cell collapse during cold stress [105,106]. The magnitude of lignification in plants is significantly associated with their potential for cold tolerance. Previous studies have shown that β-glucosidase activates several processes, including lignin precursors [107], the release of phytohormones from inactive glycosides, and the activation of several defense compounds essential for abiotic stress tolerance [108]. However, *Nicotiana benthamiana* plants over-expressing *CsBGlu12* displayed abiotic stress tolerance via the accumulation of antioxidant flavanols that played a crucial role in scavenging ROS [109]. Collectively, the upregulation of various transcripts encoding phenylpropanoid pathway genes in the present study indicates enhanced lignification-mediated cold acclimatization in maize seedlings.

A high accumulation of osmoprotectants, such as amino acids, polyamines, quaternary ammonium compounds, and sugars, mediates diverse functions in plant defense mechanisms under varying environmental conditions [110]. Herein, genes encoding β-alanine aminotransferase (*Zm00001d038453* and *Zm00001d038460*) and glutamate decarboxylase (GAD) (*Zm00001d031749*) were enhanced by cold stress in the tolerant lines (Table S7). The enzyme β-alanine aminotransferase catalyzes the biosynthesis of pyruvate and β-alanine, with the latter product being converted to an essential osmoprotective compound (β-alanine betaine) involved in plant abiotic stress tolerance [111,112]. However, GAD catalyzes the decarboxylation of L-glutamate to form γ-aminobutyric acid (GABA), which accumulates at high concentrations under abiotic stress [113]. Glutamate decarboxylation and GABA metabolism have been reported to play a crucial role in the cold acclimation of wheat and barley [114]. Thus, GABA played a vital role in the cold acclimation of the tolerant lines. Simultaneously, the enhanced expression of β-alanine aminotransferase facilitated a β-alanine-based osmoprotectant in maize during cold stress.

During cold stress, plants adjust their lipid content to retain membrane stability and integrity. Cold tolerance in peanuts was previously found to be associated with changes in membrane modifications, such as lipid metabolism and lipid signaling [115]. In the present study, 26 lipid metabolism genes were regulated by cold stress (Table S7). Among them, PLC (*Zm00001d040205*), PLA2s (*Zm00001d013461*, *Zm00001d029136*), SAD (*Zm00001d024273*), AOS (*Zm00001d028282*), α/β-hydrolase (*Zm00001d010840* and *Zm00001d012147*), KCS (*Zm00001d 046444* and *Zm00001d032728*), nsLTP (*Zm00001d027332*), and seven GDSL-like lipases were all enhanced by cold stress in the tolerant lines (Tables S6 and S7). AOS is a critical gene in the synthesis of jasmonic acid (JA), which affects the expression of cold-responsive genes and governs plant defense responses to various abiotic stressors [116]. In *Arabidopsis*, JA was found to provide cold acclimation [117]. PLC participates in signaling pathways that lead to the activation of the cold response through the CBF pathway [118]. The cold acclimation of spinach (*Spinacia oleracea*) leaves was found to be associated with the positive roles of PLA2 [119]. Higher expression of a SAD gene is linked to the total amount of unsaturated fatty acids (UFAs), which has been correlated to cold tolerance in tobacco plants [120]. On component change and permeability, the KCS gene catalyzes the biosynthesis of cuticular wax, which acts as a protective barrier against abiotic stresses [121]. GDSL lipases regulate plants' development and stress response. A previous study by Kong et al. [122] highlighted an essential role of a pepper GDSL lipase gene in regulating abiotic stress tolerance. During cold stress, nsLTPs reduce lipid fluidity and membrane solute permeability, thereby reducing solute diffusion rates across the membrane and preventing osmotic membrane rupture upon thawing [123]. A previous maize study revealed that *ZmLTPs* have a role in response to cold stress [124]. Overall, our findings highlight the putative association of multiple lipid metabolic components and nsLTP proteins in maize cold tolerance.

Membrane transport systems help maintain cellular homeostasis in environmental stressful situations by redistributing different molecules, such as phytohormones, carbohydrates, and amino acids [125]. These unique roles of plant membrane transport systems may be leveraged to enhance productivity under unfavorable stress conditions as their impact on total plant physiology [126]. The increased expression of numerous transporters and channel protein genes has been reported in the *Arabidopsis thaliana* response to various abiotic stresses [127] and rice under water stress [128]. In the present study, ABCB1 (*Zm00001d024600*, *Zm00001d025703*, *Zm00001d026041*, *Zm00001d045279*, and *Zm00001d049565*), MATE efflux (*Zm00001d031730* and *Zm00001d032971*), and polyol transporter (*Zm00001d048774*, *Zm00001d029645*) genes were enhanced by cold stress in the tolerant lines (Tables S6 and S7). Plant ABCB transporters transport molecules, such as plant hormones, lipids, metabolites, contaminants, and defense molecules, which play key roles in abiotic stress tolerance. Various environmental stresses were reported to enhance the expression of distinct ABCB transporters in maize [129]. Polyols play a crucial function in the symplastic and apoplastic transfer of carbon and energy in plants' response to salt and drought [130]. Transgenic *Arabidopsis* overexpressing the cotton MATE gene enhanced antioxidant enzyme production and abscisic acid translocation in response to cold, drought, and salt stress [131]. Therefore, the upregulation of various transporters might be associated with cold stress tolerance, the transport of plant secondary metabolites, hormones, and general growth and development in maize.

Network analysis reveals the regulatory impacts of a group of genes on target genes, revealing unique regulatory linkages that add to our understanding of abiotic stress response. The network analysis in this study had 116 nodes and 1907 connections, with 724 activation (positive) and 1183 repression (negative) connections (Figure S2, Table S8). Some of the highly connected positive regulators were found among the 116 nodes, including TFs (bHLH, MYB4, MYB8, GATA4, TALE, and WRKY53) and signaling (respiratory burst oxidase, GPCR, BAM2, RKL1, NIK3, SRF7, SRF8, and PK), antioxidant (peroxiredoxin 6, peroxidase, thioredoxin), and metabolism/biosynthesis regulators (Table S8). The cold induction of TFs regulates a set of other downstream genes. The upregulation of MYB4 (*Zm00001d041853*) in the cold-tolerant line upregulated 11 DEGs related to signaling, amino acid phosphorylation, TFs, and metabolism (Table S8). In a previous study on *Arabidopsis thaliana*, the over-expression of OsMYB4 increased cold and chilling tolerance by increasing the expression of COR genes, such as RD29A, COR15a, and PAL2 [79]. In this study, the upregulation of the WRKY53 (*Zm00001d023336*) gene in the tolerant line increased the expression of 10 additional DEGs involved in signaling, amino acid phosphorylation, and transcription factors (Table S8). In a previous study, WRKY53 was highly increased in *Arabidopsis thaliana* under cold stress, where it interacted with hub genes, such as mitogen-activated protein kinase 3 (MPK3), WRKY33, and WRKY40, all of which are implicated in plant defense [72]. As a result, WRKY53 could have influenced maize cold tolerance via the plant–pathogen interaction route. Differential expression levels of the 116 DEGs that make up the nodes in our cold studies show that various genes respond to cold stress in different ways and have varied biological functions. These DEGs could be intriguing candidates to investigate during maize seedling cold stress responses. Further research in this regard can look into the molecular specifics of any potential role of these DEGs in the adaptation of maize seedlings to cold stress.

Non-coding RNAs (ncRNAs), such as lncRNAs, have been discovered to regulate plant response to abiotic and biotic stress by controlling the expression of functional genes [132]. A previous study on cassava reported 318 lncRNAs responsive to cold and drought stress [133], while the expression of 2088 lncRNAs in grapevine (*Vitis vinifera* L.) was induced by cold stress [134]. In this study, the expression of 337 putative lncRNAs was regulated by cold stress (Figure 4). Thus, these lncRNAs might have modulated multiple biological processes involved in cold acclimation in maize by influencing gene expression at the transcriptional, post-transcriptional, and epigenetic levels. Otherwise, there is considerable interest in lncRNAs among molecular biologists, plant breeders, and geneticists, and our research may have identified crucial candidates that can aid in the development of cold-tolerant cultivars. However, more research is needed to fully understand the link between these lncRNAs and cold stress.

We developed a molecular model for cold stress tolerance in maize seedlings, as shown in Figure 11, based on our main findings of the critical cold-responsive DEGs and their associated pathways, as well as the numerous published citations in the present study.

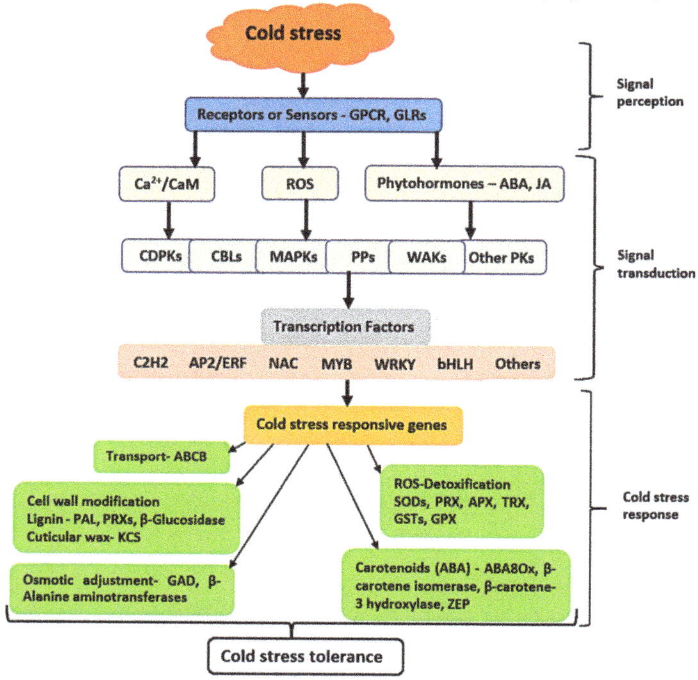

Figure 11. The schematic molecular model describing the signaling pathways involved in the acquisition of cold tolerance in maize seedlings. The model was constructed based on the main cold response components identified in this report, as well as plant abiotic stress pathway schemes previously described. The downward pointing arrows represent the sequence of events in cold tolerance in maize, from stress signal perception to acclimation mechanisms. Abbreviation key: GPCR, G-protein-coupled receptor; GLR, glutamate receptor; ROS, reactive oxygen species; ABA, abscisic acid; JA, jasmonic acid; CDPKs, calcium-dependent protein kinases; MAPK, mitogen-activated protein kinase; PPs, protein phosphatase; WAKs, wall-associated kinases; PKs, protein kinases; PRX, peroxidases; PAL, phenylalanine ammonia lyase; KCS, ketoacyl-CoA synthase; GAD, glutamate decarboxylase; SOD, superoxide dismutase; APX, ascorbate peroxidase; TRX, thioredoxin; GST, glutathione transferase; GPX, glutathione peroxidase; ZEP, zeaxanthin epoxidase.

5. Conclusions

In this study, we comprehensively compared the leaf transcriptome and phenotypic response of the maize population (24 cold-tolerant and 22 cold-sensitive lines) in response to cold stress at the seedling stage. Resultantly, the tolerant lines maintained a strong vigor with higher survival rates, while the majority of the sensitive line seedlings died and had yellow spots on the leaves. Using the RNA-seq-based approach, 2237 (1656 annotated and 581 unannotated) DEGs were identified between the tolerant and sensitive samples. Moreover, cold stress significantly enhanced 779 and 887 DEGs in the tolerant and sensitive lines, respectively. Functional annotation was carried out on the three categories (1656, 779, and 887) of DEGs. In the tolerant lines, genes associated with GPCR, Ca^{2+} signaling, protein kinases, and ROS may have played a significant role in rapid sensing and signaling, whereas genes associated with hormones, such as ABA and JA, may have played a role in signaling and cross-talk between diverse stimuli. The activation of TFs and their binding to promoter sites of certain genes results in activation of stress-responsive genes. The upregulation of several antioxidants, transport, and osmoprotectants suggested protection of the cellular machinery, whereas genes associated with the phenylpropanoid biosynthesis pathway might be involved in providing mechanical support and protection against cold stress. Moreover, genes involved in lipid metabolism may play a critical role in cold stress resistance via membrane modification. Thus, the networks involved in the function of the genes and regulators of the above-named pathways are critical in the cold acclimation of maize at the seedling stage. Moreover, genes related to ribosome, proteolysis, peroxisome, and carbon metabolism were significantly enriched in the sensitive lines. The unannotated DEGs were more inclined in the functions of long non-coding RNAs. Our findings indicate the involvement of plant signaling, transcription factors, and protective mechanisms in the molecular mechanisms underlying cold acclimation in maize at the seedling stage. Otherwise, the essential genes and metabolic pathways identified in this study may serve as valuable genetic resources or selection targets for the genetic engineering of cold-tolerant maize cultivars.

Supplementary Materials: The following are available online at https://www.mdpi.com/article/10.3390/genes12101638/s1, Figure S1: Phenotypic analysis of maize population under cold stress conditions, Figure S2: Co-expression network analysis of our DEGs, Figure S3: Validation of expression pattern of six genes identified by RNA-Seq by Ct values obtained from qRT-PCR, Table S1: The primers of six differentially expressed genes used for qRT-PCR verification, Table S2: Survival rates of the maize seedlings after cold stress treatment, Table S3: Summary of RNA sequencing results for the forty-six maize seedling leaf samples, Table S4: Expression of signaling related genes, Table S5: Expression of transcription factor-related genes, Table S6: Expression of response to cold-related genes, Table S7: Expression of metabolism/biosynthesis pathway-related genes, Table S8: The 116 critical DEGs identified by WGCNA.

Author Contributions: Conceptualization, J.L., H.W. and J.K.W.; experiments Y.S. (Ying Sun), Y.S. (Yingly Sun), C.L. and J.K.W. Analysis of the data and writing of the original draft J.K.W.; revising and editing the manuscript H.W., J.L., C.Z., Y.S. (Ying Sun), Y.S. (Yingly Sun), Q.C., C.L. and J.K.W. All authors have read and agreed to the published version of the manuscript.

Funding: This research was funded by the National Key Research and Development Program of China, grant numbers 2018YFD1000702/2018YFD1000700 and 2016YFD0101202; National Natural Science Foundation of China, grant number 31900452; key research and development program of Xinjiang province, China, grant number 2018B01006-4; and the Agricultural Science and Technology Innovation Program of CAAS.

Institutional Review Board Statement: Not applicable.

Informed Consent Statement: Not applicable.

Data Availability Statement: This manuscript includes the essential data either as figures or as Supplementary Materials. The raw sequence reads have been deposited to the Genome Sequence Archive (GSA) under accession numbers CRA003678 (https://ngdc.cncb.ac.cn/bioproject/browse/PRJCA004123) (accessed on 5 January 2021).

Acknowledgments: We thank Junjie Fu (Institute of Crop Science, Chinese Academy of Agricultural Sciences) for assisting us with B73 and CIMBL116 seeds.

Conflicts of Interest: The authors declare no conflict of interest.

Abbreviations

bHLH	Basic helix-loop-helix
DEGs	Differentially expressed genes
MYB	Myeloblastosis
KEGG	Kyoto Encyclopedia of Genes and Genomes
GO	Gene Ontology
PCA	principal component analysis
GPCR	G-protein coupled receptor
JA	jasmonic acid
PPs	protein phosphatase
WAK	wall-associated kinase

PAL	phenylalanine ammonia lyase
KCS	ketoacyl-CoA synthase
GAD	glutamate decarboxylase
SOD	Superoxide dismutase
APX	Ascorbate peroxidase
GLR	Glutamate receptor
ROS	Reactive oxygen species
ABA	Abscisic acid
PK	Protein kinase
PRX	Peroxidases
CDPK	Calcium dependent protein kinase
CaMs	Calmodulin
CBLs	Calcineurin B-like
CMLs	CaM-related proteins
MAPK	Mitogen-activated protein kinase

References

1. United States Department of Agriculture (USDA). World Agricultural Supply and Demand Estimates. 2020. Available online: https://downloads.usda.library.cornell.edu/usda-esmis/files/5q47rn72z/z890sd52n/hd76sk58s/production.pdf (accessed on 12 May 2020).
2. Gong, F.L.; Yang, F.; Tai, X.; Wang, W. "Omics" of maize stress response for sustainable food production: Opportunities and challenges. *Omics A J. Integr. Biol.* **2014**, *18*, 714–732. [CrossRef]
3. Stocker, T.F.; Qin, D.; Plattner, G.K.; Tignor, M.M.; Allen, S.K.; Boschung, J.; Nauels, A.; Xia, Y.; Bex, V.; Midgley, P.M. *Climate Change 2013: The Physical Science Basis. Contribution of Working Group I to the Fifth Assessment Report of IPCC the Intergovernmental Panel on Climate Change*; Cambridge University Press: Cambridge, UK, 2014.
4. Krasensky, J.; Jonak, C. Drought, salt, and temperature stress-induced metabolic rearrangements and regulatory networks. *J. Exp. Bot.* **2012**, *63*, 1593–1608. [CrossRef] [PubMed]
5. Peleg, Z.; Blumwald, E. Hormone balance and abiotic stress tolerance in crop plants. *Curr. Opin. Plant Biol.* **2011**, *14*, 290–295. [CrossRef]
6. Ma, S.Q.; Xi, Z.X.; Wang, Q. Risk evaluation of cold damage to corn in Northeast China. *J. Nat. Disasters* **2003**, *12*, 137–141.
7. Greaves, J.A. Improving suboptimal temperature tolerance in maize-the search for variation. *J. Exp. Bot.* **1996**, *47*, 307–323. [CrossRef]
8. Marocco, A.; Lorenzoni, C.; Fracheboud, Y. Chilling stress in maize. *Maydica* **2005**, *50*, 571–580.
9. Sobkowiak, A.; Jończyk, M.; Adamczyk, J.; Szczepanik, J.; Solecka, D.; Kuciara, I.; Hetmańczyk, K.; Trzcinska-Danielewicz, T.; Grzybowski, M.; Skoneczny, M. Molecular foundations of chilling-tolerance of modern maize. *BMC Genom.* **2016**, *17*, 125. [CrossRef] [PubMed]
10. Leipner, J.; Stamp, P. Chilling Stress in Maize Seedlings. In *Handbook of Maize: Its Biology*; Springer: New York, NY, USA, 2009; pp. 291–310.
11. Sowiński, P.; Rudzińska-Langwald, A.; Dalbiak, A.; Sowińska, A. Assimilate export from leaves of chilling-treated seedlings of maize. The path to vein. *Plant Physiol. Biochem.* **2001**, *39*, 881–889. [CrossRef]
12. Zhang, B.; Yang, L.; Li, Y. Comparison of physiological and biochemical characteristics related to cold resistance in sugarcane under field conditions. *Acta Agron. Sin.* **2011**, *37*, 496–505. [CrossRef]
13. Virdi, A.S.; Singh, S.; Singh, P. Abiotic stress responses in plants: Roles of calmodulin-regulated proteins. *Front. Plant Sci.* **2015**, *6*, 809. [CrossRef]
14. Chinnusamy, V.; Zhu, J.; Zhu, J.K. Cold stress regulation of gene expression in plants. *Trends Plant Sci.* **2007**, *12*, 444–451. [CrossRef] [PubMed]
15. Chinnusamy, V.; Zhu, J.K.; Sunkar, R. Gene Regulation During Cold Stress Acclimation in Plants. In *Plant Stress Tolerance*; Springer: Totowa, NJ, USA, 2010; Volume 639, pp. 39–55.
16. Zhao, C.; Wang, P.; Si, T.; Hsu, C.C.; Wang, L.; Zayed, O.; Yu, Z.; Zhu, Y.; Dong, J.; Tao, W.A. MAP kinase cascades regulate the cold response by modulating ICE1 protein stability. *Dev. Cell* **2017**, *43*, 618–629. [CrossRef] [PubMed]
17. Shi, Y.; Ding, Y.; Yang, S. Cold signal transduction and its interplay with phytohormones during cold acclimation. *Plant Cell Physiol.* **2015**, *56*, 7–15. [CrossRef]
18. Sah, S.K.; Reddy, K.R.; Li, J. Abscisic acid and abiotic stress tolerance in crop plants. *Front. Plant Sci.* **2016**, *7*, 571. [CrossRef]
19. Tarkowski, Ł.P.; Van den Ende, W. Cold tolerance triggered by soluble sugars: A multifaceted countermeasure. *Front. Plant Sci.* **2015**, *6*, 203. [CrossRef] [PubMed]
20. Arnholdt-Schmitt, B.; Costa, J.H.; de Melo, D.F. AOX—A functional marker for efficient cell reprogramming under stress? *Trends Plant Sci.* **2006**, *11*, 281–287. [CrossRef] [PubMed]
21. Al-Whaibi, M.H. Plant heat-shock proteins: A mini review. *J. King Saud Univ. Sci.* **2011**, *23*, 139–150. [CrossRef]

22. Alejandro, S.; Lee, Y.; Tohge, T.; Sudre, D.; Osorio, S.; Park, J.; Bovet, L.; Lee, Y.; Geldner, N.; Fernie, A.R. AtABCG29 is a monolignol transporter involved in lignin biosynthesis. *Curr. Biol.* **2012**, *22*, 1207–1212. [CrossRef]
23. Wang, B.; Guo, G.; Wang, C.; Lin, Y.; Wang, X.; Zhao, M.; Guo, Y.; He, M.; Zhang, Y.; Pan, L. Survey of the transcriptome of Aspergillus oryzae via massively parallel mRNA sequencing. *Nucleic Acids Res.* **2010**, *38*, 5075–5087. [CrossRef] [PubMed]
24. Zhang, Z.; Huang, R. Enhanced tolerance to freezing in tobacco and tomato overexpressing transcription factor TERF2/LeERF2 is modulated by ethylene biosynthesis. *Plant Mol. Biol.* **2010**, *73*, 241–249. [CrossRef]
25. Jończyk, M.; Sobkowiak, A.; Trzcinska-Danielewicz, J.; Sowiński, P. Chromatin-Level Differences Elucidate Potential Determinants of Contrasting Levels of Cold Sensitivity in Maize Lines. *Plant Mol. Biol. Report.* **2021**, *39*, 335–350. [CrossRef]
26. Grzybowski, M.; Adamczyk, J.; Jończyk, M.; Sobkowiak, A.; Szczepanik, J.; Frankiewicz, K.; Sowiński, P. Increased photosensitivity at early growth as a possible mechanism of maize adaptation to cold springs. *J. Exp. Bot.* **2019**, *70*, 2887–2904. [CrossRef] [PubMed]
27. Bolger, A.M.; Lohse, M.; Usadel, B. Trimmomatic: A flexible trimmer for Illumina sequence data. *Bioinformatics* **2014**, *30*, 2114–2120. [CrossRef]
28. Kim, D.; Langmead, B.; Salzberg, S.L. HISAT: A fast spliced aligner with low memory requirements. *Nat. Methods* **2015**, *12*, 357–360. [CrossRef] [PubMed]
29. Roberts, A.; Pimentel, H.; Trapnell, C.; Pachter, L. Identification of novel transcripts in annotated genomes using RNA-Seq. *Bioinformatics* **2011**, *27*, 2325–2329. [CrossRef]
30. Anders, S.; Pyl, P.T.; Huber, W. HTSeq—A Python framework to work with high-throughput sequencing data. *Bioinformatics* **2015**, *31*, 166–169. [CrossRef] [PubMed]
31. Anders, S.; Huber, W. Differential expression analysis for sequence count data. *Nat. Preced.* **2010**. [CrossRef]
32. Li, L.; Eichten, S.R.; Shimizu, R.; Petsch, K.; Yeh, C.T.; Wu, W.; Chettoor, A.M.; Givan, S.A.; Cole, R.A.; Fowler, J.E. Genome-wide discovery and characterization of maize long non-coding RNAs. *Genome Biol.* **2014**, *15*, R40. [CrossRef] [PubMed]
33. Kong, L.; Zhang, Y.; Ye, Z.Q.; Liu, X.Q.; Zhao, S.Q.; Wei, L.; Gao, G. CPC: Assess the protein-coding potential of transcripts using sequence features and support vector machine. *Nucleic Acids Res.* **2007**, *35*, W345–W349. [CrossRef]
34. Tian, T.; Liu, Y.; Yan, H.; You, Q.; Yi, X.; Du, Z.; Xu, W.; Su, Z. agriGO v2. 0: A GO analysis toolkit for the agricultural community, 2017 update. *Nucleic Acids Res.* **2017**, *45*, W122–W129. [CrossRef]
35. Supek, F.; Bošnjak, M.; Škunca, N.; Šmuc, T. REVIGO summarizes and visualizes long lists of gene ontology terms. *PLoS ONE* **2011**, *6*, e21800. [CrossRef] [PubMed]
36. Kanehisa, M.; Goto, S. KEGG: Kyoto encyclopedia of genes and genomes. *Nucleic Acids Res.* **2000**, *28*, 27–30. [CrossRef] [PubMed]
37. Xie, C.; Mao, X.; Huang, J.; Ding, Y.; Wu, J.; Dong, S.; Kong, L.; Gao, G.; Li, C.Y.; Wei, L. KOBAS 2.0: A web server for annotation and identification of enriched pathways and diseases. *Nucleic Acids Res.* **2011**, *39*, W316–W322. [CrossRef] [PubMed]
38. Zhang, Y.; Zhu, L.; Xue, J.; Yang, J.; Hu, H.; Cui, J.; Xu, J. Selection and Verification of Appropriate Reference Genes for Expression Normalization in Cryptomeria fortunei Under Abiotic Stress and Hormone Treatments. *Genes* **2021**, *12*, 791. [CrossRef] [PubMed]
39. Livak, K.J.; Schmittgen, T.D. Analysis of relative gene expression data using real-time quantitative PCR and the $2^{-\Delta\Delta CT}$ method. *Methods* **2001**, *25*, 402–408. [CrossRef] [PubMed]
40. Novaković, L.; Guo, T.; Bacic, A.; Sampathkumar, A.; Johnson, K.L. Hitting the wall—Sensing and signaling pathways involved in plant cell wall remodeling in response to abiotic stress. *Plants* **2018**, *7*, 89. [CrossRef] [PubMed]
41. Jin, J.; Tian, F.; Yang, D.C.; Meng, Y.Q.; Kong, L.; Luo, J.; Gao, G. PlantTFDB 4.0: Toward a central hub for transcription factors and regulatory interactions in plants. *Nucleic Acids Res.* **2016**, *45*, gkw982. [CrossRef] [PubMed]
42. Cruz, R.P.D.; Sperotto, R.A.; Cargnelutti, D.; Adamski, J.M.; de FreitasTerra, T.; Fett, J.P. Avoiding damage and achieving cold tolerance in rice plants. *Food Energy Secur.* **2013**, *2*, 96–119. [CrossRef]
43. Li, Y.; Wang, X.; Li, Y.; Zhang, Y.; Gou, Z.; Qi, X.; Zhang, J. Transcriptomic Analysis Revealed the Common and Divergent Responses of Maize Seedling Leaves to Cold and Heat Stresses. *Genes* **2020**, *11*, 881. [CrossRef]
44. Ji, C.Y.; Kim, H.S.; Lee, C.J.; Kim, S.E.; Lee, H.U.; Nam, S.S.; Kwak, S.S. Comparative transcriptome profiling of tuberous roots of two sweet potato lines with contrasting low temperature tolerance during storage. *Gene* **2020**, *727*, 144244. [CrossRef]
45. Yu, T.; Zhang, J.; Cao, J.; Cai, Q.; Li, X.; Sun, Y.; Duan, Y. Leaf transcriptomic response mediated by cold stress in two maize inbred lines with contrasting tolerance levels. *Genomics* **2021**, *113*, 782–794. [CrossRef] [PubMed]
46. Zhu, J.K. Abiotic stress signaling and responses in plants. *Cell* **2016**, *167*, 313–324. [CrossRef] [PubMed]
47. Bai, L.; Liu, Y.; Mu, Y.; Anwar, A.; He, C.; Yan, Y.; Li, Y.; Yu, X. Heterotrimeric G-protein γ subunit CsGG3. 2 positively regulates the expression of CBF genes and chilling tolerance in cucumber. *Front. Plant Sci.* **2018**, *9*, 488. [CrossRef]
48. Guo, X.; Liu, D.; Chong, K. Cold signaling in plants: Insights into mechanisms and regulation. *J. Integr. Plant Biol.* **2018**, *60*, 745–756. [CrossRef] [PubMed]
49. Xi, J.; Qiu, Y.; Du, L.; Poovaiah, B.W. Plant-specific trihelix transcription factor AtGT2L interacts with calcium/calmodulin and responds to cold and salt stresses. *Plant Sci.* **2012**, *185*, 274–280. [CrossRef]
50. Xie, Z.M.; Zou, H.F.; Lei, G.; Wei, W.; Zhou, Q.Y.; Niu, C.F.; Chen, S.Y. Soybean Trihelix transcription factors GmGT-2A and GmGT-2B improve plant tolerance to abiotic stresses in transgenic Arabidopsis. *PLoS ONE* **2009**, *4*, e6898. [CrossRef]
51. Sanders, D.; Pelloux, J.; Brownlee, C.; Harper, J.F. Calcium at the crossroads of signaling. *Plant Cell* **2002**, *14*, S401–S417. [CrossRef] [PubMed]

52. Luan, S.; Kudla, J.; Rodriguez-Concepcion, M.; Yalovsky, S.; Gruissem, W. Calmodulins and calcineurin B–like proteins: Calcium sensors for specific signal response coupling in plants. *Plant Cell* **2002**, *14*, S389–S400. [CrossRef]
53. Pareek, A.; Khurana, A.; Sharma, K.; Kumar, R. An overview of signaling regulons during cold stress tolerance in plants. *Curr. Genom.* **2017**, *18*, 498–511. [CrossRef]
54. Li, Q.; Lei, S.; Du, K.; Li, L.; Pang, X.; Wang, Z.; Xu, L. RNA-seq based transcriptomic analysis uncovers α-linolenic acid and jasmonic acid biosynthesis pathways respond to cold acclimation in Camellia japonica. *Sci. Rep.* **2016**, *6*, 36463. [CrossRef]
55. Wu, X.; Bacic, A.; Johnson, K.L.; Humphries, J. The role of *brachypodium distachyon* wall-associated kinases (WAKs) in cell expansion and stress responses. *Cells* **2020**, *9*, 2478. [CrossRef] [PubMed]
56. Kim, B.G.; Waadt, R.; Cheong, Y.H.; Pandey, G.K.; Dominguez-Solis, J.R.; Schültke, S.; Lee, S.C.; Kudla, J.; Luan, S. The calcium sensor CBL10 mediates salt tolerance by regulating ion homeostasis in Arabidopsis. *Plant J.* **2007**, *52*, 473–484. [CrossRef]
57. Lee, B.H.; Henderson, D.A.; Zhu, J.K. The Arabidopsis cold-responsive transcriptome and its regulation by ICE1. *Plant Cell* **2005**, *17*, 3155–3175. [CrossRef] [PubMed]
58. Martínez, F.A.P.; Estrada, Y.; Flores, F.B.; Ortíz-Atienza, A.; Lozano, R.; Egea, I. The Ca^{2+} Sensor Calcineurin B-like Protein 10 in Plants: Emerging New Crucial Roles for Plant Abiotic Stress Tolerance. *Front. Plant Sci.* **2020**, *11*, 2155.
59. Kamal, M.M.; Ishikawa, S.; Takahashi, F.; Suzuki, K.; Kamo, M.; Umezawa, T.; Uemura, M. Large-Scale Phosphoproteomic Study of Arabidopsis Membrane Proteins Reveals Early Signaling Events in Response to Cold. *Int. J. Mol. Sci.* **2020**, *21*, 8631. [CrossRef]
60. Kumar, M.; Gho, Y.S.; Jung, K.H.; Kim, S.R. Genome-wide identification and analysis of genes, conserved between japonica and indica rice cultivars that respond to low-temperature stress at the vegetative growth stage. *Front. Plant Sci.* **2017**, *8*, 1120. [CrossRef]
61. Zhou, H.; He, Y.; Zhu, Y.; Li, M.; Song, S.; Bo, W.; Pang, X. Comparative transcriptome profiling reveals cold stress responsiveness in two contrasting Chinese jujube cultivars. *BMC Plant Biol.* **2020**, *20*, 240. [CrossRef] [PubMed]
62. Zhang, Y.; Li, Y.; He, Y.; Hu, W.; Zhang, Y.; Wang, X.; Tang, H. Identification of NADPH oxidase family members associated with cold stress in strawberry. *FEBS Open Bio* **2018**, *8*, 593–605. [CrossRef] [PubMed]
63. Cheng, G.; Zhang, L.; Wang, H.; Lu, J.; Wei, H.; Yu, S. Transcriptomic Profiling of Young Cotyledons Response to Chilling Stress in Two Contrasting Cotton (*Gossypium hirsutum* L.) Genotypes at the Seedling Stage. *Int. J. Mol. Sci.* **2020**, *21*, 5095. [CrossRef] [PubMed]
64. Kovtun, Y.; Chiu, W.L.; Tena, G.; Sheen, J. Functional analysis of oxidative stress-activated mitogen-activated protein kinase cascade in plants. *Proc. Natl. Acad. Sci. USA* **2000**, *97*, 2940–2945. [CrossRef]
65. Schweighofer, A.; Hirt, H.; Meskiene, I. Plant PP2C phosphatases: Emerging functions in stress signaling. *Trends Plant Sci.* **2004**, *9*, 236–243. [CrossRef] [PubMed]
66. Liu, L.; Hu, X.; Song, J.; Zong, X.; Li, D.; Li, D. Over-expression of a Zea mays L. protein phosphatase 2C gene (ZmPP2C) in Arabidopsis thaliana decreases tolerance to salt and drought. *J. Plant Physiol.* **2009**, *166*, 531–542. [CrossRef] [PubMed]
67. Lee, H.G.; Seo, P.J. The MYB 96–HHP module integrates cold and abscisic acid signaling to activate the CBF–COR pathway in Arabidopsis. *Plant J.* **2015**, *82*, 962–977. [CrossRef] [PubMed]
68. Meng, J.; Hu, B.; Yi, G.; Li, X.; Chen, H.; Wang, Y.; Xu, C. Genome-wide analyses of banana fasciclin-like AGP genes and their differential expression under low-temperature stress in chilling sensitive and tolerant cultivars. *Plant Cell Rep.* **2020**, *39*, 693–708. [CrossRef] [PubMed]
69. Martín, M.L.; Busconi, L. A rice membrane-bound calcium-dependent protein kinase is activated in response to low temperature. *Plant Physiol.* **2001**, *125*, 1442–1449. [CrossRef] [PubMed]
70. Agarwal, M.; Hao, Y.; Kapoor, A.; Dong, C.H.; Fujii, H.; Zheng, X.; Zhu, J.K. A R2R3 type MYB transcription factor is involved in the cold regulation of CBF genes and in acquired freezing tolerance. *J. Biol. Chem.* **2006**, *281*, 37636–37645. [CrossRef] [PubMed]
71. Pradhan, S.K.; Pandit, E.; Nayak, D.K.; Behera, L.; Mohapatra, T. Genes, pathways and transcription factors involved in seedling stage chilling stress tolerance in indica rice through RNA-Seq analysis. *BMC Plant Biol.* **2019**, *19*, 352. [CrossRef]
72. Jiang, C.; Zhang, H.; Ren, J.; Dong, J.; Zhao, X.; Wang, X.; Yu, H. Comparative Transcriptome-Based Mining and Expression Profiling of Transcription Factors Related to Cold Tolerance in Peanut. *Int. J. Mol. Sci.* **2020**, *21*, 1921. [CrossRef] [PubMed]
73. Övernäs, E.; Sundås Larsson, A.; Söderman, E. *Two AP2 Transcription Factors, AtERF38 and AtERF39, Have Similar Function but are Differently Regulated in ABA and Stress Responses*; Uppsala University Publications: Upsala, Sweden, 2010.
74. Krishnaswamy, S.; Verma, S.; Rahman, M.H.; Kav, N.N. Functional characterization of four APETALA2-family genes (RAP2. 6, RAP2. 6L, DREB19 and DREB26) in Arabidopsis. *Plant Mol. Biol.* **2011**, *75*, 107–127. [CrossRef]
75. Wang, J.; Tao, F.; An, F.; Zou, Y.; Tian, W.; Chen, X.; Xu, X.; Hu, X. Wheat transcription factor TaWRKY70 is positively involved in high-temperature seedling plant resistance to Puccinia striiformis f. sp. tritici. *Mol. Plant Pathol.* **2017**, *18*, 649–661. [CrossRef]
76. Wei, W.; Zhang, Y.; Han, L.; Guan, Z.; Chai, T. A novel WRKY transcriptional factor from *Thlaspi caerulescens* negatively regulates the osmotic stress tolerance of transgenic tobacco. *Plant Cell Rep.* **2008**, *27*, 795–803. [CrossRef] [PubMed]
77. Wang, F.; Chen, X.; Dong, S.; Jiang, X.; Wang, L.; Yu, J.; Zhou, Y. Crosstalk of PIF4 and DELLA modulates CBF transcript and hormone homeostasis in cold response in tomato. *Plant Biotechnol. J.* **2020**, *18*, 1041–1055. [CrossRef]
78. Li, Z.; Liu, C.; Zhang, Y.; Wang, B.; Ran, Q.; Zhang, J. The bHLH family member ZmPTF1 regulates drought tolerance in maize by promoting root development and abscisic acid synthesis. *J. Exp. Bot.* **2019**, *70*, 5471–5486. [CrossRef] [PubMed]

79. Vannini, C.; Locatelli, F.; Bracale, M.; Magnani, E.; Marsoni, M.; Osnato, M.; Mattana, M.; Baldoni, E.; Coraggio, I. Overexpression of the rice Osmyb4 gene increases chilling and freezing tolerance of Arabidopsis thaliana plants. *Plant J.* **2004**, *37*, 115–127. [CrossRef] [PubMed]
80. Babitha, K.; Vemanna, R.S.; Nataraja, K.N.; Udayakumar, M. Overexpression of EcbHLH57 transcription factor from Eleusine coracana L. in tobacco confers tolerance to salt, oxidative and drought stress. *PLoS ONE* **2015**, *10*, e0137098. [CrossRef] [PubMed]
81. Khatun, K.; Robin, A.H.K.; Park, J.I.; Nath, U.K.; Kim, C.K.; Lim, K.B.; Nou, I.S.; Chung, M.Y. Molecular characterization and expression profiling of tomato GRF transcription factor family genes in response to abiotic stresses and phytohormones. *Int. J. Mol. Sci.* **2017**, *18*, 1056. [CrossRef]
82. Vogel, J.T.; Zarka, D.G.; Van Buskirk, H.A.; Fowler, S.G.; Thomashow, M.F. Roles of the CBF2 and ZAT12 transcription factors in configuring the low temperature transcriptome of Arabidopsis. *Plant J.* **2005**, *41*, 195–211. [CrossRef]
83. Tang, S.; Xian, Y.; Wang, F.; Luo, C.; Song, W.; Xie, S.; Liu, H. Comparative transcriptome analysis of leaves during early stages of chilling stress in two different chilling-tolerant brown-fiber cotton cultivars. *PLoS ONE* **2021**, *16*, e0246801. [CrossRef]
84. Agarwal, P.K.; Jha, B. Transcription factors in plants and ABA dependent and independent abiotic stress signaling. *Biol. Plant* **2010**, *54*, 201–212. [CrossRef]
85. Chen, J.; Tian, Q.; Pang, T.; Jiang, L.; Wu, R.; Xia, X.; Yin, W. Deep-sequencing transcriptome analysis of low temperature perception in a desert tree, *Populus euphratica*. *BMC Genom.* **2014**, *15*, 326. [CrossRef] [PubMed]
86. Baron, K.N.; Schroeder, D.F.; Stasolla, C. Transcriptional response of abscisic acid (ABA) metabolism and transport to cold and heat stress applied at the reproductive stage of development in *Arabidopsis thaliana*. *Plant Sci.* **2012**, *188*, 48–59. [CrossRef] [PubMed]
87. Zhang, Z.; Wang, Y.; Chang, L.; Zhang, T.; An, J.; Liu, Y.; Cao, Y.; Zhao, X.; Sha, X.; Hu, T. MsZEP, a novel zeaxanthin epoxidase gene from alfalfa (Medicago sativa), confers drought and salt tolerance in transgenic tobacco. *Plant Cell Rep.* **2016**, *35*, 439–453. [CrossRef] [PubMed]
88. Du, H.; Wang, N.; Cui, F.; Li, X.; Xiao, J.; Xiong, L. Characterization of the β-carotene hydroxylase gene DSM2 conferring drought and oxidative stress resistance by increasing xanthophylls and abscisic acid synthesis in rice. *Plant Physiol.* **2010**, *154*, 1304–1318. [CrossRef]
89. Nayyar, H.; Chander, S. Protective effects of polyamines against oxidative stress induced by water and cold stress in chickpea. *J. Agron. Crop Sci.* **2004**, *190*, 355–365. [CrossRef]
90. Li, Y.; Wang, X.; Ban, Q.; Zhu, X.; Jiang, C.; Wei, C.; Bennetzen, J.L. Comparative transcriptomic analysis reveals gene expression associated with cold adaptation in the tea plant Camellia sinensis. *BMC Genom.* **2019**, *20*, 624. [CrossRef] [PubMed]
91. Liu, T.; Li, C.X.; Zhong, J.; Shu, D.; Luo, D.; Li, Z.M.; Ma, X.R. Exogenous 1′, 4′-trans-Diol-ABA Induces Stress Tolerance by Affecting the Level of Gene Expression in Tobacco (Nicotiana tabacum L.). *Int. J. Mol. Sci.* **2021**, *22*, 2555. [CrossRef] [PubMed]
92. Filiz, E.; Ozyigit, I.I.; Saracoglu, I.A.; Uras, M.E.; Sen, U.; Yalcin, B. Abiotic stress-induced regulation of antioxidant genes in different Arabidopsis ecotypes: Microarray data evaluation. *Biotechnol. Biotechnol. Equip.* **2019**, *33*, 128–143. [CrossRef]
93. Baek, K.H.; Skinner, D.Z. Alteration of antioxidant enzyme gene expression during cold acclimation of near-isogenic wheat lines. *Plant Sci.* **2003**, *165*, 1221–1227. [CrossRef]
94. Rezaie, R.; Mandoulakani, B.A.; Fattahi, M. Cold stress changes antioxidant defense system, phenylpropanoid contents and expression of genes involved in their biosynthesis in *Ocimum basilicum* L. *Sci. Rep.* **2020**, *10*, 1–10. [CrossRef] [PubMed]
95. Wang, R.; Ma, J.; Zhang, Q.; Wu, C.; Zhao, H.; Wu, Y.; He, G. Genome-wide identification and expression profiling of glutathione transferase gene family under multiple stresses and hormone treatments in wheat (*Triticum aestivum* L.). *BMC Genom.* **2019**, *20*, 986. [CrossRef]
96. Takesawa, T.; Ito, M.; Kanzaki, H.; Kameya, N.; Nakamura, I. Overexpression of ζ glutathione S-transferase in transgenic rice enhances germination and growth at low temperature. *Mol. Breed.* **2002**, *9*, 93–101. [CrossRef]
97. Xu, J.; Yang, J.; Duan, X.; Jiang, Y.; Zhang, P. Increased expression of native cytosolic Cu/Zn superoxide dismutase and ascorbate peroxidase improves tolerance to oxidative and chilling stresses in cassava (*Manihot esculenta* Crantz). *BMC Plant Biol.* **2014**, *14*, 208. [CrossRef] [PubMed]
98. Almagro, L.; Gómez Ros, L.V.; Belchi-Navarro, S.; Bru, R.; Ros Barceló, A.; Pedreno, M.A. Class III peroxidases in plant defence reactions. *J. Exp. Bot.* **2009**, *60*, 377–390. [CrossRef] [PubMed]
99. Meyer, Y.; Siala, W.; Bashandy, T.; Riondet, C.; Vignols, F.; Reichheld, J.P. Glutaredoxins and thioredoxins in plants. *Biochim. Biophys. Acta Mol. Cell Res.* **2008**, *1783*, 589–600. [CrossRef]
100. Ortiz, D.; Hu, J.; Salas Fernandez, M.G. Genetic architecture of photosynthesis in Sorghum bicolor under non-stress and cold stress conditions. *J. Exp. Bot.* **2017**, *68*, 4545–4557. [CrossRef] [PubMed]
101. Khaledian, Y.; Maali-Amiri, R.; Talei, A. Phenylpropanoid and antioxidant changes in chickpea plants during cold stress. *Russ. J. Plant Physiol.* **2015**, *62*, 772–778. [CrossRef]
102. Srivastava, S.; Vishwakarma, R.K.; Arafat, Y.A.; Gupta, S.K.; Khan, B.M. Abiotic stress induces change in Cinnamoyl CoA Reductase (CCR) protein abundance and lignin deposition in developing seedlings of *Leucaena leucocephala*. *Physiol. Mol. Biol. Plants* **2015**, *21*, 197–205. [CrossRef] [PubMed]
103. Chun-yan, F. Research progress on 4-coumarate: Coenzyme A ligase (4CL) of plants. *Mod. Agric. Sci. Technol.* **2010**, *8*, 39–40.
104. Pandey, V.; Awasthi, M.; Singh, S.; Tiwari, S.; Dwivedi, U. A comprehensive review on function and application of plant peroxidases. *Biochem. Anal. Biochem.* **2017**, *6*, 308. [CrossRef]

105. Naikoo, M.I.; Dar, M.I.; Raghib, M.F.; Jaleel, H.; Ahmad, B.; Raina, A.; Khan, F.A.; Naushin, F. Role and regulation of plants phenolics in abiotic stress tolerance: An overview. *Plant Signal. Mol.* **2019**, 157–168. [CrossRef]
106. Vogt, T. Phenylpropanoid biosynthesis. *Mol. Plant* **2010**, *3*, 2–20. [CrossRef]
107. Dharmawardhana, D.P.; Ellis, B.E.; Carlson, J.E. A [β]-Glucosidase from lodge pole pine xylem specific for the lignin precursor coniferin. *Plant Physiol.* **1995**, *107*, 331–339. [CrossRef] [PubMed]
108. Kleczkowski, K.; Schell, J.; Bandur, R. Phytohormones conjugates: Nature and function. *Crit. Rev. Plant. Sci.* **1995**, *14*, 283–298. [CrossRef]
109. Baba, S.A.; Vishwakarma, R.A.; Ashraf, N. Functional characterization of CsBGlu12, a β-glucosidase from Crocus sativus, provides insights into its role in abiotic stress through accumulation of antioxidant flavonols. *J. Biol. Chem.* **2017**, *292*, 4700–4713. [CrossRef] [PubMed]
110. Slama, I.; Abdelly, C.; Bouchereau, A.; Flowers, T.; Savoure, A. Diversity, distribution and roles of osmoprotective compounds accumulated in halophytes under abiotic stress. *Ann. Bot.* **2015**, *115*, 433–447. [CrossRef] [PubMed]
111. Hanson, A.D.; Rathinasabapathi, B.; Chamberlin, B.; Gage, D.A. Comparative physiological evidence that β-alanine betaine and choline-O-sulfate act as compatible osmolytes in halophytic Limonium species. *Plant Physiol.* **1991**, *97*, 1199–1205. [CrossRef] [PubMed]
112. Rocha, M.; Licausi, F.; Araujo, W.L.; Nunes-Nesi, A.; Sodek, L.; Fernie, A.R.; van Dongen, J.T. Glycolysis and the tricarboxylic acid cycle are linked by alanine aminotransferase during hypoxia induced by waterlogging of Lotus japonicus. *Plant Physiol.* **2010**, *152*, 1501–1513. [CrossRef] [PubMed]
113. Renault, H.; Roussel, V.; El Amrani, A.; Arzel, M.; Renault, D.; Bouchereau, A.; Deleu, C. The Arabidopsis pop2-1 mutant reveals the involvement of GABA transaminase in salt stress tolerance. *BMC Plant Biol.* **2010**, *10*, 20. [CrossRef]
114. Mazzucotelli, E.; Tartari, A.; Cattivelli, L.; Forlani, G. Metabolism of γ-aminobutyric acid during cold acclimation and freezing and its relationship to frost tolerance in barley and wheat. *J. Exp. Bot.* **2006**, *57*, 3755–3766. [CrossRef] [PubMed]
115. Zhang, H.; Dong, J.I.; Wang, J. Research progress in membrane lipid metabolism and molecular mechanism in peanut cold tolerance. *Front. Plant Sci.* **2019**, *10*, 838. [CrossRef]
116. Hu, Y.; Jiang, Y.; Han, X.; Wang, H.; Pan, J.; Yu, D. Jasmonate regulates leaf senescence and tolerance to cold stress: Crosstalk with other phytohormones. *J. Exp. Bot.* **2017**, *68*, 1361–1369. [CrossRef] [PubMed]
117. Hu, Y.; Jiang, L.; Wang, F.; Yu, D. Jasmonate regulates the inducer of CBF expression–c-repeat binding factor/DRE binding factor1 cascade and freezing tolerance in Arabidopsis. *Plant Cell* **2013**, *25*, 2907–2924. [CrossRef]
118. Vergnolle, C.; Vaultier, M.N.; Taconnat, L.; Renou, J.P.; Kader, J.C.; Zachowski, A.; Ruelland, E. The cold-induced early activation of phospholipase C and D pathways determines the response of two distinct clusters of genes in Arabidopsis cell suspensions. *Plant Physiol.* **2005**, *139*, 1217–1233. [CrossRef]
119. Gustavsson, M.H.; Sommarin, M. Characterisation of a plasma membrane-associated phospholipase A2 activity increased in response to cold acclimation. *J. Plant Physiol.* **2002**, *159*, 1219–1227. [CrossRef]
120. Craig, W.; Lenzi, P.; Scotti, N.; De Palma, M.; Saggese, P.; Carbone, V.; Curran, N.M.; Magee, A.M.; Medgyesy, P.; Kavanagh, T.A. Transplastomic tobacco plants expressing a fatty acid desaturase gene exhibit altered fatty acid profiles and improved cold tolerance. *Transgenic Res.* **2008**, *17*, 769–782. [CrossRef] [PubMed]
121. Bernard, A.; Joubès, J. Arabidopsis Cuticular waxes: Advances in synthesis, export and regulation. *Prog. Lipid Res.* **2013**, *52*, 110–129. [CrossRef]
122. Hong, J.K.; Choi, H.W.; Hwang, I.S.; Kim, D.S.; Kim, N.H.; Choi, D.S.; Kim, Y.J.; Hwang, B.K. Function of a novel GDSL-type pepper lipase gene, CaGLIP1, in disease susceptibility and abiotic stress tolerance. *Planta* **2008**, *227*, 539–558. [CrossRef]
123. Hincha, D.K.; Neukamm, B.; Sror, H.A.; Sieg, F.; Weckwarth, W.; Rückels, M.; Schmitt, J.M. Cabbage cryoprotectin is a member of the nonspecific plant lipid transfer protein gene family. *Plant Physiol.* **2001**, *125*, 835–846. [CrossRef] [PubMed]
124. Wei, K.; Zhong, X. Non-specific lipid transfer proteins in maize. *BMC Plant Biol.* **2014**, *14*, 1–18. [CrossRef]
125. Osakabe, Y.; Yamaguchi-Shinozaki, K.; Shinozaki, K.; Tran, L.S.P. ABA control of plant macroelement membrane transport systems in response to water deficit and high salinity. *New Phytol.* **2014**, *202*, 35–49. [CrossRef]
126. Schroeder, J.I.; Delhaize, E.; Frommer, W.B.; Guerinot, M.L.; Harrison, M.J.; Herrera-Estrella, L.; Sanders, D. Using membrane transporters to improve crops for sustainable food production. *Nature* **2013**, *497*, 60–66. [CrossRef]
127. Seki, M.; Narusaka, M.; Ishida, J.; Nanjo, T.; Fujita, M.; Oono, Y.; Shinozaki, K. Monitoring the expression profiles of 7000 Arabidopsis genes under drought, cold and high-salinity stresses using a full-length cDNA microarray. *Plant J.* **2002**, *31*, 279–292. [CrossRef]
128. Chai, C.; Subudhi, P.K. Comprehensive analysis and expression profiling of the OsLAX and OsABCB auxin transporter gene families in rice (Oryza sativa) under phytohormone stimuli and abiotic stresses. *Front. Plant Sci.* **2016**, *7*, 593. [CrossRef]
129. Yue, R.; Tie, S.; Sun, T.; Zhang, L.; Yang, Y.; Qi, J.; Yan, S.; Han, X.; Wang, H.; Shen, C. Genome-wide identification and expression profiling analysis of ZmPIN, ZmPILS, ZmLAX and ZmABCB auxin transporter gene families in maize (*Zea mays* L.) under various abiotic stresses. *PLoS ONE* **2015**, *10*, e0118751. [CrossRef] [PubMed]
130. Noiraud, N.; Maurousset, L.; Lemoine, R. Transport of polyols in higher plants. *Plant Physiol. Biochem.* **2001**, *39*, 717–728. [CrossRef]
131. Lu, P.; Magwanga, R.O.; Kirungu, J.N.; Hu, Y.; Dong, Q.; Cai, X.; Liu, F. Overexpression of cotton a DTX/MATE gene enhances drought, salt, and cold stress tolerance in transgenic Arabidopsis. *Front. Plant Sci.* **2019**, *10*, 299. [CrossRef] [PubMed]

132. Di, C.; Yuan, J.; Wu, Y.; Li, J.; Lin, H.; Hu, L.; Zhang, T.; Qi, Y.; Gerstein, M.B.; Guo, Y. Characterization of stress-responsive lncRNAs in Arabidopsis thaliana by integrating expression, epigenetic and structural features. *Plant J.* **2014**, *80*, 848–861. [CrossRef] [PubMed]
133. Li, S.; Yu, X.; Lei, N.; Cheng, Z.; Zhao, P.; He, Y.; Wang, W.; Peng, M. Genome-wide identification and functional prediction of cold and/or drought-responsive lncRNAs in cassava. *Sci. Rep.* **2017**, *7*, 45981. [CrossRef]
134. Wang, P.; Dai, L.; Ai, J.; Wang, Y.; Ren, F. Identification and functional prediction of cold-related long non-coding RNA (lncRNA) in grapevine. *Sci. Rep.* **2019**, *9*, 6638. [CrossRef]

genes

MDPI

Review

HD-ZIP Gene Family: Potential Roles in Improving Plant Growth and Regulating Stress-Responsive Mechanisms in Plants

Rahat Sharif [1,2], Ali Raza [3,4], Peng Chen [5], Yuhong Li [2,*], Enas M. El-Ballat [6], Abdur Rauf [7], Christophe Hano [8] and Mohamed A. El-Esawi [6,*]

1 Department of Horticulture, College of Horticulture and Plant Protection, Yangzhou University, Yangzhou 225009, China; rahatsharif2016@nwafu.edu.cn
2 College of Horticulture, Northwest A&F University, Yangling 712100, China
3 Fujian Provincial Key Laboratory of Crop Molecular and Cell Biology, Oil Crops Research Institute, Center of Legume Crop Genetics and Systems Biology, College of Agriculture, Fujian Agriculture and Forestry University, Fuzhou 350002, China; alirazamughal143@gmail.com
4 Key Laboratory of Biology and Genetic Improvement of Oil Crops, Oil Crops Research Institute, Chinese Academy of Agriculture Science (CAAS), Wuhan 430062, China
5 College of Life Science, Northwest A&F University, Yangling 712100, China; pengchen@nwsuaf.edu.cn
6 Botany Department, Faculty of Science, Tanta University, Tanta 31527, Egypt; enas.elballat@science.tanta.edu.eg
7 Department of Chemistry, University of Swabi, Anbar 23430, Pakistan; mashaljcs@yahoo.com
8 Laboratoire de Biologie des Ligneux et des Grandes Cultures (LBLGC), INRAE USC1328, Université d'Orléans, 28000 Chartres, France; hano@univ-orleans.fr
* Correspondence: liyuhong73@nwsuaf.edu.cn (Y.L.); mohamed.elesawi@science.tanta.edu.eg (M.A.E.-E.)

Citation: Sharif, R.; Raza, A.; Chen, P.; Li, Y.; El-Ballat, E.M.; Rauf, A.; Hano, C.; El-Esawi, M.A. HD-ZIP Gene Family: Potential Roles in Improving Plant Growth and Regulating Stress-Responsive Mechanisms in Plants. *Genes* **2021**, *12*, 1256. https://doi.org/10.3390/genes12081256

Academic Editor: Patrizia Galeffi

Received: 6 July 2021
Accepted: 12 August 2021
Published: 17 August 2021

Abstract: Exploring the molecular foundation of the gene-regulatory systems underlying agronomic parameters or/and plant responses to both abiotic and biotic stresses is crucial for crop improvement. Thus, transcription factors, which alone or in combination directly regulated the targeted gene expression levels, are appropriate players for enlightening agronomic parameters through genetic engineering. In this regard, homeodomain leucine zipper (HD-ZIP) genes family concerned with enlightening plant growth and tolerance to environmental stresses are considered key players for crop improvement. This gene family containing HD and LZ domain belongs to the homeobox superfamily. It is further classified into four subfamilies, namely HD-ZIP I, HD-ZIP II, HD-ZIP III, and HD-ZIP IV. The first HD domain-containing gene was discovered in maize cells almost three decades ago. Since then, with advanced technologies, these genes were functionally characterized for their distinct roles in overall plant growth and development under adverse environmental conditions. This review summarized the different functions of HD-ZIP genes in plant growth and physiological-related activities from germination to fruit development. Additionally, the HD-ZIP genes also respond to various abiotic and biotic environmental stimuli by regulating defense response of plants. This review, therefore, highlighted the various significant aspects of this important gene family based on the recent findings. The practical application of HD-ZIP biomolecules in developing bioengineered plants will not only mitigate the negative effects of environmental stresses but also increase the overall production of crop plants.

Keywords: abiotic stress; biotic stress; crop improvement; HD-ZIP; plant development

1. Introduction

The genes containing the homeobox domain were discovered for the first time in *Drosophila*. This was due to the homeotic mutation, which transformed one part into another part in the *Drosophila* body [1]. Homeobox domain genes are mainly involved in controlling the growth and developmental processes such as transition through phases in an organism by encoding a certain transcription factor [2]. The additional presence of homeodomain (HD), which comprises 60 amino acid sequences and later makes a three-helix tertiary structure, supports the promoter regions to interact with specific target genes [3]. In plants, the first HD-containing gene was reported in maize (*Zea mays*), where a *Knotted1* gene was observed to control the leaf differentiation mechanism. Due to this phenotypic characteristic, the name *Knotted1* was given to this gene and perhaps the first HD family gene from plant genomes [4]. Following that, a series of discoveries reported a large set of genes-possessing HD domain and different other additional domains in a single copy of a gene [5]. These different homeobox gene families exhibit structure and functional similarities [2]. The functional importance of HD-ZIP genes has been documented in a wide range of plant species. For instance, HD-ZIP genes are involved in regulating plant architecture, organogenesis, and reproductive processes [6–8]. The aided importance of HD-ZIP genes in curbing environmental stresses is also well highlighted. For instance, most of the HD-ZIP genes in transgenic research showed pronounced effects against drought and salinity [9,10]. Apart from that, these genes respond to various other adverse conditions, including heat, heavy metals, and biotic stresses [11,12]. Therefore, the present review documented several aspects of the homeodomain leucine zipper (HD-ZIP) gene family, such as structural characteristics, interaction with other gene families, and potential in regulating plant growth, development, and responses to environmental cues.

2. Structural Characteristics of HD-ZIP Gene Family

The HD-ZIP gene family is composed of two functional domains, i.e., HD and leucine zipper (LZ). Based on their sequence conservation and functional properties, HD-ZIP is further divided into four subfamilies (HD-Zip I, HD-Zip II, HD-Zip III, and HD-Zip IV) [13,14]. The subfamily I and II genes encode a small transcription factor (TF) with a similar structure. Both the subfamily I and II consist of a highly conserved HD domain and a contrasting less conserved LZ domain [15,16]. Both class I and class II shared structure similarities; however, some elements are still varied, which differentiate between them. Such as, the HD region of class II contains two introns and three exons and encodes alpha-helixes 2 and 3, whereas, genes in class I comprised one intron at the LZ domain region or alpha-helix 1 [15]. Moreover, an additional Cys, Pro, Ser, Cys, and Glu (CPSCE) motif on the C-terminal differentiates class I from II (Figure 1). The extra motif facilitates the formation of multimeric proteins responsible for the Cys-Cys inter-molecular bond [17]. Further, the class I and II genes showed differences for their specific target sites. For example, the pseudopalindromic sequences CAATNATTG have different central nucleotides A/T and C/G in class I and II genes, respectively. According to an earlier study [18], this class-based target specificity is caused by various amino acids. The amino acids at the alpha-helix 3 (ranging between 46 and 56 nucleotides) are different for class I (ala and trip) and II (Glu and Thr). The changes in these amino acids coupled with Arg55 play a pivotal role during their interaction with DNA molecules [18]. The genes from class I and class II both interact with DNA only in the form of dimers. The strength of the interaction between HD-ZIP proteins and DNA molecules largely depends on the loop region between the first and the second α-helixes and the structure of the N-terminal [19,20].

Likewise, class III and IV genes comprised an additional steroidogenic acute regulatory protein-related lipid transfer (START) domain and a conserved SAD (START-associated domain along with HD and LZ domains). The class III family genes also contain an additional highly conserved methionine-glutamic-lysine-histidine-leucine-alanine (MEKHLA) domain. The MEKHLA domain is unique to class III subfamily genes in plants HD-ZIP gene family [2,21]. The class III MEKHLA domain shares a high similarity with the PAS domain. However, studies are limited over the potential role of the MEKHLA domain in plants [22] besides their involvement in embryo patterning and transportation of auxin [23]. The START domain (~200 amino acid residues) is involved in lipid and sterol transport in animals; however, no study reported their interaction with DNA molecules [24]. On the other hand, no clear evidence of the function of START domain in the plant genome was found. However, the protein-containing START domain could be regulated in plants by lipid/sterol-associated proteins (Figure 1). This regulation could be the outcome of the direct interaction of START domain-containing proteins with lipid/sterol proteins or by third mediator protein [25]. A study supported this notion by reporting that an HD-ZIP class IV gene regulates the phospholipid signaling in arabidopsis roots [26]. Another report concluded that the START domain is essential for the proper functioning of HD-ZIP genes in the cotton plant [27]. The research body is limited regarding class III and IV genes interaction with DNA molecules due to their polymorphic nature. The common distinctive feature of these genes is due to the presence of TAAA sequence in their target sites [2].

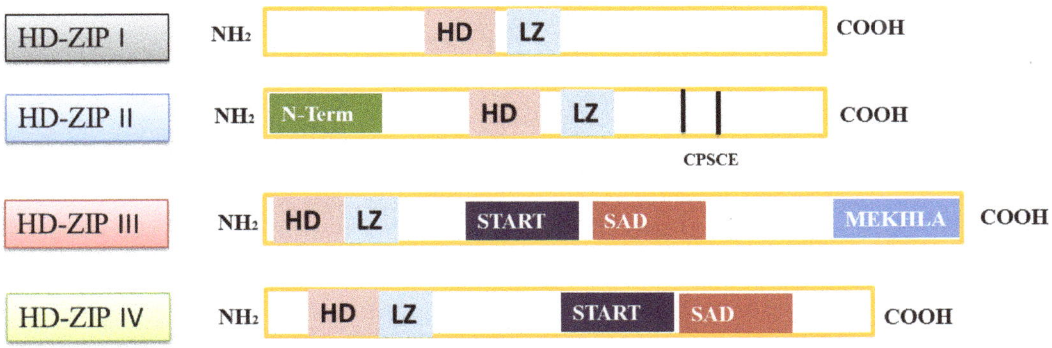

Figure 1. Schematic representation of HD-ZIP genes and their structural distribution split them into four classes (HD-ZIP I, HD-ZIP II, HD-ZIP III, and HD-ZIP IV). HD, LZ, START, SAD, and MEKHLA can be seen in the decoded form in the text.

3. Role of HD-ZIP Genes Family in Plant Growth and Regulation

Numerous HD-ZIP I genes that have evolutionary resemblance generally show the same expression pattern in various plant tissues. For instance, *ATHB1* plays a crucial role in the developmental processes of tobacco (*Nicotiana tabacum*) leaf cells [21]. The transgenic plant overexpressing *ATHB23* or *ATHB3*, *ATHB13*, and *ATHB20* fine-tuned the cotyledon and leaf development processes significantly [15,22]. The ectopic expression of the tomato (*Solanum lycopersicum*) *LeHB-1* gene disrupts the normal flowering process in the transgenic plant [23]. The study also reported that the transgenic plants also resulted in multiple flower production, an abnormal transformation of sepals into carpel and regulates the floral morphogenesis, and triggered the fruit ripening process [23]. Similarly, the grape (*Vitis*) *VvHB58* controls the fruit size, reduced the number of seeds, and hindered the pericarp expansion in the tomato fruit by modulating the multiple-hormones pathway [24].

Additionally, this HD-ZIP I TF regulates the growth and development of plants under various adverse conditions. For example, the *HDZI-4* promoter drives DREB/CBF expression under severe drought conditions, which mitigates the negative effects of drought stress and restricts the declination in yield and other growth attributes in wheat and barley [25]. Recently, Ma et al. [26] addressed the crucial role of *ATHB13* in floral induction. Flower induction at an appropriate time is crucial for seed setting, survival, and germination [27,28]. The citrus *PtHB13* is homologous to *Arabidopsis ATHB13*. The ectopic expression of *PtHB13* in *Arabidopsis* inhibited the floral induction process and could regulate the flowering-related genes [26]. Majority of the reports available on HD-ZIP I TFs suggested that they are mostly induced under abiotic stresses and thus crucial for maintaining plant growth under unfavorable environments.

There are nine genes in the *Arabidopsis* HD-ZIP II subfamily. The main role of this class in plant development is their shade-avoiding mechanism during the photosynthetic process [29–31]. For example, one member of class II subfamily *ATHB2*, when overexpressed in *Arabidopsis*, unfolded its role in plant development under illumination conditions [32]. On the other hand, microarray analysis revealed that *HAT2*, a member of the class II HD-ZIP gene family, was significantly influenced by the auxin during the seedlings stage [33]. To confirm that, *Arabidopsis* plants overexpressing the *HAT2* gene produced epinastic cotyledons, long hypocotyls, long petioles, and small leaves. All these traits resembled to the mutants, generating auxin in high quantity [33,34]. Fruit ripening is an important qualitative factor that defines the fate market value of postharvest produces. Ethylene is generally considered a potent regulator of the fruit ripening process. In this regard, the overexpression of *PpHB.G7*, a class II HD-ZIP family gene in peach (*Prunus persica*), mediates the ripening process by altering the expression and production of ethylene biosynthesis genes and ethylene, respectively [35]. In a recent study, the rice (*Oryza sativa*) *sgd2* gene was found responsible for small grain size and dwarf plant phenotype. The study further showed that the *sgd2* gene is a transcriptional suppresser of GA biosynthetic genes, particularly suppressing the generation of endogenous GA_1 [36]. The majority of the class II HD-ZIP genes that are differentially expressed in various plant tissues confer their importance in regulating plant developmental activities.

Arabidopsis genome contains five members of the class III HD-ZIP gene family. Numerous mutants of these genes have been reported previously. Most of the class III genes are responsible for sustaining the normal organ polarity and shoot apical meristem (SAM) [37]. Single loss of HD-ZIP III protein function does not display any obvious phenotypic changes. However, a double or triple mutant of class III genes such as *phb-6/phv-5/rev-9* lacked SAM along with single abaxialized cotyledon, suggesting their overlapping nature [38]. Additionally, overexpression of *Arabidopsis ATHB8* hastened the xylem formation because of the ectopic production of procambial cells [39]. In contrast, loss of function of *ATHB8* failed to show any physiological and morphological changes [39]. The *ATHB15* gained the *icu4-1* function allele, resulting in an abnormal arrangement of root meristem and more number of lateral roots production than the wild type (WT) [40]. Taken together, the aforementioned statements elucidated the crucial role of class III genes in root formation and vascular development. Another study [41] supported the notion by reporting the role of class III gene in nodule formation, root development, and vascular activities regulation. The results highlighted that *GmHD-ZIP III 2* demonstrated strong interaction with *GmZPR3d*, ensuing in the ectopic formation of secondary root xylem and also a dominant expression of soybean (*Glycine max*) vessel-specific genes [41].

The class IV HD-ZIP gene family has been previously characterized in various plants such as *Arabidopsis*, maize, and rice. These genes generally show a dominant expression trend in the outer layer of SAM and the epidermal cells [42,43]. Additionally, these genes are mainly involved in the developmental processes of stomata, trichome and epidermis, cuticle, and root hairs [2]. In line with that, two functionally redundant class IV genes, *ARABIDOPSIS THALIANA MERISTEM LAYER1* (*ATML1*) and *PROTODERMAL FACTOR2* (*PDF2*) in *Arabidopsis*, were reported for their crucial role in regulating the epidermis and embryo development and also in the patterning of floral identity [44,45]. The *TRICHOMELESS1* (GL2) gene in *Arabidopsis*, a member of the class IV gene family, has been recognized for fine-tuning the trichome and root hair development [46]. Anthocyanins are potent regulators of leaf pigments and mainly responsible for protecting chloroplast against deleterious environmental effects [47]. The *Arabidopsis ANTHOCYANINLESS2* (*AtANL2*) controls the deposition of anthocyanins, root growth and ectopic root hairs development, and also epidermal cells proliferation [48,49]. Improved root growth is significant in providing support to the plant in water-scarce conditions. The class IV gene *ATHDG11* led to the overall improvement root system in the overexpressed *Arabidopsis* transgenic plants [50,51]. Apart from *Arabidopsis*, the function of class IV HD-ZIP genes have been in other economically important crops such as rice and maize. The maize *ZmOCL1* and *ZmOCL4* have been reported to regulate cuticle deposition, kernel development, and trichome formation [42,52]. In rice, the *Roc4* gene, a member of the class IV HD-ZIP gene family, manipulates flowering time by regulating the expression of *Ghd7* gene. The results revealed that the overexpressed *Roc4* rice transgenic plants showed repressed expression of *Ghd7* under long days and thus hastened the flower induction processes [53]. Altogether, the aforementioned evidence highlighted that the class IV HD-ZIP gene family has an imposing role in plant growth and developmental activities.

4. The Crucial Role of HD-ZIP Gene Family in Regulating Abiotic Stress

4.1. Role of HD-ZIP I Subfamily in Abiotic Stress Control

Plants adopt various mechanisms to cope with numerous abiotic stresses [54,55]. The HD-ZIP class I genes are generally known for assisting with abiotic stress responses and tolerance, particularly drought, salinity, and cold stress. Thus, in the subsequent sections, we have explained the vital role of HD-ZIP genes-regulating stress-responsive mechanisms under numerous abiotic and biotic cues. Apart from the textual explanation, a large amount of literature has been tabulated and presented in Table 1.

4.1.1. Drought Stress

Drought is a major stress suffered by plants. It impairs plant physiological and biochemical functions and is considered a major threat to food security in the current time [56,57]. The *AtHB7* and *AtHB12*, two paralogous genes, induced significantly under ABA and water stress conditions by regulating stomata closure [58,59]. The *Oshox4* interacted with DELLA-like genes and further regulated the gibberellic acid (GA)-signaling pathway that confers drought stress tolerance in rice [60]. Additionally, the rice *Oshox22* showed dominant transcriptional activities under the prolonged drought stress [16]. The sunflower (*Helianthus annuus*) *Hah-4* gene was overexpressed in the maize plants to elucidate its role in mitigating the drought stress. The study revealed the crucial role of *Hah-4* gene in increasing the resistance of maize plants against drought stress without hindering the agronomic traits and colonization of root Arbuscular mycorrhizal fungi activity [61]. The accumulation of ABA in the leaf is significant and plays a key role in maintaining normal plant growth under drought stress [62]. The *Nicotiana attenuata* class I HD-ZIP gene *NaHD20,* when overexpressed, facilitates the ABA accumulation in leaf under water-scarce conditions, which also triggered the expression level of dehydration responsive genes such as *NaOSM* [62]. On the contrary, the *NaHDZ20* gene-silenced plants displayed increased susceptibility to drought stress. The reduction in the NaHDZ20-silenced plants' drought tolerance could be attributed to the suppressed expression level of dehydration responsive genes [62]. The wheat (*Triticum aestivum*) gene *TaHDZ5-6A* was overexpressed in *Arabidopsis.* The transgenic *Arabidopsis* plants generated high proline contents, better water holding capacity, and a good survival rate under drought stress than the wild-type plants [9]. This growing evidence confirmed the role of HD-ZIP I subfamily genes in maintaining plant growth under water deficit conditions.

4.1.2. Salinity Stress

Around 40 million hectares of world irrigated arable land are affected by salinity, which causes massive economic losses to the countries with the worst sodic soil [63]. Salt stress or salinity affects the plants when the soil NaCl content is more than the required amount [64,65]. The HD-ZIP I subfamily genes have been reported for their mitigatory role against salt stress in plants [59]. For example, the *AtHB1* induced strongly under salinity stress in *Arabidopsis* [15]. Similarly, the rice *OsHOX22* gene restored resistance significantly against prolonged NaCl stress by mediating the ABA signaling machinery [66]. Two genes from *Craterostigma plantagineum* (*CpHB6* and *CpHB7*) simultaneously curb the drought and salinity stress by showing an induced expression trend in roots and leaves [67]. The *GhHB1* gene has been functionally characterized in cotton (*Gossypium hirsutum*) plants. A remarkable increase in the expression activity of *GhHB1* gene was observed under 1% NaCl stress [68]. The results further revealed that the transgenic cotton plants showed enhanced resistance to salinity stress by modulating the root developmental processes [68]. The maize *ZmHDZ10* was overexpressed in rice. The transgenic rice plants hastened their resistance against salinity by triggering the production of proline while alleviated the malondialdehyde (MDA) activities in comparison to that of wild type [69]. In a recent study, the *JcHDZ07* gene was isolated from physic nut (*Jatropha curcas*) and overexpressed in the *Arabidopsis*. The transgenic *Arabidopsis* plants showed increased sensitivity to salinity stress by exhibiting higher electrolyte leakage activities, lower proline content, and hindered antioxidant activities [70]. Taken together, these results suggested the important regulatory role of the HD-ZIP I subfamily in plants against salinity stress.

4.1.3. Low-Temperature Stress

Low-temperature stress alters the photosynthetic, ions transport, and metabolic activities by directly targeting the cell fluidity [71–73]. Plants use different mechanisms and signaling pathways to deal with low-temperature stress. In this regard, HD-ZIP I subfamily genes have been characterized in various plants and yielded significant results. For example, the wheat *TaHDZipI-2* was overexpressed in barley resulted in the acclimatization of barley plants to cold conditions. The overexpressed transgenic plants also exhibited better flowering under low temperatures than the wild type [74]. Similarly, the *TaHDZipI-5* showed upregulated expression trends in flowers and grains. Further, under low temperature, *TaHDZipI-5* indicated its role in cold tolerance during the reproductive stage [75]. To confirm that, transgenic wheat plants overexpressing the *TaHDZipI-5* restore the normal flowering activities under cold stress; however, compromised agronomic and yield-related traits were observed [75]. Overexpression of the *AtHB13* gene confers cold stress tolerance by maintaining cellular stability in *Arabidopsis* plants [59]. The expression level of several glucanase, anti-freezing proteins (AFP), pathogenesis-related proteins, glucanase, and chitinase enhanced significantly in the *HaHB1* sunflower and soybean transgenic plants showed improved resistance to cold stress [76]. Therefore, it is confirmed that the HD-ZIP I subfamily genes facilitate the resistance mechanism against cold stress by triggering the expression of the cell membrane-related proteins and AFP (Figure 2).

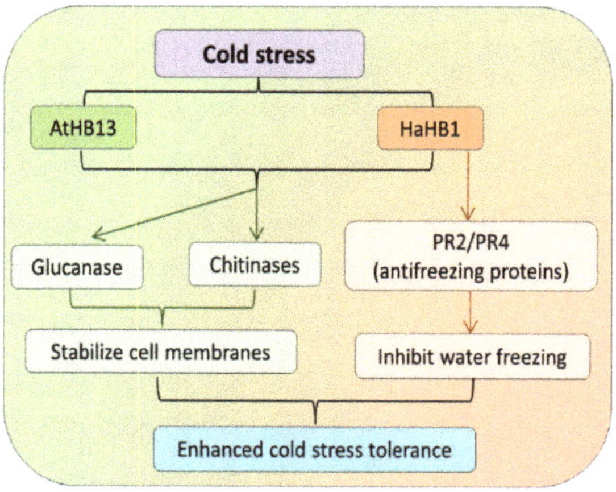

Figure 2. Role of HD-ZIP I subfamily in regulating low-temperature stress. The cold stress induces *AtHB13* and *HaHB1* gene, which further activates the transcription of chitinases, glucanase, and PR2 genes. These genes help stabilize the water transport and inhibit it from freezing inside cell membrane.

4.1.4. Heavy Metal Stress

The increasing soil pollution with heavy metals, such as cadmium, chromium, iron, lead, nickel, selenium, etc., causes toxic reactions that hamper the physiological and morphological activities of plants [77–79]. Recent studies have reported the involvement of HD-ZIP I genes in regulating heavy metals stress. In *Citrus sinensis*, for example, the cDNA-AFLP methodology revealed that two genes from HD-ZIP I subfamily (*TDF #170-1* and *170-1k*) enhanced significantly under manganese (Mn) toxicity, suggesting their possible role in Mn stress tolerance [80]. Based on this, it could be of high interest to elucidate the role of these genes under various important toxic heavy metals.

4.1.5. Heat Stress

The rise in global temperature is becoming increasingly challenging to crop scientists as heat stress causes early maturity of the plants and subsequent manifold reduction in overall yield [55,81,82]. Expression-based analysis in cucumber (*Cucumis sativus*) suggested that two members (*CsHDZ02* (*Csa1G045550*) and *CsHDZ33* (*Csa6G499720*)) of HD-ZIP subfamily I showed induced expression pattern under heat stress [83]. The sunflower *HaHB4* gene has been functionally characterized in soybean plants under field conditions [11]. The transgenic soybean plants overexpressing *HaHB4* genes exhibited better tolerance capacity to heat stress by triggering the transcriptional activity of heat shock proteins (*AT-HSC70-1*, *AT-HSFB2A*, and *Hsp81.4*) [11]. On the other hand, *HaHB4* transgenic plants recorded better yield by reducing heat stress damage during seed setting in soybean pods [11]. The perennial ryegrass (*Lolium perenne*) is generally regarded as heat-sensitive because of its temperate growth nature [84]. Perennial ryegrass is mostly grown for turf or forage purposes; however, increasing temperature due to global warming hampered its production manifold [84]. The HD-ZIP I subfamily gene *LpHOX21* possessed upregulated expression in the heat-tolerant cultivar of perennial ryegrass, which suggested its possible involvement in enhancing resistance to heat stress [84]. Although the HD-ZIP subfamily I genes are well characterized under other abiotic stresses, still, relatively less research is available regarding their role in mitigating heat stress.

4.1.6. Flooding Stress

Flooding stress refers to the plant's submergence, which creates an anaerobic condition in the surroundings and affects plant productivity [85,86]. The HD-ZIP I subfamily gene *HaHB11* was overexpressed in the *Arabidopsis* and exposed to flooding stress [87]. The transgenic *Arabidopsis* plants carrying gain of function *HaHB11* gene induced the tolerance to flood stress and increased the biomass and yielded more seeds than control [87]. Flooding is becoming a serious threat due to climate change, and therefore the HD-ZIP TFs could be utilized to generate flooding resistance cultivars.

4.1.7. Nutrient Stress

Excess or deficiency of plant nutrients in the soil is generally regarded as nutrient stress. The transition heavy metals such as manganese, zinc, copper, and iron are essential micronutrients for regulating the plant's growth and developmental activities [88,89]. Iron in a relatively small amount is considered an important nutrient and involves key regulatory processes (chlorophyll biosynthesis and photosynthesis) of plant development [90]. Higher plants solubilize the ferric iron in the rhizosphere region, facilitating the uptake of iron efficiently [91]. The HD-ZIP I subfamily member gene *AtHB1* was previously reported for its involvement in iron homeostasis [92]. The lack of function *athb1* gene showed strong tolerance to iron deficiency by upregulating the expression of *Iron-Regulated Transporter1* (*IRT1*) and exhibited higher chlorophyll contents than the control [92]. In contrast to that, the overexpression of *AtHB1* genes suppressed the transcription activity of *IRT1* genes, which hampered the plant's iron regulation, indicating a crucial role of the *AtHB1* gene in iron homeostasis [92]. This suggested the importance of HD-ZIP I subfamily genes in maintaining the uptake and translocation of iron and other essential nutrients and could be used as a genetic tool to improve crop productivity and nutrient efficiency.

4.2. Role of HD-ZIP II Subfamily in Abiotic Stress Control

4.2.1. Drought Stress

The HD-ZIP II subfamily is renowned for providing resistance against important abiotic stresses such as drought, cold, and salinity stress. The *SiHDZ13* and *SiHDZ42* showed upregulated transcriptional activity under prolonged drought stress in the sesame

(*Sesamum indicum*) plant [93]. In another study, the expression of wheat *Tahdz4-A* strongly increased under drought stress, conferring its responsive nature to this important abiotic stress [94]. Similarly, in *Arabidopsis*, increased mRNA level of *HAT2* and *HAT22* genes was observed under water deficit conditions [95]. Eucalyptus is an industrial plant and generally used for paper and timber production [96]. However, its production has been affected and reduced significantly by water scarcity [97]. The gain of function *EcHB1* gene significantly boosted the photosynthetic capacity, which increased the number of chloroplast unit per leaf area under drought stress in transgenic eucalyptus plants [98]. Numerous expression studies suggested the importance of HD-ZIP II subfamily genes in regulating drought stress. However, the smaller number of functional studies encourages future research over HD-ZIP II subfamily genes in various important plants.

4.2.2. Light Stress

The vast number of HD-ZIP II subfamily genes across different plant species has been reported to respond to light stress, and shade avoidance in particular [99]. The *AtHB2/HAT4* is strongly induced under the dark condition in the etiolated seedlings [100]. To confirm their function, transgenic lines overexpressing *AtHB2/HAT4* produced longer hypocotyls [101]. This indicated that *AtHB2/HAT4* is responsible for controlling the growth of seedlings in fluctuating light conditions. Additionally, the *AtHB2* (protein) directly interacts with PIF proteins because the expression was completely lost in *pif4, pif5,* and *pifq* [102–104]. The ectopic overexpression of various HD-ZIP II subfamily members could phenocopies the positive shade avoidance effects over other organs such as flowers [31]. Light stress is controlled by multiple pathways and, therefore, further studies are required to unfold the potent role of HD-ZIP II subfamily genes in plants.

4.2.3. Salinity Stress

The HD-ZIP II subfamily has been examined extensively in different plants under salinity stress. However, functional characterization of these genes under NaCl stress is far little compared with subfamily I genes. The tea (*Camellia sinensis*) *CsHDZ15* and *CsHDZ16* increased significantly throughout the stress period, suggesting that they are involved in responding to salinity stress [105]. The *StHOX17, StHOX20,* and *StHOX27* genes possessed dominant expression under the saline condition in potato (*Solanum tuberosum*) plants [106]. Additionally, the *Capsicum annum* (*CaHB1*) showed an enhanced expression trend under various stresses, including salt. To verify its role, the *CaHB1* gene was overexpressed in tomato plants. The transgenic tomato plants displayed improved resistance against NaCl stress. Moreover, the transgenic tomato plants developed better agronomic traits than the wild type [10]. These results imply the beneficial roles of HD-ZIP II subfamily genes in mitigating the salinity stress and could be useful in future crop-breeding programs.

4.3. Role of HD-ZIP III Subfamily in Abiotic Stress Control

4.3.1. Drought Stress

The members of HD-ZIP III subfamily are mainly involved in the leaf-rolling mechanism of plants. Leaf rolling is an important factor that provides assistance to plants under water deficit conditions. The HD-ZIP III subfamily genes are the major target genes of *miRNA165/166*. In line with that, rice miRNA166 loss-of-function mutant (STTM166) developed rolled leaf phenotype because of damaged sclenrenchymatous cells along with abnormal bulliform cells [107]. The molecular dissection of the STTM166 mutant revealed that the *OsHB4* gene is targeting the *miRNA166*, a member of the class III HD-ZIP gene family [107]. The *miRNA166* STTM166 mutant lines showed enhanced resistance to drought stress. To validate that, *OsHB4* overexpressed transgenic rice plants influenced the expression of polysaccharide synthesis genes, which facilitates the cell wall and vascular developmental activities and also imposed the rolled leaf phenotype conferred tolerance to drought stress [107]. In another study, the *Arabidopsis miRNA160/166* double mutant displayed enhanced resistance under water-scarce conditions by influencing the expression of auxin-related genes and exhibiting rolled leaf phenotype [108]. Based on the above-mentioned shortcomings, it can be concluded that the HD-ZIP III subfamily genes also possessed drought stress-responsive factors and, therefore, can be considered to characterize in different plants apart from model species.

4.3.2. Salinity Stress

Studies based on expression analysis suggested that HD-ZIP III subfamily genes are responsive to salinity stress. For example, the wheat HD-ZIP III genes *Tahdz1* and *Tahdz23* both induced under NaCl stress [94]. The *MtHDZ5, MtHDZ13,* and *MtHDZ22* showed differential expression under 180 mM and 200 mM NaCl stress in 2-week old seedlings of *Medicago truncatula* [109]. However, research is required to elucidate the functional role of these genes under saline conditions in numerous plant species.

4.3.3. Heavy Metal Stress

Heavy metal toxicity is consistently hampering plant productivity due to the increasing environmental pollution. They generally compromised plants' physiological and molecular pathways and caused irreparable damage [78,79]. Among them, cadmium (Cd) is a highly toxic metal that has been reported for causing yield losses in various plant species [78,110–112]. The application of Cd induced the expression of rice *OsHB4,* whereas the miRNA166 was deduced under Cd treatment [12]. This suggested the possible involvement of this gene in regulating Cd stress in rice. To confirm this, an overexpression assay was performed for miRNA166. The overexpression of miRNA significantly reduced the transcriptional activity of *OsHB4* in root and leaf tissue. On the other hand, the miRNA166 was strongly induced in both the root and leaf and hindered the Cd translocation from root to stem [12]. Additionally, the accumulation of Cd in the rice grain was also arrested in the miRNA166 overexpressed transgenic lines. On the contrary bases, the overexpression of *OsHB4* made the root and leaf more sensitive to Cd toxicity, whereas RNAi silencing of *OsHB4* made the transgenic plants tolerant to Cd stress [12]. These evidences clearly suggest that the induced expression of *OsHB4* increased the rice plant's sensitivity to Cd stress. Furthermore, the majority of the HD-ZIP III subfamily genes showed pronounced expression during root development-related activities [113,114]. Therefore, this could be vital in providing stress response to heavy metals.

4.4. Role of HD-ZIP IV Subfamily in Abiotic Stress Control

4.4.1. Drought Stress

The HD-ZIP IV subfamily genes have been recently characterized in many plants to induce drought stress tolerance in plants. The gain of function *OsHDG11* gene (a member of HD-ZIP IV) enhanced the overall yield and drought stress tolerance mechanism in rice plants. The transgenic rice plants overexpressing *OsHDG11* significantly influenced the root system, improved water holding capacity, triggered the proline content, and enhanced endogenous ABA production [51]. Similar results were found when the *AtEDT1/HDG11* gene was overexpressed in Chinese kale; however, altered endogenous ABA imposed stomatal closure [115]. Lignification in the plant is associated with an array of abiotic stress tolerance. The *Oryza sativa transcription factor I-like* (*OsTFIL*) gene has been reported for its beneficial role in providing tolerance against drought stress in rice. The transgenic rice plants carrying *OsTFIL* gene showed enhanced lignin accumulation in shoot tissue than the RNAi or WT plants [116]. Additionally, the high transcriptional activity of lignin biosynthesis genes also facilitates stomatal closure under drought stress [116]. This *HDG11* gene was further reported in cotton to induce water use efficiency (WUE) and improved stress tolerance [117]. A recent study investigated the genetic pathway of the *HDG11* gene on how it facilitates the WUE and tolerance of a plant to water stress conditions. The study unfolded that the genetic pathway consists of *EDT1/HDG11, ERECTA,* and *E2Fa* loci. Initially, *ERECTA* become transcriptionally activated by binding with HD element in its promoter region. *ERECTA* then modulates the transcription of cell-cycle pathway genes, which further helps in the transition of mitosis into endocycle. This mechanism positively affected the leaf cell size by triggering the ploidy level, which in turn altered the stomatal density [118]. The reduced density of stomata modulates the WUE system of plants and thus provides resistance against drought stress. Other members of this subfamily also showed a response to drought stress, such as that in *Nicotiana tabacum*. The *NtHD-ZIP IV 4* and *NtHD-ZIP IV 10* displayed a dominant expression trend under prolonged drought stress conditions [119].

4.4.2. Salt Stress

Cotton crop, although known as a moderate salt-tolerant crop, is still affected by salinity [63]. Salt stress causes a substantial delay in flowering, which implies less fruiting and decreased cotton ball weight [63,120]. The effect of salt stress is more pronounced during the germination and seedling stage of cotton [63,120]. The HD-ZIP IV gene *AtEDT1/HDG11* restored the cotton plant resistance to salt stress by the induction of proline and soluble sugar contents along with an improved antioxidant enzymes system [117]. Remarkably, the transgenic cotton plants showed no compromised agronomic traits and thus yielded more numbers of cotton balls per plant than the wild type [117]. Exogenous application of jasmonic acid (JA) on plants ameliorates the deleterious effects of many abiotic stresses, including salt stress [121,122]. The *EDT1/HDG11* gene was overexpressed in *Arabidopsis,* which hastened the transcriptional level of numerous JA biosynthetic genes and influenced the formation of lateral root significantly by activating the auxin signaling pathway [123]. The endogenous JA level was also high in the roots of transgenic plants [123]. The above statement suggested that the *EDT1/HDG11* transgenic plants could be resistant to multiple environmental stresses, including salt stress.

4.4.3. Osmotic Stress

Osmotic stress dysfunction affects plants' normal physiological processes by disturbing the transport of ion and water [124]. The cotton *GaHDG11* gene was overexpressed in the *Arabidopsis* plant. The transgenic *Arabidopsis* plants showed better performance under osmotic stress because of the high generation of osmoprotectants such as proline, enhanced antioxidant activities, and elongated roots [125]. The elongation of primary roots supports the plant by lowering the rate of water loss [125]. Due to these noticeable functional characteristics, more research is required to functionally elucidate the role of HD-ZIP IV subfamily genes under osmotic stress.

Table 1. The HD-ZIP family genes and their potential role in providing resistance against abiotic stresses.

Stress Control	Plant Species	Gene	Functions	References
			Subfamily I	
Drought	*Arabidopsis thaliana*	*AtHB7*	Overexpression of *AtHB7* regualte the expression of drought stress-specific genes.	[126]
	Arabidopsis thaliana	*AtHB13*	The *AtHB13* work upstream of the *JUB1* gene to confer drought stress.	[127]
	Arabidopsis thaliana	*HaHB11*	The *HaHB11* transgenic plants closed their stomata faster and lost less water than controls.	[128]
	Alfalfa	*HaHB11*	Longer roots and rolled leaves in *HaHB11* transgenic *alfalfa* plant.	[128]
	Arabidopsis thaliana	*AtHB12*	Nullify the negative effects of ABA signaling genes (*PYL5* and *PYL8*).	[129]
	Oryza sativa	*OsHOX4*	The *OsHOX4* modulate GA signaling by interacting with DELLA-like genes and GA oxidase genes.	[60]
	Oryza Sativa	*OsHOX22*	Higher expression of *OsHOX22* gene under drought stress.	[16]
	Nicotiana attenuate	*NaHD20*	Augmented ABA accumualtion in leaf.	[62]

Table 1. *Cont.*

Stress Control	Plant Species	Gene	Functions	References
	Triticum aestivum	*TaHDZ5-6A*	*TaHDZ5-6A* transgenic plants displayed enhanced drought tolerance by lowering the water loss rates, higher survival rates, and higher proline contents.	[9]
Salinity	*Arabidopsis thaliana*	*HaHB11*	Higher expression of salt stress-related genes.	[128]
	Alfalfa	*HaHB11*	Strong root activities.	[128]
	Oryza sativa	*OSHOX22*	Regualted ABA signaling.	[66]
	Physic nut	*JcHDZ07*	Overexpression of *JcHDZ07*-induced sensitivity to salinity stress.	[70]
	Zea mays	*ZmHDZ10*	Lower relative electrolyte leakage (REL), lowee MDA and increased proline content in overexpressed *ZmHDZ10* transgenic plant.	[69]
Heat stress	*Soybean*	*HaHB4*	*HaHB4* transgenic plant possesses larger xylem area, and increased water use efficiency under high temperature stress.	[11]
	Perennial ryegrass	*LpHOX21*	Higher expression of *LpHOX21* gene was recorded in heat-tolerant cultivar.	[84]
Heavy metal (manganese)	*Citrus sinensis*	*TDF #170-1, 170-1k*	Induced expression of these genes were observed under heavy metal stress.	[80]
Cold stress	*Triticum aestivum*	*TaHDZipI-2*	Frost toelrance-related genes were upregualted in *TaHDZip1-2* overxpressed plants.	[74]
	Triticum aestivum	*TaHDZipI-5*	Induction in lipid biosynthesis genes induced cold tolerance.	[75]
	Arabidopsis thaliana	*AtHB13*	Higher antioxidant activities of *AtHB13* transgenic plants.	[59]
	Arabidopsis thaliana	*AtHB1/HaHB1*	Induction of pathogenesis-related and glucanase proteins.	[76]
Flooding stress	*Arabidopsis thaliana*	*HaHB11*	Modulation of genes genes involved in glycolisis and fermentative pathways.	[87]
Nutrient stress (iron deficiency)	*Arabidopsis thaliana*	*AtHB1*	Overexpression of *AtHB1* regualtes iron homeostasis.	[92]
			Subfamily II	
Drought	*Sesame*	*SiHDZ13, SiHDZ42*	Higher expression under drought stress.	[93]
	Triticum aestivum	*Tahdz4-A*	Upregualted mRNA level under drought stress.	[94]
	Eucalyptus	*EcHB1*	Increased the leaf photosynthesis.	[98]
	Arabidopsis thaliana	*HAT2, HAT22*	High response to hormonal treatment.	[95]
Salinity	*Camellia sinensis*	*CsHDZ15, CsHDZ16*	Augmented expression under salinity stress.	[105]
	Solanum tuberosum	*StHOX17, StHOX20, StHOX27*	Higher expression under salinity stress.	[106]
	Capsicum annum	*CaHB1*	Upregulation of multiple genes involved in plant osmotic stress resistance.	[10]
Light stress	*Arabidopsis thaliana*	*AtHB2/HAT4*	Stimualted expression of phytochrome genes in overexpressed *AtHAT4* gene transgenic plant.	[100,130]
			Subfamily III	
Drought	*Oryza sativa*	*OsHB4*	LeafrRolling and altering stem xylem development.	[107]
	Triticum aestivum	*Tahdz1, Tahdz23*	Induced mRNA level under salinity stress.	[94]
Salinity	*Medicago truncatula*	*MtHDZ5, MtHDZ13, MtHDZ22*	Higher expression under salinity stress.	[109]
Cadmium stress	*Oryza sativa*	*OsHB4*	Silencing of *OsHB4* gene reduced Cd accumulation in the leaves and grains.	[12]

Table 1. *Cont.*

Stress Control	Plant Species	Gene	Functions	References
			Subfamily IV	
Drought	*Oryza sativa*	*OsHDG11*	Transgenic rice plants had higher levels of abscisic acid, proline, soluble sugar, and reactive oxygen species-scavenging enzyme activities under stress.	[51]
	Chinese kale	*AtEDT1/HDG11*	Induced stomatal closure.	[115]
	Gossypium herbaceum	*HDG11*	Augmeneted proline content, soluble sugar content, and activities of reactive oxygen species-scavenging enzymes.	[117]
	Nicotiana tobaccum	*NtHD-ZIP IV 4, NtHD-ZIP IV 10*	Higher expression under drought stress.	[119]
	Nicotiana tobaccum	*NtHDG2*	Induced flavonoid biosynthesis.	[131]
	Oryza sativa	*OsTFIL*	Promotes lignin biosynthesis and stomatal closure.	[116]
Salinity	*Gossypium herbaceum*	*AtEDT1/HDG11*	Better proline content, soluble sugar content.	[117]
	Arabidopsis thaliana	*EDT1/HDG11*	Promotes lateral root formation in *Arabidopsis* mutant edt1 by upregulating jasmonate biosynthesis.	[123]
	Nicotiana tobaccum	*NtHDG2*	Higher antioxidant activities.	[131]
Osmotic	*Arabidopsis thaliana*	*GaHDG11*	Upregualted expression level was observed under osmotic stress.	[125]

5. Role of HD-ZIP Gene Family in Regulating Biotic Stress

Climate change made not only abiotic stresses but also biotic stresses more challenging for plant scientists. Often, fluctuations in temperature or water stress directly trigger biotic stressors' negative response and do irreversible damage to the plants [132,133]. The positive roles of HD-ZIP genes in mitigating abiotic stresses have been discussed above. Besides, the HD-ZIP genes could play a powerful role in amending the deleterious effects of biotic stresses (Table 2). In this context, these myriad biomolecules could be utilized to curb the simultaneous stresses (biotic and abiotic). Below, we discussed the roles of HD-ZIP genes in arming the plants against biotic stresses.

5.1. HD-ZIP I: Role in Coping Biotic Stress

Biotic stresses generally affect the plant morphologically and physiologically, which can be challenging to control at times [134–136]. For example, the powdery disease infecting numerous crops worldwide cost millions of dollar to the economy [137]. The HD-ZIP I subfamily member *AtHB13* increased *Arabidopsis* plants' resistance to powdery mildew fungi by regulating the expression of many stress-specific TFs. In contrast, the silencing of *AtHB13* increased the sensitivity of *Arabidopsis* to powdery mildew disease [138]. These results supported the notion that *AtHB13* might be involved in providing resistance against simultaneous abiotic and biotic stresses [138]. The *HAHB4* expression is strongly induced under the herbivores attack or jasmonic acid (JA) treatment. The induced expression produced green leaf volatiles and trypsin protease inhibitors (TPI). The overexpression of *HAHB4* in *Zea mays* and *Arabidopsis* triggered the transcript level of stress-related genes such as lipoxygenase and TPI. The lipoxygenase and TPI genes in plants provide a protective response to *Spodoptera littoralis* or *Spodoptera frugiperda* larvae [139]. Additionally, the transgenic plants overexpressing *HAHB4* generated a higher amount of JA, JA-isoleucine, and ethylene (ET), which lead us to assume that this gene could enhance the resistance against biotic stress casual agents [139]. The *Verticillium dahlia* is a fungal pathogen that is responsible for vascular wilt disease in a plethora of plant species, including cotton. JA has been previously reported for enhancing the resistance of cotton plants to *Verticillium dahlia* [140]. In line with that, the overexpression of the *GhHB12* gene suppressed the transcriptional activities of JA biosynthesis and responsive genes (*GhJAZ2, GhPR3*). It thus made the cotton plant more susceptible to *Verticillium dahlia* fungus [141]. Minimal research is available on the role of HD-ZIP I subfamily genes in mitigating biotic stresses in comparison to abiotic stresses. However, it could be of great interest to functionally characterize these genes under various biotic stresses.

5.2. HD-ZIP II: Role in Coping Biotic Stress

The HD-ZIP II subfamily members are investigated under various biotic stress and showed differential expression patterns in several plant species. The *Phytophthora infestans* (*P. infestans*) is a bacterial pathogen, which causes the late blight disease particularly in potato and tomato, and becomes a major challenge for many crop producers around the world [142,143]. The potato *StHOX28* and *StHOX30* exhibited high expression under the P. *infestans* stress. This suggested their responsive behavior toward biotic stresses [106]. The *Phytophthora capsici* (*P. capsici*) is a multi-host fungus pathogen with more drastic effects on Solanaceae (pepper and tomato) and Cucurbitaceae (cucumber and pumpkin) [144,145]. The overexpression of *Capsicum annuum* HD-ZIP II gene *CaHB1* in tomato increased the thickness of cell wall and cuticle layer, enhanced expression of defense genes (*SlPR1, SlGluA, SlChi3,* and *SlPR23*), and larger cell size than the control plants conferred tolerance to *P. capsici* [10]. Therefore, HD-ZIP II subfamily genes could be considered for potential crop improvement in the future.

5.3. HD-ZIP III: Role in Coping Biotic Stress

Expression analysis-based studies revealed that the HD-ZIP III genes of potato *StHOX7, StHOX16, StHOX26,* and *StHOX38* showed upregulated expression trend under *P. infestans* stress [106]. The *Arabidopsis AtHB8* genes induced significantly at 5 and 7 days post-inoculation (dpi) of root-knot nematode (RKN) *Meloidogyne incognita* [146]. The *AtHB8* plays an important role in the root developmental activities and, therefore, could be a potential candidate gene in providing a gateway to RKN to form gall around the root [146]. The *PHB* and *PHV* genes of the *Arabidopsis* class III family are responsible for the upward curled leaf phenotype. Similar characteristics were shown by the plants when treated with *Tomato yellow leaf curl China virus* (TYLCCNV) [147]. The results were confirmed in βC1 (pathogenesis protein) overexpressing transgenic plants, which showed an increase in the mRNA level of *PHB* and *PHV* genes while suppressed the expression of miRNA166 [147]. Therefore, it can be suggested that *PHB* and *PHV* play a crucial role in regulating the response of plants to TYLCCNV. However, no conclusive evidence is available to confirm the role of *PHB, PHV,* and other members of HD-ZIP III genes under TYLCCNV or other biotic stress casual agents.

5.4. HD-ZIP IV: Role in Coping Biotic Stress

The cuticle layer in plants provides support against many abiotic stresses. Several reports also highlighted that these cuticle layer films around plant cells serve as the first line of defense against pathogen attack [148–150]. The activation of the HD-ZIP IV gene *AaHD8* strongly induced the expression of cuticle development-related genes and significantly affected cuticle formation processes in the Artemisia annua plant [151]. The study also revealed that *AaHD8* interacts with the *AaMIXTA1* gene (regulator of cuticle formation), modulating the *AaHD1* transcription and regulating a network of other cuticle developmental genes [151]. The phenols present inside a trichome generally provide a chemical barrier to the invading pathogen and protect the plant from drastic damage, particularly from chewing pests, such as herbivores [152]. The HD-ZIP IV gene *AaHD8* and *CmGL* were reported for their potent role in trichome formation and development in Artemisia annua and melon plants, respectively [151,153]. The *ZmOCL1* (member of HD-ZIP IV) gene was overexpressed in *Zea mays*. The transgenic maize plants showed induction in the expression of *LIPID TRANSFER PROTEIN TYPE 2* (*nsLTPII*), *CARBOXYLESTERASE* (*AtCXE-18*), and *PHOSPHATIDYL INOSITOL TRASNPORT PROTEIN* (*SEC14*) [42]. Among them, the *LTPII* is crucial in the transportation of cuticle lipids across the cell wall [154]. These

LTP genes were also reported to increase resistance against a plethora of biotic stresses. These proteins belong to the plant defensins family and exhibit remarkable antifungal and antibacterial ability [155,156]. These genes are generally expressed in the outer layer or epidermis [156,157], the same as the HD-ZIP IV subfamily genes. The durum wheat *TdGL7* gene under wounding stress elevated significantly in the grain tissue, similarly to the defensins genes [158]. This provides potential grounds for the biotechnological manipulation of the *TdGL7* gene in wheat to protect the grain from chewing insects or fungi [158]. Therefore, it can be assumed that the HD-ZIP IV subfamily genes could be indirectly involved in regulating biotic stresses (Figure 3). Moreover, the potato *StHOX21* and *StHOX42* increased manifold under *P. infestans* stress [106]. However, no functional study is available to confirm the direct involvement of HD-ZIP IV subfamily genes in increasing tolerance against biotic factors.

Table 2. List of functionally characterized HD-ZIP family genes under biotic stress.

Subfamilies	Plant	Gene	Pathogen	Functions	Reference
Subfamily I	*Arabidopsis thaliana*	*AtHB13*	Powdery mildew (*Odium neolycopersici*), downy mildew (*Hyaloperonospora arabidopsidis*)	Overexpression of *AtHB13* stimualted the expression of various defense related genes.	[138]
	Zea mays	*HaHB4*	*Spodoptera littoralis*	Modulate signals from the jasmonic acid and ethylene pathways.	[139]
	Gossypium hirsutum	*GhHB12*	*Verticillium dahliae*	Increased susceptibility of the cotton plant via suppression of the jasmonic acid (JA)-response genes *GhJAZ2* and *GhPR3*.	[141]
Subfamily II	*Solanum tuberosum*	*StHOX28*, *StHOX30*	*Phytophthora infestans*	Induced expression pattern under *Phytophora infestans*.	[106]
	Capsicum annuum	*CaHB1*	*Phytophthora capsici*	Overexpression of *CaHB1* in tomato resulted in a thicker cell wall.	[10]
Subfamily III	*Solanum tuberosum*	*StHOX7*, *StHOX16*, *StHOX26*, *StHOX38*	*Phytophthora infestans*	Induced expression pattern under *Phytophora infestans*.	[106]
	Arabidopsis thaliana	*AtHB8*	*Meloidogyne incognita*	The promoters of procambial marker gene *ATHB8* were activated in *M. incognita*-induced galls.	[146]
	Arabidopsis thaliana	*PHB*, *PHV*	TYLCCNV	Suppress selective jasmonic acid responses.	[147]
Subfamily IV	*Zea mays*	*ZmOCL1*	*Pseudomonas syringae*	Overexpression of *ZmOCL1* induced antifungal activity of a lipid transfer proteins.	[42,159,160]
	Solanum tuberosum	*StHOX21*, *StHOX42*	*Phytophthora infestans*	Induced expression pattern under *Phytophora infestans*.	[106]

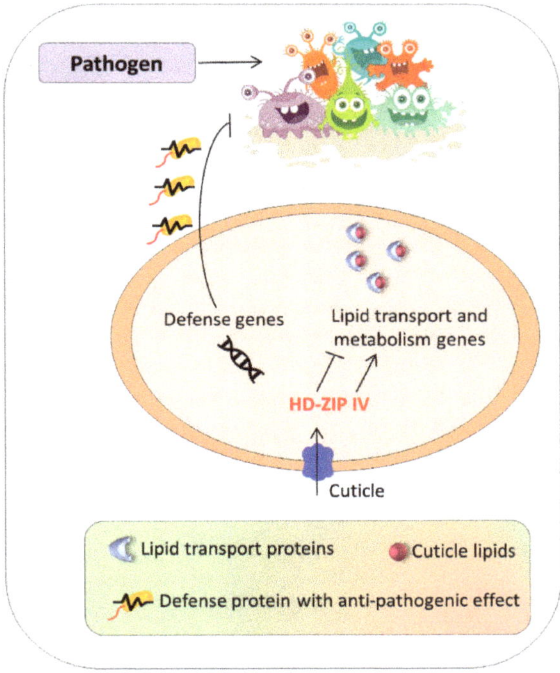

Figure 3. Indirect involvement of HD-ZIP IV subfamily in enhancing the resistance to biotic stress. The HD-ZIP IV genes participate in the activation of cuticle formation and defensins genes. The induction/suppression of lipid transport and metabolism genes largely depends on the HD-ZIP IV genes. The majorities of these genes reside in the epidermis and work synergistically in responding to pathogens.

6. Conclusions and Future Perspective

The HD-ZIP is an important gene family involved in the diverse roles of plant growth and developmental activities. Apart from their role in plant growth, several studies proved the potential of HD-ZIP genes in enhancing plant tolerance to various abiotic and biotic stresses. For example, the HD-ZIP I subfamily genes are involved in responding to drought and salinity stress in particular, whereas significant results were achieved in transgenic plants against various biotic stresses. The HD-ZIP II subfamily genes protect the plants from the deleterious effects of a shade. A member gene of HD-ZIP II subfamily CaHB1 was also reported for providing resistance against *P. capsici* and salt stress. The HD-ZIP III subfamily genes are characterized mainly under drought stress; meanwhile, another study [12] showed that the silencing of the *OsHB4* gene induced the plant immunity against Cd stress. Additionally, the HD-ZIP IV subfamily genes are mainly expressed in the outer cell membrane and provide the first line of defense against different environmental stresses.

Further investigations are still required to characterize the function of these important TFs under numerous abiotic (heat, heavy metals, flooding, and nutrient imbalance) and biotic (powdery mildew) stresses. Expression-based studies suggested their responsive role against heat and powdery mildew stress [87]. Moreover, another study [96] highlighted the crucial role of the *AtHB1* gene in iron homeostasis. Therefore, it could be of great importance to examine the role of other members under nutrient starvation and homeostasis. Heavy metals such as Cd are increasing in the soil due to the massive industrial waste and mineralization of rocks. The uptakes of these heavy metals by major food crops are harmful not only to plant but also to human health. The HD-ZIP III subfamily gene OsHB4, when silenced, significantly reduced the Cd accumulation in rice grain. Therefore, it can be used as a potential biomarker to curb the toxic effects of Cd on plants and humans. Altogether, the genetic manipulation of HD-ZIP genes could be a handful strategy to maximize the crop yield under the looming threat of climate change using state-of-the-art genome-editing tools like the CRISPR/Cas system.

Author Contributions: Y.L. conceived the idea and supervised the manuscript. R.S., A.R. (Ali Raza), and M.A.E.-E. collected the literature. R.S. composed and wrote the manuscript. A.R. (Ali Raza), E.M.E.-B., and M.A.E.-E. contributed in writing and prepared the figures and tables. Y.L., P.C., E.M.E.-B., A.R. (Abdur Rauf), C.H. and M.A.E.-E. critically reviewed, revised, and edited the manuscript. All authors have read and agreed to the published version of the manuscript.

Funding: This research was funded by the National Natural Science Foundation of China (31772300 and 31471891), the Shaanxi Province's Major research and Development Projects (2019TSLNY01-04) and the National Key R&D Project, China (2016YFD0101705). This research work was also supported by Tanta University, Egypt.

Institutional Review Board Statement: Not applicable.

Informed Consent Statement: Not applicable.

Data Availability Statement: Not applicable.

Acknowledgments: We thank Abdullah Shalmani and Izhar Muhammad (State Key Laboratory of Crop Stress Biology in Arid Areas, College of Life Sciences, Northwest A&F University) for critically reviewing the manuscript.

Conflicts of Interest: The authors declare no conflict of interest.

References

1. Garber, R.; Kuroiwa, A.; Gehring, W.J. Genomic and cDNA clones of the homeotic locus Antennapedia in Drosophila. *EMBO J.* **1983**, *2*, 2027–2036. [CrossRef]
2. Ariel, F.D.; Manavella, P.A.; Dezar, C.A.; Chan, R.L. The true story of the HD-Zip family. *Trends Plant Sci.* **2007**, *12*, 419–426. [CrossRef]
3. Moens, C.B.; Selleri, L. Hox cofactors in vertebrate development. *Dev. Biol.* **2006**, *291*, 193–206. [CrossRef]
4. Vollbrecht, E.; Veit, B.; Sinha, N.; Hake, S. The developmental gene *Knotted-1* is a member of a maize homeobox gene family. *Nature* **1991**, *350*, 241–243. [CrossRef] [PubMed]
5. Bharathan, G.; Janssen, B.-J.; Kellogg, E.A.; Sinha, N. Did homeodomain proteins duplicate before the origin of angiosperms, fungi, and metazoa? *Proc. Natl. Acad. Sci. USA* **1997**, *94*, 13749–13753. [CrossRef] [PubMed]
6. Xie, Q.; Gao, Y.; Li, J.; Yang, Q.; Qu, X.; Li, H.; Zhang, J.; Wang, T.; Ye, Z.; Yang, C. The HD-Zip IV transcription factor *SlHDZIV8* controls multicellular trichome morphology by regulating the expression of Hairless-2. *J. Exp. Bot.* **2020**, *71*, 7132–7145. [CrossRef] [PubMed]
7. Hu, T.; Ye, J.; Tao, P.; Li, H.; Zhang, J.; Zhang, Y.; Ye, Z. The tomato HD-Zip I transcription factor *SlHZ24* modulates ascorbate accumulation through positive regulation of the d-mannose/l-galactose pathway. *Plant J.* **2016**, *85*, 16–29. [CrossRef]
8. Jiang, C.-Z.; Chang, X.; Donnelly, L.; Reid, M.S.; Jiang, C.-Z. A HD-ZIP Transcription Factor Regulates Flower Senescence via Ethylene and ABA Cross-talks in Petunia. In *Hortscience;* 113 S WEST ST, STE 200; American Society Horticultural Science: Alexandria, VA, USA, 2014; p. S138.
9. Li, S.; Chen, N.; Li, F.; Mei, F.; Wang, Z.; Cheng, X.; Kang, Z.; Mao, H. Characterization of wheat homeodomain-leucine zipper family genes and functional analysis of TaHDZ5-6A in drought tolerance in transgenic *Arabidopsis*. *BMC Plant Biol.* **2020**, *20*, 50. [CrossRef]
10. Oh, S.-K.; Yoon, J.; Choi, G.J.; Jang, H.A.; Kwon, S.-Y.; Choi, D. Capsicum annuum homeobox 1 (*CaHB1*) is a nuclear factor that has roles in plant development, salt tolerance, and pathogen defense. *Biochem. Biophys. Res. Commun.* **2013**, *442*, 116–121. [CrossRef]
11. Ribichich, K.F.; Chiozza, M.; Ávalos-Britez, S.; Cabello, J.V.; Arce, A.L.; Watson, G.; Arias, C.; Portapila, M.; Trucco, F.; Otegui, M.E. Successful field performance in warm and dry environments of soybean expressing the sunflower transcription factor HaHB4. *J. Exp. Bot.* **2020**, *71*, 3142–3156. [CrossRef]
12. Ding, Y.; Gong, S.; Wang, Y.; Wang, F.; Bao, H.; Sun, J.; Cai, C.; Yi, K.; Chen, Z.; Zhu, C. MicroRNA166 modulates cadmium tolerance and accumulation in rice. *Plant Physiol.* **2018**, *177*, 1691–1703. [CrossRef]
13. Elhiti, M.; Stasolla, C. Structure and function of homodomain-leucine zipper (HD-Zip) proteins. *Plant Signal. Behav.* **2009**, *4*, 86–88. [CrossRef] [PubMed]
14. Harris, J.C.; Hrmova, M.; Lopato, S.; Langridge, P. Modulation of plant growth by HD-Zip class I and II transcription factors in response to environmental stimuli. *New Phytol.* **2011**, *190*, 823–837. [CrossRef]

15. Henriksson, E.; Olsson, A.S.; Johannesson, H.; Johansson, H.; Hanson, J.; Engström, P.; Söderman, E. Homeodomain leucine zipper class I genes in Arabidopsis. Expression patterns and phylogenetic relationships. *Plant Physiol.* **2005**, *139*, 509–518. [CrossRef]
16. Agalou, A.; Purwantomo, S.; Övernäs, E.; Johannesson, H.; Zhu, X.; Estiati, A.; de Kam, R.J.; Engström, P.; Slamet-Loedin, I.H.; Zhu, Z. A genome-wide survey of HD-Zip genes in rice and analysis of drought-responsive family members. *Plant Mol. Biol.* **2008**, *66*, 87–103. [CrossRef] [PubMed]
17. Tron, A.E.; Bertoncini, C.W.; Chan, R.L.; Gonzalez, D.H. Redox regulation of plant homeodomain transcription factors. *J. Biol. Chem.* **2002**, *277*, 34800–34807. [CrossRef]
18. Tron, A.E.; Comelli, R.N.; Gonzalez, D.H. Structure of homeodomain-leucine zipper/DNA complexes studied using hydroxyl radical cleavage of DNA and methylation interference. *Biochemistry* **2005**, *44*, 16796–16803. [CrossRef]
19. Palena, C.M.; Tron, A.E.; Bertoncini, C.W.; Gonzalez, D.H.; Chan, R.L. Positively charged residues at the N-terminal arm of the homeodomain are required for efficient DNA binding by homeodomain-leucine zipper proteins. *J. Mol. Biol.* **2001**, *308*, 39–47. [CrossRef]
20. Tron, A.E.; Welchen, E.; Gonzalez, D.H. Engineering the loop region of a homeodomain-leucine zipper protein promotes efficient binding to a monomeric DNA binding site. *Biochemistry* **2004**, *43*, 15845–15851. [CrossRef]
21. Aoyama, T.; Dong, C.-H.; Wu, Y.; Carabelli, M.; Sessa, G.; Ruberti, I.; Morelli, G.; Chua, N.-H. Ectopic expression of the *Arabidopsis* transcriptional activator *Athb-1* alters leaf cell fate in tobacco. *Plant Cell* **1995**, *7*, 1773–1785. [PubMed]
22. Mattsson, J.; Ckurshumova, W.; Berleth, T. Auxin signaling in *Arabidopsis* leaf vascular development. *Plant Physiol.* **2003**, *131*, 1327–1339. [CrossRef]
23. Lin, Z.; Hong, Y.; Yin, M.; Li, C.; Zhang, K.; Grierson, D. A tomato HD-Zip homeobox protein, *LeHB-1*, plays an important role in floral organogenesis and ripening. *Plant J.* **2008**, *55*, 301–310. [CrossRef]
24. Li, Y.; Zhang, S.; Dong, R.; Wang, L.; Yao, J.; van Nocker, S.; Wang, X. The grapevine homeobox gene *VvHB58* influences seed and fruit development through multiple hormonal signaling pathways. *BMC Plant Biol.* **2019**, *19*, 523. [CrossRef]
25. Yang, Y.; Al-Baidhani, H.H.J.; Harris, J.; Riboni, M.; Li, Y.; Mazonka, I.; Bazanova, N.; Chirkova, L.; Sarfraz Hussain, S.; Hrmova, M. DREB/CBF expression in wheat and barley using the stress-inducible promoters of HD-Zip I genes: Impact on plant development, stress tolerance and yield. *Plant Biotechnol. J.* **2020**, *18*, 829–844. [CrossRef]
26. Ma, Y.-J.; Li, P.-T.; Sun, L.-M.; Zhou, H.; Zeng, R.-F.; Ai, X.-Y.; Zhang, J.-Z.; Hu, C.-G. HD-ZIP I Transcription Factor (*PtHB13*) Negatively Regulates Citrus Flowering through Binding to FLOWERING LOCUS C Promoter. *Plants* **2020**, *9*, 114. [CrossRef]
27. Song, Y.H.; Ito, S.; Imaizumi, T. Flowering time regulation: Photoperiod-and temperature-sensing in leaves. *Trends Plant Sci.* **2013**, *18*, 575–583. [CrossRef]
28. Gaudinier, A.; Blackman, B.K. Evolutionary processes from the perspective of flowering time diversity. *New Phytol.* **2020**, *225*, 1883–1898. [CrossRef] [PubMed]
29. Sessa, G.; Carabelli, M.; Sassi, M.; Ciolfi, A.; Possenti, M.; Mittempergher, F.; Becker, J.; Morelli, G.; Ruberti, I. A dynamic balance between gene activation and repression regulates the shade avoidance response in Arabidopsis. *Genes Dev.* **2005**, *19*, 2811–2815. [CrossRef] [PubMed]
30. Morelli, G.; Ruberti, I. Light and shade in the photocontrol of Arabidopsis growth. *Trends Plant Sci.* **2002**, *7*, 399–404. [CrossRef]
31. Sessa, G.; Carabelli, M.; Possenti, M.; Morelli, G.; Ruberti, I. Multiple pathways in the control of the shade avoidance response. *Plants* **2018**, *7*, 102. [CrossRef] [PubMed]
32. Rueda, E.C.; Dezar, C.A.; Gonzalez, D.H.; Chan, R.L. *Hahb-10*, a sunflower homeobox-leucine zipper gene, is regulated by light quality and quantity, and promotes early flowering when expressed in Arabidopsis. *Plant Cell Physiol.* **2005**, *46*, 1954–1963. [CrossRef]
33. Sawa, S.; Ohgishi, M.; Goda, H.; Higuchi, K.; Shimada, Y.; Yoshida, S.; Koshiba, T. The *HAT2* gene, a member of the HD-Zip gene family, isolated as an auxin inducible gene by DNA microarray screening, affects auxin response in *Arabidopsis*. *Plant J.* **2002**, *32*, 1011–1022. [CrossRef]
34. Delarue, M.; Prinsen, E.; Va, H.; Caboche, M.; Bellini, C. Sur2 mutations of *Arabidopsis thaliana* define a new locus involved in the control of auxin homeostasis. *Plant J.* **1998**, *14*, 603–611. [CrossRef]
35. Gu, C.; Guo, Z.-H.; Cheng, H.-Y.; Zhou, Y.-H.; Qi, K.-J.; Wang, G.-M.; Zhang, S.-L. A HD-ZIP II HOMEBOX transcription factor, PpHB.G7, mediates ethylene biosynthesis during fruit ripening in peach. *Plant Sci.* **2019**, *278*, 12–19. [CrossRef]
36. Chen, W.; Cheng, Z.; Liu, L.; Wang, M.; You, X.; Wang, J.; Zhang, F.; Zhou, C.; Zhang, Z.; Zhang, H. Small Grain and Dwarf 2, encoding an HD-Zip II family transcription factor, regulates plant development by modulating gibberellin biosynthesis in rice. *Plant Sci.* **2019**, *288*, 110208. [CrossRef]
37. Prigge, M.J.; Otsuga, D.; Alonso, J.M.; Ecker, J.R.; Drews, G.N.; Clark, S.E. Class III homeodomain-leucine zipper gene family members have overlapping, antagonistic, and distinct roles in Arabidopsis development. *Plant Cell* **2005**, *17*, 61–76. [CrossRef]
38. Emery, J.F.; Floyd, S.K.; Alvarez, J.; Eshed, Y.; Hawker, N.P.; Izhaki, A.; Baum, S.F.; Bowman, J.L. Radial patterning of Arabidopsis shoots by class III HD-ZIP and KANADI genes. *Curr. Biol.* **2003**, *13*, 1768–1774. [CrossRef] [PubMed]
39. Baima, S.; Possenti, M.; Matteucci, A.; Wisman, E.; Altamura, M.M.; Ruberti, I.; Morelli, G. The Arabidopsis *ATHB-8* HD-zip protein acts as a differentiation-promoting transcription factor of the vascular meristems. *Plant Physiol.* **2001**, *126*, 643–655. [CrossRef]

40. Hawker, N.P.; Bowman, J.L. Roles for class III HD-Zip and *KANADI* genes in Arabidopsis root development. *Plant Physiol.* **2004**, *135*, 2261–2270. [CrossRef] [PubMed]
41. Damodaran, S.; Dubois, A.; Xie, J.; Ma, Q.; Hindié, V.; Subramanian, S. *GmZPR3d* Interacts with *GmHD-ZIP* III Proteins and Regulates Soybean Root and Nodule Vascular Development. *Int. J. Mol. Sci.* **2019**, *20*, 827. [CrossRef] [PubMed]
42. Javelle, M.; Vernoud, V.; Depege-Fargeix, N.; Arnould, C.; Oursel, D.; Domergue, F.; Sarda, X.; Rogowsky, P.M. Overexpression of the epidermis-specific homeodomain-leucine zipper IV transcription factor Outer Cell Layer1 in maize identifies target genes involved in lipid metabolism and cuticle biosynthesis. *Plant Physiol.* **2010**, *154*, 273–286. [CrossRef]
43. Chew, W.; Hrmova, M.; Lopato, S. Role of homeodomain leucine zipper (HD-Zip) IV transcription factors in plant development and plant protection from deleterious environmental factors. *Int. J. Mol. Sci.* **2013**, *14*, 8122–8147. [CrossRef]
44. Kamata, N.; Okada, H.; Komeda, Y.; Takahashi, T. Mutations in epidermis-specific HD-ZIP IV genes affect floral organ identity in *Arabidopsis thaliana*. *Plant J.* **2013**, *75*, 430–440. [CrossRef]
45. Ogawa, E.; Yamada, Y.; Sezaki, N.; Kosaka, S.; Kondo, H.; Kamata, N.; Abe, M.; Komeda, Y.; Takahashi, T. *ATML1* and *PDF2* play a redundant and essential role in *Arabidopsis* embryo development. *Plant Cell Physiol.* **2015**, *56*, 1183–1192. [CrossRef]
46. Wang, S.; Kwak, S.-H.; Zeng, Q.; Ellis, B.E.; Chen, X.-Y.; Schiefelbein, J.; Chen, J.-G. *TRICHOMELESS1* regulates trichome patterning by suppressing *GLABRA1* in *Arabidopsis*. *Development* **2007**, *134*, 3873–3882. [CrossRef]
47. Landi, M.; Tattini, M.; Gould, K.S. Multiple functional roles of anthocyanins in plant-environment interactions. *Environ. Exp. Bot.* **2015**, *119*, 4–17. [CrossRef]
48. Kubo, H.; Peeters, A.J.; Aarts, M.G.; Pereira, A.; Koornneef, M. *ANTHOCYANINLESS2*, a homeobox gene affecting anthocyanin distribution and root development in Arabidopsis. *Plant Cell* **1999**, *11*, 1217–1226. [CrossRef]
49. Kubo, H.; Hayashi, K. Characterization of root cells of anl2 mutant in *Arabidopsis thaliana*. *Plant Sci.* **2011**, *180*, 679–685. [CrossRef] [PubMed]
50. Yu, H.; Chen, X.; Hong, Y.-Y.; Wang, Y.; Xu, P.; Ke, S.-D.; Liu, H.-Y.; Zhu, J.-K.; Oliver, D.J.; Xiang, C.-B. Activated expression of an Arabidopsis HD-START protein confers drought tolerance with improved root system and reduced stomatal density. *Plant Cell* **2008**, *20*, 1134–1151. [CrossRef] [PubMed]
51. Yu, L.; Chen, X.; Wang, Z.; Wang, S.; Wang, Y.; Zhu, Q.; Li, S.; Xiang, C. Arabidopsis enhanced drought tolerance1/HOMEODOMAIN GLABROUS11 confers drought tolerance in transgenic rice without yield penalty. *Plant Physiol.* **2013**, *162*, 1378–1391. [CrossRef]
52. Vernoud, V.; Laigle, G.; Rozier, F.; Meeley, R.B.; Perez, P.; Rogowsky, P.M. The HD-ZIP IV transcription factor *OCL4* is necessary for trichome patterning and anther development in maize. *Plant J.* **2009**, *59*, 883–894. [CrossRef]
53. Wei, J.; Choi, H.; Jin, P.; Wu, Y.; Yoon, J.; Lee, Y.-S.; Quan, T.; An, G. GL2-type homeobox gene *Roc4* in rice promotes flowering time preferentially under long days by repressing Ghd7. *Plant Sci.* **2016**, *252*, 133–143. [CrossRef] [PubMed]
54. Raza, A. Eco-physiological and biochemical responses of rapeseed (*Brassica napus* L.) to abiotic stresses: Consequences and mitigation strategies. *J. Plant Growth Regul.* **2020**, *40*, 1368–1388. [CrossRef]
55. Raza, A.; Ashraf, F.; Zou, X.; Zhang, X.; Tosif, H. Plant Adaptation and Tolerance to Environmental Stresses: Mechanisms and Perspectives. In *Plant Ecophysiology and Adaptation under Climate Change: Mechanisms and Perspectives I*; Springer: Berlin/Heidelberg, Germany, 2020; pp. 117–145.
56. Zandalinas, S.I.; Mittler, R.; Balfagón, D.; Arbona, V.; Gómez-Cadenas, A. Plant adaptations to the combination of drought and high temperatures. *Physiol. Plant* **2018**, *162*, 2–12. [CrossRef]
57. Gupta, A.; Rico-Medina, A.; Caño-Delgado, A.I. The physiology of plant responses to drought. *Science* **2020**, *368*, 266–269. [CrossRef] [PubMed]
58. Perotti, M.F.; Ribone, P.A.; Chan, R.L. Plant transcription factors from the homeodomain-leucine zipper family I. Role in development and stress responses. *IUBMB Life* **2017**, *69*, 280–289. [CrossRef] [PubMed]
59. Gong, S.; Ding, Y.; Hu, S.; Ding, L.; Chen, Z.; Zhu, C. The role of HD-Zip class I transcription factors in plant response to abiotic stresses. *Physiol. Plant* **2019**, *167*, 516–525. [CrossRef]
60. Zhou, W.; Malabanan, P.B.; Abrigo, E. *OsHox4* regulates GA signaling by interacting with DELLA-like genes and GA oxidase genes in rice. *Euphytica* **2015**, *201*, 97–107. [CrossRef]
61. Colombo, R.P.; Ibarra, J.G.; Bidondo, L.F.; Silvani, V.A.; Bompadre, M.J.; Pergola, M.; Lopez, N.I.; Godeas, A.M. Arbuscular mycorrhizal fungal association in genetically modified drought-tolerant corn. *J. Environ. Qual.* **2017**, *46*, 227–231. [CrossRef]
62. Re, D.A.; Dezar, C.A.; Chan, R.L.; Baldwin, I.T.; Bonaventure, G. *Nicotiana attenuata NaHD20* plays a role in leaf ABA accumulation during water stress, benzylacetone emission from flowers, and the timing of bolting and flower transitions. *J. Exp. Bot.* **2011**, *62*, 155–166. [CrossRef] [PubMed]
63. Sharif, I.; Aleem, S.; Farooq, J.; Rizwan, M.; Younas, A.; Sarwar, G.; Chohan, S.M. Salinity stress in cotton: Effects, mechanism of tolerance and its management strategies. *Physiol. Mol. Biol. Plants* **2019**, *25*, 807–820. [CrossRef]
64. Isayenkov, S.V.; Maathuis, F.J. Plant salinity stress: Many unanswered questions remain. *Front. Plant Sci.* **2019**, *10*, 80. [CrossRef]
65. Negrão, S.; Schmöckel, S.; Tester, M. Evaluating physiological responses of plants to salinity stress. *Ann. Bot.* **2017**, *119*, 1–11. [CrossRef]
66. Zhang, S.; Haider, I.; Kohlen, W.; Jiang, L.; Bouwmeester, H.; Meijer, A.H.; Schluepmann, H.; Liu, C.-M.; Ouwerkerk, P.B. Function of the HD-Zip I gene *Oshox22* in ABA-mediated drought and salt tolerances in rice. *Plant Mol. Biol.* **2012**, *80*, 571–585. [CrossRef]
67. Deng, X.; Phillips, J.; Meijer, A.H.; Salamini, F.; Bartels, D. Characterization of five novel dehydration-responsive homeodomain leucine zipper genes from the resurrection plant *Craterostigma plantagineum*. *Plant Mol. Biol.* **2002**, *49*, 601–610. [CrossRef]

68. Ni, Y.; Wang, X.; Li, D.; Wu, Y.; Xu, W.; Li, X. Novel cotton homeobox gene and its expression profiling in root development and in response to stresses and phytohormones. *Acta Biochim. Biophys. Sin.* **2008**, *40*, 78–84. [CrossRef]
69. Zhao, Y.; Ma, Q.; Jin, X.; Peng, X.; Liu, J.; Deng, L.; Yan, H.; Sheng, L.; Jiang, H.; Cheng, B. A novel maize homeodomain–leucine zipper (HD-Zip) I gene, *Zmhdz10*, positively regulates drought and salt tolerance in both rice and arabidopsis. *Plant Cell Physiol.* **2014**, *55*, 1142–1156. [CrossRef]
70. Tang, Y.; Bao, X.; Wang, S.; Liu, Y.; Tan, J.; Yang, M.; Zhang, M.; Dai, R.; Yu, X. A physic nut stress-responsive HD-Zip transcription factor, *JcHDZ07*, confers enhanced sensitivity to salinity stress in transgenic *Arabidopsis*. *Front. Plant Sci.* **2019**, *10*, 942. [CrossRef]
71. Ye, Y.; Ding, Y.; Jiang, Q.; Wang, F.; Sun, J.; Zhu, C. The role of receptor-like protein kinases (RLKs) in abiotic stress response in plants. *Plant Cell Rep.* **2017**, *36*, 235–242. [CrossRef]
72. Mehmood, S.S.; Lu, G.; Luo, D.; Hussain, M.A.; Raza, A.; Zafar, Z.; Zhang, X.; Cheng, Y.; Zou, X.; Lv, Y. Integrated Analysis of Transcriptomics and Proteomics provides insights into the molecular regulation of cold response in *Brassica napus*. *Environ. Exp. Bot.* **2021**, *187*, 104480. [CrossRef]
73. He, H.; Lei, Y.; Yi, Z.; Raza, A.; Zeng, L.; Yan, L.; Xiaoyu, D.; Yong, C.; Xiling, Z. Study on the mechanism of exogenous serotonin improving cold tolerance of rapeseed (*Brassica napus* L.) seedlings. *Plant Growth Regul.* **2021**, *94*, 161–170. [CrossRef]
74. Kovalchuk, N.; Chew, W.; Sornaraj, P.; Borisjuk, N.; Yang, N.; Singh, R.; Bazanova, N.; Shavrukov, Y.; Guendel, A.; Munz, E. The homeodomain transcription factor *TaHDZipI-2* from wheat regulates frost tolerance, flowering time and spike development in transgenic barley. *New Phytol.* **2016**, *211*, 671–687. [CrossRef]
75. Yang, Y.; Luang, S.; Harris, J.; Riboni, M.; Li, Y.; Bazanova, N.; Hrmova, M.; Haefele, S.; Kovalchuk, N.; Lopato, S. Overexpression of the class I homeodomain transcription factor *TaHDZipI-5* increases drought and frost tolerance in transgenic wheat. *Plant Biotechnol. J.* **2018**, *16*, 1227–1240. [CrossRef] [PubMed]
76. Cabello, J.V.; Arce, A.L.; Chan, R.L. The homologous HD-Zip I transcription factors *HaHB1* and *AtHB13* confer cold tolerance via the induction of pathogenesis-related and glucanase proteins. *Plant J.* **2012**, *69*, 141–153. [CrossRef] [PubMed]
77. Raza, A.; Hussain, S.; Javed, R.; Hafeez, M.B.; Hasanuzzaman, M. Antioxidant Defense Systems and Remediation of Metal Toxicity in Plants. In *Approaches to the Remediation of Inorganic Pollutants*; Springer: Singapore, 2021; pp. 91–124.
78. Raza, A.; Habib, M.; Kakavand, S.N.; Zahid, Z.; Zahra, N.; Sharif, R.; Hasanuzzaman, M. Phytoremediation of Cadmium: Physiological, Biochemical, and Molecular Mechanisms. *Biology* **2020**, *9*, 177. [CrossRef] [PubMed]
79. Raza, A.; Habib, M.; Charagh, S.; Kakavand, S.N. Genetic engineering of plants to tolerate toxic metals and metalloids. In *Handbook of Bioremediation*; Elsevier: Amsterdam, The Netherlands, 2021; pp. 411–436.
80. Zhou, C.-P.; Li, C.-P.; Liang, W.-W.; Guo, P.; Yang, L.-T.; Chen, L.-S. Identification of manganese-toxicity-responsive genes in roots of two citrus species differing in manganese tolerance using cDNA-AFLP. *Trees* **2017**, *31*, 813–831. [CrossRef]
81. Hu, S.; Ding, Y.; Zhu, C. Sensitivity and responses of chloroplasts to heat stress in plants. *Front. Plant Sci.* **2020**, *11*, 375. [CrossRef]
82. Raza, A.; Tabassum, J.; Kudapa, H.; Varshney, R.K. Can omics deliver temperature resilient ready-to-grow crops? *Crit. Rev. Biotechnol.* **2021**, *94*, 1–24. [CrossRef] [PubMed]
83. Sharif, R.; Xie, C.; Wang, J.; Cao, Z.; Zhang, H.; Chen, P.; Yuhong, L. Genome wide identification, characterization and expression analysis of HD-ZIP gene family in *Cucumis sativus* L. under biotic and various abiotic stresses. *Int. J. Biol. Macromol.* **2020**, *158*, 502–520. [CrossRef]
84. Wang, J.; Zhuang, L.; Zhang, J.; Yu, J.; Yang, Z.; Huang, B. Identification and characterization of novel homeodomain leucine zipper (HD-Zip) transcription factors associated with heat tolerance in perennial ryegrass. *Environ. Exp. Bot.* **2019**, *160*, 1–11. [CrossRef]
85. Striker, G.G. Flooding stress on plants: Anatomical, morphological and physiological responses. *Botany* **2012**, *1*, 3–28.
86. Pan, J.; Sharif, R.; Xu, X.; Chen, X. Waterlogging Response Mechanisms in Plants: Research Progress and Prospects. *Front. Plant Sci.* **2020**, *11*, 2319.
87. Cabello, J.V.; Giacomelli, J.I.; Piattoni, C.V.; Iglesias, A.A.; Chan, R.L. The sunflower transcription factor *HaHB11* improves yield, biomass and tolerance to flooding in transgenic Arabidopsis plants. *J. Biotechnol.* **2016**, *222*, 73–83. [CrossRef] [PubMed]
88. Gong, Z.; Xiong, L.; Shi, H.; Yang, S.; Herrera-Estrella, L.R.; Xu, G.; Chao, D.-Y.; Li, J.; Wang, P.-Y.; Qin, F. Plant abiotic stress response and nutrient use efficiency. *Sci. China Life Sci.* **2020**, *63*, 635–674. [CrossRef] [PubMed]
89. Gulzar, S.; Hassan, A.; Nawchoo, I.A. A Review of Nutrient Stress Modifications in Plants, Alleviation Strategies, and Monitoring through Remote Sensing. In *Plant Micronutrients*; Springer: Berlin/Heidelberg, Germany, 2020; pp. 331–343.
90. Aznar, A.; Chen, N.W.; Thomine, S.; Dellagi, A. Immunity to plant pathogens and iron homeostasis. *Plant Sci.* **2015**, *240*, 90–97. [CrossRef]
91. Kobayashi, T.; Nozoye, T.; Nishizawa, N.K. Iron transport and its regulation in plants. *Free Radic. Biol. Med.* **2019**, *133*, 11–20. [CrossRef]
92. Romani, F.; Capella, M.; Ribone, P.; Chan, R.L. The Arabidopsis transcription factor *AtHB1* plays a role in iron homeostasis. In Proceedings of the 11th International Congress of Plant Molecular Biology, Iguazú Falls, Brazil, 25–30 October 2015.
93. Wei, M.; Liu, A.; Zhang, Y.; Zhou, Y.; Li, D.; Dossa, K.; Zhou, R.; Zhang, X.; You, J. Genome-wide characterization and expression analysis of the HD-Zip gene family in response to drought and salinity stresses in sesame. *BMC Genom.* **2019**, *20*, 748. [CrossRef]
94. Yue, H.; Shu, D.; Wang, M.; Xing, G.; Zhan, H.; Du, X.; Song, W.; Nie, X. Genome-Wide Identification and Expression Analysis of the HD-Zip Gene Family in Wheat (*Triticum aestivum* L.). *Genes* **2018**, *9*, 70. [CrossRef] [PubMed]

95. Huang, D.; Wu, W.; Abrams, S.R.; Cutler, A.J. The relationship of drought-related gene expression in *Arabidopsis thaliana* to hormonal and environmental factors. *J. Exp. Bot.* **2008**, *59*, 2991–3007. [CrossRef]
96. Stape, J.L.; Binkley, D.; Ryan, M.G. Eucalyptus production and the supply, use and efficiency of use of water, light and nitrogen across a geographic gradient in Brazil. *For. Ecol. Manag.* **2004**, *193*, 17–31. [CrossRef]
97. Almeida, A.C.; Soares, J.V.; Landsberg, J.J.; Rezende, G.D. Growth and water balance of *Eucalyptus grandis* hybrid plantations in Brazil during a rotation for pulp production. *For. Ecol. Manag.* **2007**, *251*, 10–21. [CrossRef]
98. Sasaki, K.; Ida, Y.; Kitajima, S.; Kawazu, T.; Hibino, T.; Hanba, Y.T. Overexpressing the HD-Zip class II transcription factor *EcHB1* from Eucalyptus camaldulensis increased the leaf photosynthesis and drought tolerance of Eucalyptus. *Sci. Rep.* **2019**, *9*, 14121. [CrossRef] [PubMed]
99. Ciarbelli, A.R.; Ciolfi, A.; Salvucci, S.; Ruzza, V.; Possenti, M.; Carabelli, M.; Fruscalzo, A.; Sessa, G.; Morelli, G.; Ruberti, I. The Arabidopsis homeodomain-leucine zipper II gene family: Diversity and redundancy. *Plant Mol. Biol.* **2008**, *68*, 465–478. [CrossRef]
100. Carabelli, M.; Morelli, G.; Whitelam, G.; Ruberti, I. Twilight-zone and canopy shade induction of the *Athb-2* homeobox gene in green plants. *Proc. Natl. Acad. Sci. USA* **1996**, *93*, 3530–3535. [CrossRef] [PubMed]
101. Steindler, C.; Matteucci, A.; Sessa, G.; Weimar, T.; Ohgishi, M.; Aoyama, T.; Morelli, G.; Ruberti, I. Shade avoidance responses are mediated by the *ATHB-2* HD-zip protein, a negative regulator of gene expression. *Development* **1999**, *126*, 4235–4245. [CrossRef]
102. Lorrain, S.; Allen, T.; Duek, P.D.; Whitelam, G.C.; Fankhauser, C. Phytochrome-mediated inhibition of shade avoidance involves degradation of growth-promoting bHLH transcription factors. *Plant J.* **2008**, *53*, 312–323. [CrossRef] [PubMed]
103. Hornitschek, P.; Kohnen, M.V.; Lorrain, S.; Rougemont, J.; Ljung, K.; López-Vidriero, I.; Franco-Zorrilla, J.M.; Solano, R.; Trevisan, M.; Pradervand, S. Phytochrome interacting factors 4 and 5 control seedling growth in changing light conditions by directly controlling auxin signaling. *Plant J.* **2012**, *71*, 699–711. [CrossRef]
104. Leivar, P.; Tepperman, J.M.; Cohn, M.M.; Monte, E.; Al-Sady, B.; Erickson, E.; Quail, P.H. Dynamic antagonism between phytochromes and PIF family basic helix-loop-helix factors induces selective reciprocal responses to light and shade in a rapidly responsive transcriptional network in Arabidopsis. *Plant Cell* **2012**, *24*, 1398–1419. [CrossRef]
105. Shen, W.; Li, H.; Teng, R.; Wang, Y.; Wang, W.; Zhuang, J. Genomic and transcriptomic analyses of HD-Zip family transcription factors and their responses to abiotic stress in tea plant (*Camellia sinensis*). *Genomics* **2019**, *111*, 1142–1151. [CrossRef]
106. Li, W.; Dong, J.; Cao, M.; Gao, X.; Wang, D.; Liu, B.; Chen, Q. Genome-wide identification and characterization of HD-ZIP genes in potato. *Gene* **2019**, *697*, 103–117. [CrossRef]
107. Zhang, J.; Zhang, H.; Srivastava, A.K.; Pan, Y.; Bai, J.; Fang, J.; Shi, H.; Zhu, J.-K. Knockdown of Rice MicroRNA166 Confers Drought Resistance by Causing Leaf Rolling and Altering Stem Xylem Development. *Plant Physiol.* **2018**, *176*, 2082–2094. [CrossRef] [PubMed]
108. Yang, T.; Wang, Y.; Teotia, S.; Wang, Z.; Shi, C.; Sun, H.; Gu, Y.; Zhang, Z.; Tang, G. The interaction between miR160 and miR165/166 in the control of leaf development and drought tolerance in Arabidopsis. *Sci. Rep.* **2019**, *9*, 2832. [CrossRef] [PubMed]
109. Li, Z.; Gao, Z.; Li, R.; Xu, Y.; Kong, Y.; Zhou, G.; Meng, C.; Hu, R. Genome-wide identification and expression profiling of HD-ZIP gene family in *Medicago truncatula*. *Genomics* **2020**, *112*, 3624–3635. [CrossRef] [PubMed]
110. DalCorso, G.; Farinati, S.; Furini, A. Regulatory networks of cadmium stress in plants. *Plant Signal. Behav.* **2010**, *5*, 663–667. [CrossRef]
111. Andresen, E.; Küpper, H. Cadmium toxicity in plants. In *Cadmium: From Toxicity to Essentiality*; Springer: Berlin/Heidelberg, Germany, 2013; pp. 395–413.
112. Ismael, M.A.; Elyamine, A.M.; Moussa, M.G.; Cai, M.; Zhao, X.; Hu, C. Cadmium in plants: Uptake, toxicity, and its interactions with selenium fertilizers. *Metallomics* **2019**, *11*, 255–277. [CrossRef] [PubMed]
113. Franco, D.M.; Silva, E.M.; Saldanha, L.L.; Adachi, S.A.; Schley, T.R.; Rodrigues, T.M.; Dokkedal, A.L.; Nogueira, F.T.S.; de Almeida, L.F.R. Flavonoids modify root growth and modulate expression of SHORT-ROOT and HD-ZIP III. *J. Plant Physiol.* **2015**, *188*, 89–95. [CrossRef]
114. Singh, A.; Roy, S.; Singh, S.; Das, S.S.; Gautam, V.; Yadav, S.; Kumar, A.; Singh, A.; Samantha, S.; Sarkar, A.K. Phytohormonal crosstalk modulates the expression of miR166/165s, target Class III HD-ZIPs, and *KANADI* genes during root growth in Arabidopsis thaliana. *Sci. Rep.* **2017**, *7*, 3408. [CrossRef]
115. Zhu, Z.; Sun, B.; Xu, X.; Chen, H.; Zou, L.; Chen, G.; Cao, B.; Chen, C.; Lei, J. Overexpression of *AtEDT1/HDG11* in Chinese kale (*Brassica oleracea* var. alboglabra) enhances drought and osmotic stress tolerance. *Front. Plant Sci.* **2016**, *7*, 1285. [CrossRef]
116. Bang, S.W.; Lee, D.-K.; Jung, H.; Chung, P.J.; Kim, Y.S.; Choi, Y.D.; Suh, J.-W.; Kim, J.-K. Overexpression of *OsTF1L*, a rice HD-Zip transcription factor, promotes lignin biosynthesis and stomatal closure that improves drought tolerance. *Plant Biotechnol. J.* **2019**, *17*, 118–131. [CrossRef]
117. Yu, L.H.; Wu, S.J.; Peng, Y.S.; Liu, R.N.; Chen, X.; Zhao, P.; Xu, P.; Zhu, J.B.; Jiao, G.L.; Pei, Y. Arabidopsis *EDT1/HDG11* improves drought and salt tolerance in cotton and poplar and increases cotton yield in the field. *Plant Biotechnol. J.* **2016**, *14*, 72–84. [CrossRef]
118. Guo, X.Y.; Wang, Y.; Zhao, P.X.; Xu, P.; Yu, G.H.; Zhang, L.Y.; Xiong, Y.; Xiang, C.B. *AtEDT1/HDG11* regulates stomatal density and water-use efficiency via ERECTA and E2Fa. *New Phytol.* **2019**, *223*, 1478–1488. [CrossRef] [PubMed]
119. Zhang, H.; Ma, X.; Li, W.; Niu, D.; Wang, Z.; Yan, X.; Yang, X.; Yang, Y.; Cui, H. Genome-wide characterization of NtHD-ZIP IV: Different roles in abiotic stress response and glandular Trichome induction. *BMC Plant Biol.* **2019**, *19*, 444. [CrossRef]

120. Ahmad, S.; Khan, N.; Iqbal, M.Z.; Hussain, A.; Hassan, M. Salt tolerance of cotton (*Gossypium hirsutum* L.). *Asian J. Plant Sci.* **2002**, *1*, 715–719.
121. Raza, A.; Charagh, S.; Zahid, Z.; Mubarik, M.S.; Javed, R.; Siddiqui, M.H.; Hasanuzzaman, M. Jasmonic acid: A key frontier in conferring abiotic stress tolerance in plants. *Plant Cell Rep.* **2021**, *40*, 1513–1541. [CrossRef]
122. Mir, M.A.; John, R.; Alyemeni, M.N.; Alam, P.; Ahmad, P. Jasmonic acid ameliorates alkaline stress by improving growth performance, ascorbate glutathione cycle and glyoxylase system in maize seedlings. *Sci. Rep.* **2018**, *8*, 2831. [CrossRef]
123. Cai, X.T.; Xu, P.; Wang, Y.; Xiang, C.B. Activated expression of *AtEDT1/HDG11* promotes lateral root formation in *Arabidopsis* mutant edt1 by upregulating jasmonate biosynthesis. *J. Integr. Plant Biol.* **2015**, *57*, 1017–1030. [CrossRef]
124. Lin, Z.; Li, Y.; Zhang, Z.; Liu, X.; Hsu, C.-C.; Du, Y.; Sang, T.; Zhu, C.; Wang, Y.; Satheesh, V. A RAF-SnRK2 kinase cascade mediates early osmotic stress signaling in higher plants. *Nat. Commun.* **2020**, *11*, 613. [CrossRef]
125. Chen, E.; Zhang, X.; Yang, Z.; Wang, X.; Yang, Z.; Zhang, C.; Wu, Z.; Kong, D.; Liu, Z.; Zhao, G. Genome-wide analysis of the HD-ZIP IV transcription factor family in *Gossypium arboreum* and *GaHDG11* involved in osmotic tolerance in transgenic *Arabidopsis*. *Mol. Genet. Genom.* **2017**, *292*, 593–609. [CrossRef]
126. Romani, F.; Ribone, P.A.; Capella, M.; Miguel, V.N.; Chan, R.L. A matter of quantity: Common features in the drought response of transgenic plants overexpressing HD-Zip I transcription factors. *Plant Sci.* **2016**, *251*, 139–154. [CrossRef] [PubMed]
127. Ebrahimian-Motlagh, S.; Ribone, P.A.; Thirumalaikumar, V.P.; Allu, A.D.; Chan, R.L.; Mueller-Roeber, B.; Balazadeh, S. *JUNG-BRUNNEN1* confers drought tolerance downstream of the HD-Zip I transcription factor *AtHB13*. *Front. Plant Sci.* **2017**, *8*, 2118. [CrossRef]
128. Cabello, J.V.; Giacomelli, J.I.; Gómez, M.C.; Chan, R.L. The sunflower transcription factor *HaHB11* confers tolerance to water deficit and salinity to transgenic *Arabidopsis* and alfalfa plants. *J. Biotechnol.* **2017**, *257*, 35–46. [CrossRef] [PubMed]
129. Valdés, A.E.; Övernäs, E.; Johansson, H.; Rada-Iglesias, A.; Engström, P. The homeodomain-leucine zipper (HD-Zip) class I transcription factors *ATHB7* and *ATHB12* modulate abscisic acid signalling by regulating protein phosphatase 2C and abscisic acid receptor gene activities. *Plant Mol. Biol.* **2012**, *80*, 405–418. [CrossRef]
130. Steindler, C.; Carabelli, M.; Borello, U.; Morelli, G.; Ruberti, I. Phytochrome A, phytochrome B and other phytochrome (s) regulate ATHB-2 gene expression in etiolated and green Arabidopsis plants. *Plant Cell Environ.* **1997**, *20*, 759–763. [CrossRef]
131. Wang, Z.; Wang, S.; Xiao, Y.; Li, Z.; Wu, M.; Xie, X.; Li, H.; Mu, W.; Li, F.; Liu, P. Functional characterization of a HD-ZIP IV transcription factor *NtHDG2* in regulating flavonols biosynthesis in *Nicotiana tabacum*. *Plant Physiol. Biochem.* **2020**, *146*, 259–268. [CrossRef]
132. Ramegowda, V.; Senthil-Kumar, M. The interactive effects of simultaneous biotic and abiotic stresses on plants: Mechanistic understanding from drought and pathogen combination. *J. Plant Physiol.* **2015**, *176*, 47–54. [CrossRef] [PubMed]
133. You, M.P.; Rui, T.; Barbetti, M.J. Plant genotype and temperature impact simultaneous biotic and abiotic stress-related gene expression in Pythium-infected plants. *Plant Pathol.* **2020**, *69*, 655–668. [CrossRef]
134. Gull, A.; Lone, A.A.; Wani, N.U.I. Biotic and abiotic stresses in plants. In *Abiotic and Biotic Stress in Plants*; IntechOpen: London, UK, 2019; pp. 1–19. [CrossRef]
135. Sharif, R.; Mujtaba, M.; Ur Rahman, M.; Shalmani, A.; Ahmad, H.; Anwar, T.; Tianchan, D.; Wang, X. The Multifunctional Role of Chitosan in Horticultural Crops; A Review. *Molecules* **2018**, *23*, 872. [CrossRef] [PubMed]
136. Sharif, R.S.; Xie, C.; Zhang, H.; Arnao, M.B.; Ali, M.; Ali, Q.; Muhammad, I.; Shalmani, A.; Nawaz, M.A.; Chen, P.; et al. Melatonin and Its Effects on Plant Systems. *Molecules* **2018**, *23*, 2352. [CrossRef] [PubMed]
137. Micali, C.; Göllner, K.; Humphry, M.; Consonni, C.; Panstruga, R. The powdery mildew disease of Arabidopsis: A paradigm for the interaction between plants and biotrophic fungi. *Am. Soc. Plant Biol.* **2008**, *6*, e0115. [CrossRef]
138. Gao, D.; Appiano, M.; Huibers, R.P.; Chen, X.; Loonen, A.E.; Visser, R.G.; Wolters, A.-M.A.; Bai, Y. Activation tagging of *ATHB13* in *Arabidopsis thaliana* confers broad-spectrum disease resistance. *Plant Mol. Biol.* **2014**, *86*, 641–653. [CrossRef]
139. Manavella, P.A.; Dezar, C.A.; Bonaventure, G.; Baldwin, I.T.; Chan, R.L. *HAHB4*, a sunflower HD-Zip protein, integrates signals from the jasmonic acid and ethylene pathways during wounding and biotic stress responses. *Plant J.* **2008**, *56*, 376–388. [CrossRef] [PubMed]
140. Tian, L.; Yu, J.; Wang, Y.; Tian, C. The C_2H_2 transcription factor VdMsn2 controls hyphal growth, microsclerotia formation, and virulence of *Verticillium dahliae*. *Fungal Biol.* **2017**, *121*, 1001–1010. [CrossRef]
141. He, X.; Wang, T.; Zhu, W.; Wang, Y.; Zhu, L. *GhHB12*, a HD-ZIP I Transcription Factor, Negatively Regulates the Cotton Resistance to *Verticillium dahliae*. *Int. J. Mol. Sci.* **2018**, *19*, 3997. [CrossRef]
142. Mulugeta, T.; Abreha, K.; Tekie, H.; Mulatu, B.; Yesuf, M.; Andreasson, E.; Liljeroth, E.; Alexandersson, E. Phosphite protects against potato and tomato late blight in tropical climates and has varying toxicity depending on the Phytophthora infestans isolate. *Crop. Prot.* **2019**, *121*, 139–146. [CrossRef]
143. Ren, Y.; Armstrong, M.; Qi, Y.; McLellan, H.; Zhong, C.; Du, B.; Birch, P.R.; Tian, Z. Phytophthora infestans RXLR effectors target parallel steps in an immune signal transduction pathway. *Plant Physiol.* **2019**, *180*, 2227–2239. [CrossRef] [PubMed]
144. Hausbeck, M.K.; Lamour, K.H. Phytophthora capsici on vegetable crops: Research progress and management challenges. *Plant Dis.* **2004**, *88*, 1292–1303. [CrossRef] [PubMed]
145. Wang, W.; Liu, X.; Han, T.; Li, K.; Qu, Y.; Gao, Z. Differential Potential of Phytophthora capsici Resistance Mechanisms to the Fungicide Metalaxyl in Peppers. *Microorganisms* **2020**, *8*, 278. [CrossRef]

146. Yamaguchi, Y.L.; Suzuki, R.; Cabrera, J.; Nakagami, S.; Sagara, T.; Ejima, C.; Sano, R.; Aoki, Y.; Olmo, R.; Kurata, T. Root-knot and cyst nematodes activate procambium-associated genes in *Arabidopsis* roots. *Front. Plant Sci.* **2017**, *8*, 1195. [CrossRef]
147. Yang, J.-Y.; Iwasaki, M.; Machida, C.; Machida, Y.; Zhou, X.; Chua, N.-H. βC1, the pathogenicity factor of TYLCCNV, interacts with AS1 to alter leaf development and suppress selective jasmonic acid responses. *Genes Dev.* **2008**, *22*, 2564–2577. [CrossRef]
148. Lewandowska, M.; Keyl, A.; Feussner, I. Wax biosynthesis in response to danger: Its regulation upon abiotic and biotic stress. *New Phytol.* **2020**, *227*, 698–713. [CrossRef]
149. Ziv, C.; Zhao, Z.; Gao, Y.G.; Xia, Y. Multifunctional roles of plant cuticle during plant-pathogen interactions. *Front. Plant Sci.* **2018**, *9*, 1088. [CrossRef]
150. Tafolla-Arellano, J.C.; Báez-Sañudo, R.; Tiznado-Hernández, M.E. The cuticle as a key factor in the quality of horticultural crops. *Sci. Hortic.* **2018**, *232*, 145–152. [CrossRef]
151. Yan, T.; Li, L.; Xie, L.; Chen, M.; Shen, Q.; Pan, Q.; Fu, X.; Shi, P.; Tang, Y.; Huang, H. A novel HD-ZIP IV/MIXTA complex promotes glandular trichome initiation and cuticle development in Artemisia annua. *New Phytol.* **2018**, *218*, 567–578. [CrossRef] [PubMed]
152. Karabourniotis, G.; Liakopoulos, G.; Nikolopoulos, D.; Bresta, P. Protective and defensive roles of non-glandular trichomes against multiple stresses: Structure–function coordination. *J. For. Res.* **2019**, *31*, 1–12. [CrossRef]
153. Zhu, H.; Sun, X.; Zhang, Q.; Song, P.; Hu, Q.; Zhang, X.; Li, X.; Hu, J.; Pan, J.; Sun, S. GLABROUS (*CmGL*) encodes a HD-ZIP IV transcription factor playing roles in multicellular trichome initiation in melon. *Theor. Appl. Genet.* **2018**, *131*, 569–579. [CrossRef] [PubMed]
154. DeBono, A.; Yeats, T.H.; Rose, J.K.; Bird, D.; Jetter, R.; Kunst, L.; Samuels, L. Arabidopsis LTPG is a glycosylphosphatidylinositol-anchored lipid transfer protein required for export of lipids to the plant surface. *Plant Cell* **2009**, *21*, 1230–1238. [CrossRef] [PubMed]
155. Zou, H.-W.; Tian, X.-H.; Ma, G.-H.; Li, Z.-X. Isolation and functional analysis of *ZmLTP3*, a homologue to *Arabidopsis* LTP3. *Int. J. Mol. Sci.* **2013**, *14*, 5025–5035. [CrossRef]
156. Hairat, S.; Baranwal, V.K.; Khurana, P. Identification of *Triticum aestivum* nsLTPs and functional validation of two members in development and stress mitigation roles. *Plant Physiol. Biochem.* **2018**, *130*, 418–430. [CrossRef]
157. Kovalchuk, N.; Li, M.; Wittek, F.; Reid, N.; Singh, R.; Shirley, N.; Ismagul, A.; Eliby, S.; Johnson, A.; Milligan, A.S. Defensin promoters as potential tools for engineering disease resistance in cereal grains. *Plant Biotechnol. J.* **2010**, *8*, 47–64. [CrossRef]
158. Kovalchuk, N.; Wu, W.; Bazanova, N.; Reid, N.; Singh, R.; Shirley, N.; Eini, O.; Johnson, A.A.; Langridge, P.; Hrmova, M. Wheat wounding-responsive HD-Zip IV transcription factor GL7 is predominantly expressed in grain and activates genes encoding defensins. *Plant Mol. Biol.* **2019**, *101*, 41–61. [CrossRef]
159. Zottich, U.; Da Cunha, M.; Carvalho, A.O.; Dias, G.B.; Silva, N.C.; Santos, I.S.; do Nacimento, V.V.; Miguel, E.C.; Machado, O.L.; Gomes, V.M. Purification, biochemical characterization and antifungal activity of a new lipid transfer protein (LTP) from Coffea canephora seeds with α-amylase inhibitor properties. *Biochim. Biophys. Acta (BBA)-Gen. Subj.* **2011**, *1810*, 375–383. [CrossRef] [PubMed]
160. Boutrot, F.; Chantret, N.; Gautier, M.-F. Genome-wide analysis of the rice and Arabidopsis non-specific lipid transfer protein (*nsLtp*) gene families and identification of wheat nsLtp genes by EST data mining. *BMC Genom.* **2008**, *9*, 86. [CrossRef] [PubMed]

genes

MDPI

Article

Biochemical and Molecular Effects Induced by Triacontanol in Acquired Tolerance of Rice to Drought Stress

Basmah M. Alharbi [1], Awatif Mahfouz Abdulmajeed [2] and Heba Hassan [3,*]

1 Biology Department, Faculty of Science, University of Tabuk, Tabuk 71421, Saudi Arabia; b.alharbi@ut.edu.sa
2 Biology Department, Faculty of Science, University of Tabuk, Umluj 46429, Saudi Arabia; awabdulmajeed@ut.edu.sa
3 Botany Department, Faculty of Science, Ain Shams University, Cairo 11566, Egypt
* Correspondence: hebametwally@sci.asu.edu.eg

Abstract: To assess the effect of triacontanol (TRIA) on rice plants grown under normal or drought conditions, rice seeds were presoaked in TRIA (35 ppm) for two hours. After 20 days of sowing, rice seedlings developed from TRIA-treated or untreated seeds were subjected to drought stress. After 10 days of plant exposure to drought stress, data of major growth attributes and the content of photosynthetic pigments were recorded. Moreover, the effect of drought stress on stomatal conductance and the photochemical efficiency of PSII (Fv/Fm) were followed. The data obtained indicated that the species of rice (*Oryza sativa* L.) cultivar Giza 177 under investigation was sensitive to drought stress where there were significant decreases in the fresh and dry weights of shoots and roots and in stomatal conductance, as well as in the content of chlorophyll a, chlorophyll b, and carotenoids. Seed priming with TRIA enhanced both growth and acquired plant tolerance to drought stress. Thus, TRIA via the enhancement of stomatal conductance through the regulation of stomatal closure, the rate of water loss, ABA metabolism, the accumulation of osmolytes, and the regulation of aquaporins genes improved the water status of plants grown under water scarcity. Moreover, TRIA via increasing the content of free amino acids and sugars under drought stress may increase the chance of plant tissues to retain more water under scarcity conditions.

Keywords: triacontanol; drought; rice; aquaporins; *PIP1,1*, *PIP1,2*, *PIP2,4* and *PIP2,5* genes

Citation: Alharbi, B.M.; Abdulmajeed, A.M.; Hassan, H. Biochemical and Molecular Effects Induced by Triacontanol in Acquired Tolerance of Rice to Drought Stress. *Genes* **2021**, *12*, 1119. https://doi.org/10.3390/genes12081119

Academic Editor: Patrizia Galeffi

Received: 2 June 2021
Accepted: 20 July 2021
Published: 23 July 2021

1. Introduction

Drought is one of the major environmental constrictions limiting plant development and productivity [1]. Drought menaces about 70% of arable land worldwide. Consequently, the major crops will exhibit over 65% reduction in their yield by 2050 because of drought all over the world [2,3]. Rice is one of the major staple food crops for most of world population, and belongs to semi-aquatic plants, so it requires a high soil moisture level [4]. Thus, rice plants are susceptible to water scarcity, which induces a variety of morphological, molecular, and physiological changes [5]. It was reported that major growth attributes of important crops are severely affected by drought stress [6]. It was also reported that drought stress disturbs the leaf water potential, transpiration rate, and stomatal conductance [7]. Moreover, drought induces oxidative stress that destroys various macromolecules as proteins, lipids, and nucleic acids concomitant with cell membranes damage [8]. It was recorded that seedling growth, dry weight, and vegetative growth were reduced under drought stress in various important crops including pea (*Pisum sativum* L.), alfalfa (*Medicago sativa* L.), and rice (*Oryza sativa* L.) [6,9,10].

Plants can tolerate drought stress by developing different structural, biochemical, and molecular strategies including accumulation of certain osmolytes and proteins [11]. Indeed, drought stress accelerates abscisic acid (ABA) biosynthesis, which plays a crucial role in stomatal conductance [12,13]. In this connection, the accumulation of root ABA under drought stress was reported in many plants such as rice, beans, and potato [14–16].

Furthermore, plants can cope with drought stress by stimulating the expression of various genes as genes of some protective proteins, water channel proteins (aquaporins), enzymes catalyzing osmolyte biosynthesis, proteases, and detoxification enzymes. Similarly, genes encode various proteins such as kinases, transcription factors, phosphatases, and enzymes that regulate certain pathways including ABA biosynthesis and phospholipid metabolism were also regulated under drought condition [11,17]. Aquaporins (AQPs) are a class of intrinsic proteins that play an important role in regulating the transmembrane transport of water [18,19] and small molecules like glycerol, urea, and CO_2 [20,21]. Many types of AQPs are known in plants for their importance in stabilizing cell membrane homeostasis and keeping movement of water through the plant body under drought conditions. Plasma membrane intrinsic proteins (PIP) belong to one of the subfamilies of AQPs, and PIPs are further subdivided into two phylogenetic subgroups: PIP1s and PIP2s [22].

Triacontanol (TRIA), the material used as a seed primer in the present work, is a saturated primary alcohol classified as a plant growth regulator (PGR) that stimulates many physiological and biochemical processes in crop plants [23,24]. Triacontanol at relatively low concentrations enhances the growth of most crops such as rice (*O. sativa* L.) and maize (*Zea mays* L.) [25,26]. Currently, TRIA has been used to improve plant tolerance against abiotic stresses such as chilling, drought, and heavy metal and salt stresses [25,27,28]. Notably, under abiotic stresses, exogenously applied TRIA stimulates growth, increases the content of photosynthetic pigments, and increases compatible osmolyte accumulation [29,30]. Additionally, it enhances enzymatic and non-enzymatic antioxidant defense systems [27,30–33]. TRIA can also mitigate stress hazards via the regulation of the expression of some genes [28,32]. The present work investigates the efficiency of TRIA in enhancing drought tolerance of rice plants.

2. Materials and Methods

2.1. Materials and Growth Conditions

Grains of rice (*O. sativa* L.) cultivar Giza 177 were obtained from the Agriculture Research Center, Rice Research Institute in Giza, Egypt. Triacontanol (TRIA) was obtained from Sigma-Aldrich (Lot 637238, St. Louis, MO, USA). This research was conducted at Faculty of Science, Ain Shams University, Egypt using two controlled growth chambers, model V3-DM, Vision scientific company, Korea. The grains were surface sterilized by immersion in 1% (*w/v*) sodium hypochlorite solution for 5 min, then washed three times with sterile distilled water prior to an experimental procedure to prevent fungal contamination.

2.2. Imposition of Treatments

The sterilized grains were divided into two groups, which were soaked either in water or TRIA (35 ppm) for 2 h. The experiment was conducted in a naturally lit greenhouse (day/night temperatures about 27/32 ± 2 °C and a 14 h photoperiod) of the Botany Department, Faculty of Science, Ain Shams University. This experiment was carried out in a complete randomized design with three replicates. The sterilized rice grains of the two groups were sown in plastic pots (25 × 25 cm^2) filled with homogenous soil (50 pots for each group). The physical and chemical analysis of soil are given in Table 1.

Table 1. Some physical and chemical properties of soil.

pH	Electrical Conductivity (EC) (dS m^{-1})	Clay (%)	Silt (%)	Sand (%)	Texture
8.2	2.8	53.3	31.4	15.3	Clayey

The irrigation of all pots was carried out using the same volume of water based on the maximum water-holding capacity of the soil used in the present work. After 15 days of sowing, thinning was done so that 10 uniform seedlings were left in each pot. After 20 days of sowing, drought stress was imposed on half of each group by withholding irrigation.

After 10 days of rice seedling exposure to drought stress, the experiment was terminated as severe growth retardation was observed, compared with the control or with seedlings developed from pre-soaked grains in TRIA. Both shoots and roots were collected directly frozen in liquid nitrogen and then stored at −80 °C for biochemical analyses.

2.3. Methods

2.3.1. Measurement of Number of Stomata

Direct microscopic measurements of number of stomata were carried out following the method described by [34]. Leaf epidermal strips were obtained from a fully expanded leaf and immediately immersed in absolute alcohol for fixation and preservation. The epidermal strip was prepared on a slide and then covered with a cover slip. The total number of stomata as well as number of open stomata per μm^2 on the upper and lower epidermis were counted using an eye-piece graticule, which is calibrated by using stage micrometer scale.

2.3.2. Chlorophyll Fluorescence Measurements

The chlorophyll a fluorescence measurement was assessed in leaves in the morning hours. The intact flag leaves were dark adapted for 30 min using light-withholding clips. Leaf chlorophyll fluorescence was measured simultaneously using a pulse amplitude modulation portable fluorometer (Handy PEA, Hansatech, Norfolk, UK). After the adaptation of leaves to darkness, a single, strong, 1 s light pulse (3500 $\mu mol\ m^2\ s$) was applied. Three replicates were used for each treatment. The fast fluorescence kinetics (F_0 to Fm) value was recorded during 10 μs to 1 s [35]. The maximum quantum efficiency of PSII photochemistry (Fv/Fm) was calculated according to the equation:

$$Fv/Fm = (Fm - F_0)/Fm$$

where F_0 means fluorescence intensity at 50 μs, Fm represents maximal fluorescence intensity, and Fv represents variable fluorescence.

2.3.3. Measurement of Photosynthetic Pigments

The photosynthetic pigments chlorophyll (Chl) a, Chl b, and carotenoids were extracted and determined according to the method of [36]. Fresh leaves (1 g) were homogenized in 85% aqueous acetone for 5 min. Then, the homogenate was centrifuged, and the supernatant was made up to 100 mL with 85% acetone. The extinction was measured against a blank of pure 85% aqueous acetone at three different wave lengths (452.5, 644, and 663 nm) by using spectrophotometer (Spectronic 601, Milton Roy Company, Ivyland, PA, USA)

2.3.4. Measurement of Total Soluble Sugars

Total soluble sugars were analyzed by reacting 0.1 mL of the ethanolic extract with 3 mL of freshly prepared anthrone reagent (150 mg anthrone + 100 mL 72% H_2SO_4) in boiling water bath for 10 min. After cooling, the absorbance was measured at 620 nm by using a spectrophotometer [37].

2.3.5. Measurement of Free Amino Acids

Free amino acids were determined according to [38] by grinding the plant tissue (0.5 g) in water; then, 0.1 mL of the water extract was added to 1.5 mL (ethanol/acetone) of a 1:1 (v/v) mixture of 0.1 mL phosphate buffer (pH 6.5) and 2 mL ninhydrin reagent (0.5% in n-butanol). Then, the mixture was placed in a boiling water bath for 20 min, and then cooled immediately in ice water, and methanol was added to 10 mL. The absorbance was measured directly at 580 nm by using spectrophotometer.

2.3.6. Determination of Free Proline

The total free proline was assessed by the method described by [39] using ninhydrin reagent. The plant tissue (0.5 g) was grinded in 6 mL of 3% (*w*/*v*) sulfosalicylic acid solution. Then, the filtrate (2 mL) was reacted with 2 mL ninhydrin reagent and 2 mL glacial acetic acid, and the mixture was kept in boiling water bath for 1 h. Then, the mixture was cooled in ice and was separated using a separating funnel. The absorbance of the upper phase was read at 520 nm by using spectrophotometer.

2.3.7. Determination of Electrolyte Leakage (EL)

The stress injury was measured by electrolyte leakage as described by [40]. Leaf samples (0.5 g) were incubated with 20 mL of deionized water for 24 h at 25 °C. Then, the electrical conductivity of the solution (L_1) was quantified using a conductivity meter (HI 8733, Hanna Instruments, Woonsocket, RI, USA). Samples were then autoclaved at 120 °C for 20 min and then the final conductivity (L_2) was assessed after equilibration at 25 °C. The EL was determined according to the following equation:

$$EL\% = (L_1/L_2) \times 100$$

2.3.8. Lipid Peroxidation

Lipid peroxidation was determined by measuring the amount of malondialdehyde (MDA) produced by the thiobarbituric acid reaction as described by [41]. The plant tissue (0.5 g) was grinded in water then the crude extract was mixed with the same volume of a 0.5% (*w*/*v*) thiobarbituric acid solution containing 20% (*w*/*v*) trichloroacetic acid. The mixture was heated at 95 °C for 30 min and then quickly cooled in an ice-bath. The mixture was centrifuged at 3000× *g* for 5 min and the absorbance of the supernatant was measured at 532 and 600 nm by using a spectrophotometer.

2.3.9. Relative Water Content (RWC)

The relative water content was measured following the method described by [42]. Leaf discs from the fully expanded and uniform leaves were taken. The fresh mass (FM) of leaf discs was measured, and then samples were placed in a Petri dish with distilled water for 4 h. The water saturated mass (WSM) was then measured, and the leaf samples were placed in an oven at 80 °C for 48 h to determine the dry mass (DM). Leaf RWC was calculated as:

$$RWC\,[\%] = [(FM - DM)/(WSM - DM] \times 100$$

2.3.10. Extraction, Separation, and Determination of Abscisic Acid (ABA)

The method of hormones extraction was essentially similar to that adopted by [43]. The frozen tissue was homogenized in cold 85% ethanol by an electric automixer and then extracted by an electric stirrer with 85% ethanol at about 0 °C. The solvent was changed three times. After filtration, the three extracts were combined together and concentrated under a vacuum at 20–25 °C to a few mL. The concentrated aqueous phase was adjusted to pH 8.8 by using 1% NaOH. The alkaline aqueous phase was shaken three times with equal quantities of ethyl acetate using a separating funnel. The combined ethyl acetate fraction was evaporated to dryness and held for further purification. The aqueous fraction was acidified to pH 2.8 with 1% HCI and shaken three times with equal volumes of ethyl acetate. The remaining aqueous phase was discarded. The combined acidic ethyl acetate phase was reduced to a certain volume to determine the abscisic acid (ABA) by using gas chromatography (GC). The dried basic ethyl acetate fraction was dissolved in 80% methanol.

2.3.11. Quantitative Real-Time PCR (qRT-PCR) Analysis

The total RNA was extracted from rice tissue (100 mg) of all treatments with 30% PEG6000 using the RNeasy Plant Mini Kit (Qiagen, Amsterdam, The Netherlands). The total RNA (1 μg) from each sample was transformed into cDNA by the reverse transcription using the c.DNA Kit (TaKaRa) according to the manufacturer's instructions. The qRT-PCR was conducted on an ABI 7500 system (Applied Biosystems, New York, NY, USA) using a TransStart™ Green qRT-PCR Super Mix Kit (TransGen, Beijing, China). OsActin rRNA was used as a reference gene to standardize the relative transcriptional abundance and to minimize different copy numbers of cDNA templates [44]. All data were calculated from three replicates based on the $2^{-\Delta\Delta Ct}$ method [45]. The primers of the *PIP1,1*, *PIP1,2*, *PIP2,4*, and *PIP2,5* genes (Table 2) used in the qRT-PCR excluded the highly conserved protein domain and had high efficiency and specificity.

Table 2. The primers used for real-time PCR analysis.

Primer Name	Primer Sequence 5′-3′	Gene Accession Number
PIP1,1 F′	TGCGCAGCCGACGACATG	AK061769
PIP1,1 R	CATACAGTGACTGAGTACTGGATTAC	
PIP1,2 F	CTGTCAAGATGCCAATCCAGAG	AK098849
PIP1,2 R	GAACCGAACTCCAATAGGAGGA	
PIP2,4 F	GAGCTCGTCTGGTGATATCC	AK072632
PIP2,4 R	CATGAAGACAACAGAGGGACAG	
PIP2,5 F	GCTTAAGCCGCAATCAAATGTGC	AK107700
PIP2,5 R	CGATCGAACAATGTCACACTTGC	
OsActin F	CTGGGTTCGCCGGAGATGAT	XM_015774830.2
OsActin R	TGAGATCACGCCCAGCAAGG	

2.4. Statistical Analysis

The experimental data presented in this work were statistically analyzed by the one-way analysis of variance (ANOVA) using SPSS v20.0 (SPSS Inc., Chicago, IL, USA) analyzing software. Statistical significances of the means were compared with Duncan's test at $p \leq 0.05$ levels and the standard error (SE) of the means are shown in tables and figures as mean ± SE, with the number of degrees of freedom (n) = 3.

3. Results

TRIA treatment of unstressed rice seedlings led to significant increases in the fresh and dry weights of shoots and roots as compared with the unstressed control (Figure 1a,b). Meanwhile, the imposition of drought stress induced a significant decrease in the fresh and dry weights of both shoots and roots as compared with the unstressed control, while stressed plants treated with TRIA showed an increase in the fresh and dry weights as compared with the stressed control (Figure 1a,b).

Moreover, the results obtained showed that the relative water content was decreased in the leaves of rice seedlings exposed to drought. On the other hand, the pretreatment with TRIA significantly increased the relative water content of stressed leaves (Figure 1c).

Likewise, drought stress induced a significant decrease in the leaf content of Chl a, Chl b, and carotenoids (Table 3). Notably, TRIA pretreatment induced a significant increase in Chl a, Chl b, and the contents of leaves of drought-stressed seedlings, compared with the untreated stressed controls (Table 3). TRIA increased the photosynthetic pigments concomitant with increments in the Fv/Fm values of the leaves of stressed rice seedlings (Table 3).

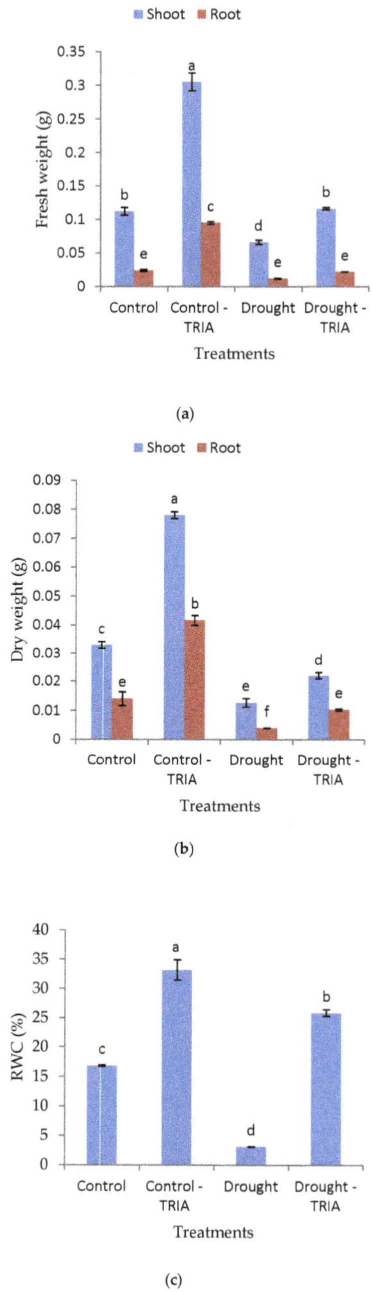

Figure 1. Effect of TRIA (35 ppm) treatment on (**a**) fresh weight, (**b**) dry weight, and (**c**) relative water content of drought-stressed rice seedlings. Each value is the mean of three replicates ± SE. Columns with different letters are significantly different at $p \leq 0.05$.

Table 3. Effects of TRIA (35 ppm) on photosynthetic pigment contents (μg/g FW) and the maximal photochemical efficiency of the primary photochemistry (Fv/Fm) of the leaves of rice seedlings exposed to drought stress. Data are means of three replicates ± SE.

Treatments	Chl (a)	Chl (b)	Carotenoid	Fv/Fm
Control	0.59 ± 0.0058 b	0.26 ± 0.013 c	0.39 ± 0.00 b	0.78 ± 0.012 c
Control-TRIA	0.79 ± 0.023 a	0.31 ± 0.0135 b	2.4 ± 0.205 a	0.93 ± 0.09 a
Drought	0.29 ± 0.012 d	0.25 ± 0.015 c	0.24 ± 0.012 b	0.04 ± 00.00 d
Drought-TRIA	0.52 ± 0.015 c	0.39 ± 0.006 a	0.5 ± 0.006 b	0.7 ± 0.006 b

Columns with different letters are significantly different at $p \leq 0.05$.

Moreover, in this study, drought obviously induced a significant decrease in percentage of open stomata (Figure 2) as compared with the unstressed control. The pretreatment of rice with TRIA significantly reduced the percentage of stomatal openings on both the upper and lower surface of rice leaves by 26% and 23.9%, respectively, as compared with the stressed control.

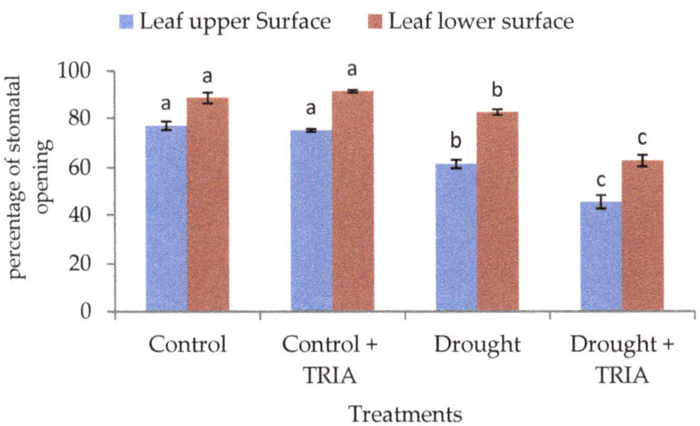

Figure 2. Effect of TRIA (35 ppm) treatment on percentage of stomatal openings on both the upper and lower surface of leaves of drought-stressed rice. Each value is the mean of three replicates ± SE. Columns with different letters are significantly different at $p \leq 0.05$.

The current data also revealed that exposure to drought stress significantly increased and decreased the ABA content of the roots and shoots, respectively (Figure 3a). In addition, proline and amino acids showed a significant increase in both the shoots and roots of drought-stressed seedlings as compared with the unstressed control (Figure 3b,c).

Meanwhile, the total soluble sugars recorded a non-significant decrease and a significant increase in the shoots and roots of drought-stressed seedlings, respectively (Figure 3d). On the other hand, TRIA treatment of drought-stressed rice seedlings led to a significant increase in ABA, proline, amino acids, and total soluble sugar contents in both the shoots and roots (Figure 3a–d).

Furthermore, drought stress induced a significant increase in the lipid peroxidation product, concomitant with a significant increase in the electrolyte leakage value (Figure 4a,b), respectively, as compared with those of the unstressed plants. The pretreatment with TRIA significantly reduced the lipid peroxidation product and electrolyte leakage of drought-stressed rice seedlings (Figure 4a,b).

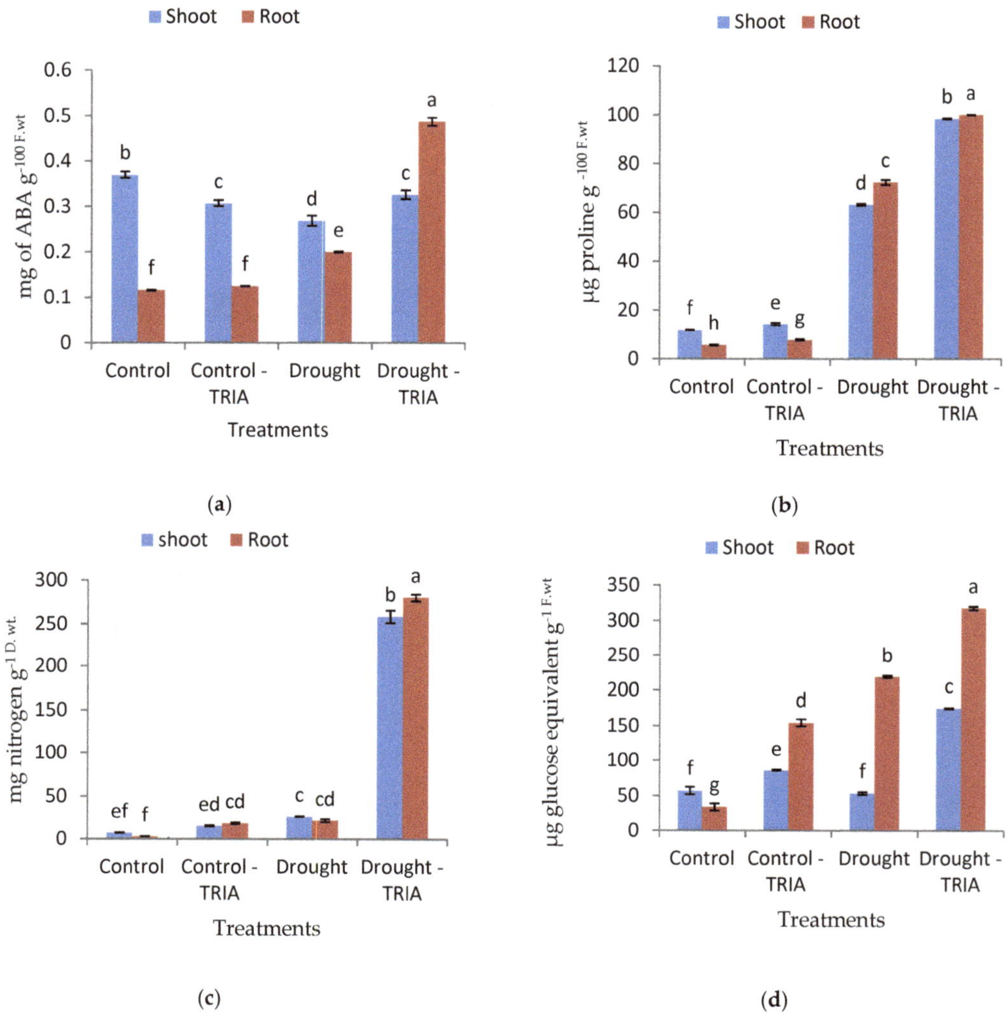

Figure 3. Effects of TRIA (35 ppm) treatment on (**a**) abscisic acid, (**b**) free proline, (**c**) free amino acids, and (**d**) total soluble sugars of drought-stressed rice seedlings. Each value is the mean of three replicates ± SE. Columns with different letters are significantly different at $p \leq 0.05$.

Notably, the RT-PCR analysis showed that drought downregulated *PIP1,1*, *PIP1,2*, *PIP2,4*, and *PIP2,5* expressions (Figure 5a–d). Meanwhile, TRIA pretreatment stimulated the overexpression of *PIP1,1*, *PIP1,2*, *PIP2,4*, and *PIP2,5* of drought-stressed rice shoots and roots (Figure 5a–d) as compared with the stressed, untreated plants. The maximum expressions of *PIP1,1*, *PIP2,4*, and *PIP2,5* were observed in TRIA-primed stressed leaves; however, the greatest expression of *PIP1,2*, was assayed in TRIA-primed stressed roots.

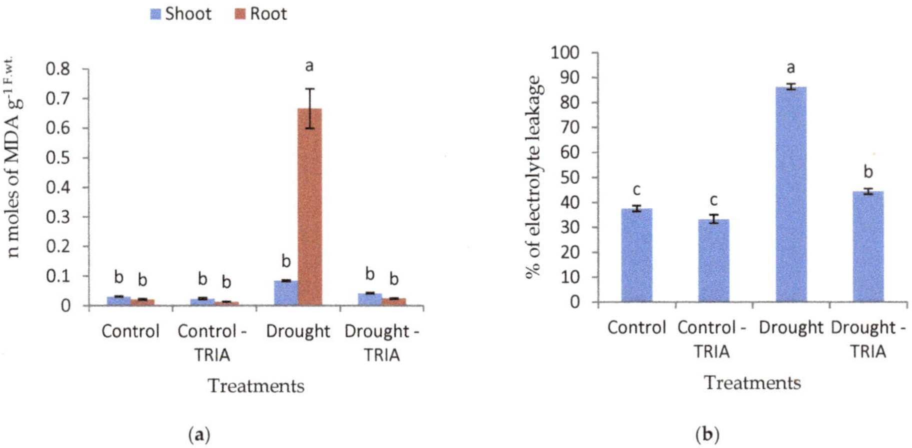

Figure 4. Effects of TRIA (35 ppm) treatment on (**a**) malondialdehyde (MDA) content and (**b**) the percentage of electrolytes leakage (EL) of drought-stressed rice seedlings. Each value is the mean of three replicates ± SE. Columns with different letters are significantly different at $p \leq 0.05$.

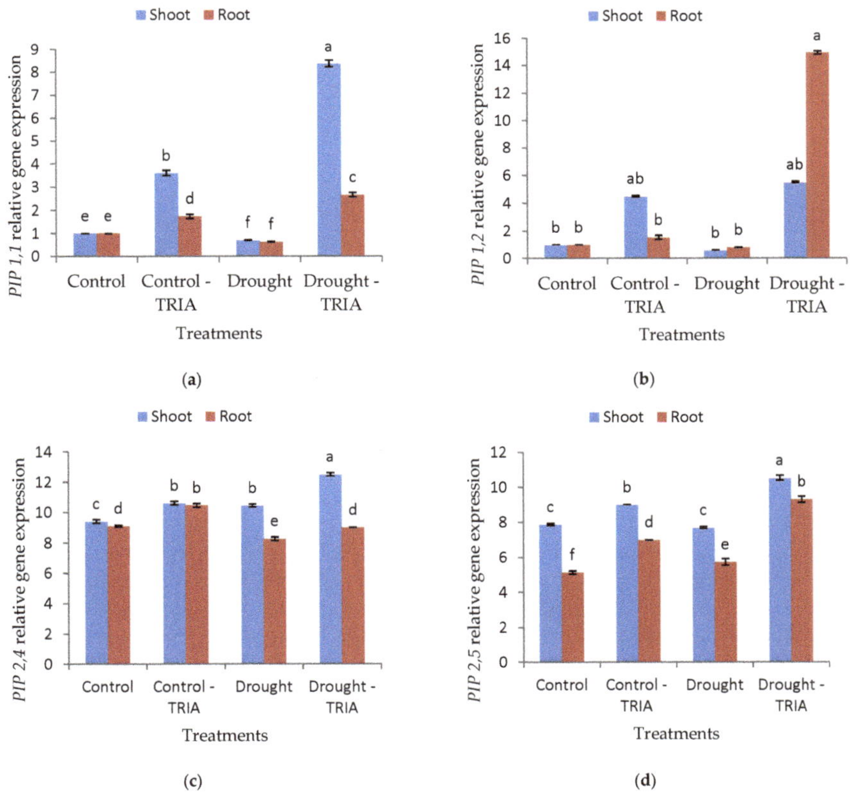

Figure 5. Effects of TRIA (35 ppm) treatment on mRNA expression of plasma membrane intrinsic protein (**a**) PIP1,1, (**b**) PIP1,2, (**c**) PIP2,4, and (**d**) PIP2,5 of drought-stressed rice seedlings. Each value is the mean of three replicates ± SE. Columns with different letters are significantly different at $p \leq 0.05$.

4. Discussion

Drought stress adversely affects plant growth and development. On the other hand, the ability to survive the drought state is a result of adaptations that prevent or decrease cellular destruction that occurs with the scarcity of water. The results presented in Figure 1a,b showed a significant decrease in both the fresh and dry weighs of shoots and roots of rice exposed to drought conditions. Such an effect was concomitant with a significant decrease in the relative water content (RWC) of rice leaves (Figure 1c), which may reduce the cell turgor pressure and, thus, cause growth retardation [46,47]. The drop in RWC of drought-stressed seedlings may be attributed to a decline in water uptake by roots, which was recorded by [48]. Moreover, drought adversely affects the photosynthetic process, as it caused a decline in the photosynthetic pigments content in our study (Table 3). This effect may be attributed to destruction or photooxidation of chlorophyll and/or inhibition of Chl synthesis, or to an increase in the activity of chlorophyllase [48,49].

Triacontanol (TRIA) is a promising plant growth regulator, as it plays an active role in the upregulation of major physiological activities required in different stages of plant growth [50,51]. In the present study, the priming of rice grains with TRIA (35 ppm) increased the fresh and dry weights of the shoots and roots of rice grown under well-watered conditions (Figure 1a,b) compared with plants exposed to drought stress. Similarly, TRIA enhanced the growth of several crops like ginger [23], tomato [52], rice [53], and viviparous [54]. The significant promoting effects of TRIA on the fresh and dry weights of TRIA-treated rice seedlings concomitant with increases in the chlorophyll a, chlorophyll b, and carotenoid contents of leaves (Table 3) demonstrated the motivating effect of TRIA on the photosynthetic efficiency, which, sequentially, improved the assimilation rate and the accumulation of photosynthates. In this regard, it has been reported by many researchers that TRIA stimulated photosynthesis in several crops such as *Papaver somniferum* L. [55], *Vigna radiata* L. [56], and *Lablab purpureus* L. [57]. Given this connection, it was reported that TRIA may enhance the photosynthetic process via increasing the rate of chlorophyll synthesis, as well as the number and size of chloroplasts [58–60]. Moreover, Fv/Fm values were markedly increased in TRIA-treated rice seedlings (Table 3), and thereby, may contribute to improving the photosynthetic efficiency of PSII and to lessening the degree of photoinhibition [59,61]. In addition, TRIA may increase photosynthesis via enhancing the activity of the Rubisco enzyme, photosynthetic pigments, and the upregulation of many photosynthetic genes [23,62].

Now it has been well documented that TRIA functions as a signaling molecule and accelerates plant tolerance against various abiotic stresses [25,63]. TRIA clearly improved seedling growth, as measured by both the fresh and dry weights of rice seedlings exposed to drought stress as compared with the stressed control ones (Figure 1a,b). Similar results were obtained by [64,65] on drought-stressed *V. radiata* L and rice seedlings, respectively. The increments in the previous parameters may be attributed to the significant increase obtained in the RWC (Figure 1c) and the significant decrease obtained in the percentage of stomatal opening (Figure 2), which may have played a crucial role in controlling water loss. The stomatal closure is an important strategy to avoid water loss and help plants tolerate drought conditions [7,66]. TRIA as a constituent of wax in the cuticle of plants [67] might have a role in controlling the rate of water loss and might have counteracted the drought-induced disturbance in the leaf water potential. Moreover, TRIA treatment increased the abscisic acid content (ABA) of the shoots and roots of drought-stressed rice seedlings as compared with the control (Figure 3a). ABA is among the hormones that affect the water status of plants via regulation of the stomatal function [68,69].

In addition, the results obtained suggested that TRIA priming helped plants to improve the water status under drought stress through osmotic adjustment, as attained by the accumulation of some osmolytes, including proline, free amino acids, and total soluble sugars (Figure 3b–d). The accumulation of proline, total soluble sugars, and free amino acids helps in reducing the cell osmotic potential, thereby diminishing water loss under water scarcity [70]. In the present investigation, the priming of rice grains with

TRIA stimulated the accumulation of high levels of proline under stressed conditions (Figure 3b). Proline is a significant indicator for finding out how tolerant plants are under water-restrictive conditions [71]. Proline protects plants from the hazards of dehydration stress via maintaining osmotic adjustment, maintains membrane integrity, and enhances the antioxidant defense system [72,73]. Proline accumulation in TRIA-treated seedlings exposed to either drought or salt stress has been reported by [31,74].

In the present work, rice exposure to drought stress led to oxidative imbalance as indicated from the marked increase in the percentage of electrolyte leakage (EL) and malondialdehyde (MDA) content (Figure 4), as well as the significant decrease in photosynthetic pigments, which may be a result of its photo-oxidation and degradation under the effect of accumulated free radicals, induced by oxidative stress [75]. In response to TRIA priming, the reverse was true.

Hence, it was reported that a major requirement for plants to tolerate drought stress is the ability to withstand and/or counteract the oxidative imbalance associated with the decrease in available water. TRIA, in this regard, played a crucial role in lessening the hazards of oxidative stress via increasing the content of free amino acids in stressed plants, which among other effects, could mitigate oxidative stress via reducing reactive oxygen species [76]. The increase in total soluble sugars and free amino acids may be attributed to the acceleration of the photosynthesis process via increasing the photosynthetic pigment contents. In this context, the current study showed that TRIA increased the Chl a and Chl b contents compared with the stressed plants. Such an effect of TRIA may be attributed to its role in protecting the chlorophyll from oxidation by increasing the carotenoid content in TRIA-treated plants (Table 3). Carotenoids act as an important antioxidant protecting pigments from the oxidation induced by the stressful condition [77]. Chlorophyll fluorescence is a good indicator of stress tolerance [78], and in the current study, TRIA treatment showed an increment in Fv/Fm values under drought stress conditions, which refers to a higher photochemical efficiency of PSII [59]. It was reported that TRIA has also increased the Fv/Fm values under various abiotic stresses, such as salt and chilling stresses [29,30].

Predicted functions of sugars and amino acids, particularly of those which are hydrophilic, for the improvement of the water status of plants subjected to drought stress include the following: water replacement molecules, when acting as protectants, and stabilizing subcellular structures in drought conditions [79]. Moreover, some of these amino acids and sugars, which have polar groups within their structures, may coat intracellular macromolecules with a cohesive water layer providing preferential hydration to these molecules and, hence, more retention of water under its scarcity [80].

The alleviation of oxidative stress by TRIA in drought-stressed plants could be achieved via reducing membrane injury [81]. Likewise, the present study showed that TRIA notably decreased EL in association with a decrease in the lipid peroxidation product MDA content as compared with drought-stressed plants (Figure 4a,b). In this context, TRIA has been observed to reduce membrane permeability and MDA contents in maize seedlings under salinity stress [32]. Such an effect of TRIA may be attributed to the activation of some antioxidant enzymes that contribute to buffering the excess reactive oxygen species (ROS), which results in alleviating the stress damage effects on plants. It was reported in many studies that TRIA stimulated the activity of some antioxidant enzymes under drought and salinity stresses [31,65,82].

Aquaporins (AQPs) play a crucial role in regulation of water transport through plants; hence, they take part in drought stress tolerance. Plasma membrane intrinsic proteins (PIP) belong to one of the subfamilies of AQPs, and PIPs are further subdivided into two phylogenetic subgroups: PIP1s and PIP2s. In the current study, the increase in water content in TRIA-treated plants was accompanied with the upregulation of *PIP1,1*, *PIP1,2*, *PIP2,4*, and *PIP2,5* genes (Figure 5a–d) in both the shoots and roots of plants, either under normal or drought conditions. In this context, it was reported that TRIA can alleviate the toxic effects of stress by regulating the gene expression [28,32]. Generally, the downregulation of specific *PIP* isoforms leads to a decrease in water permeability of protoplasts, and consequently increases susceptibility to drought and osmotic stress [83,84]. Hence, *PIP* isoform overexpression participates in the increments of root osmotic hydraulic conductivity [85,86]. Moreover, the overexpression of *PIPs* genes in TRIA-treated plants may be attributed to the accumulation of ABA [87,88]. It was reported in many studies that the application of exogenous ABA increased *PIP* gene expression under normal water supply [89–91]. In addition, the accumulation of ABA under drought stress plays a crucial role in regulating AQP gene expression [87,88].

5. Conclusions

The current results provide molecular and physiological evidence supporting the vital roles of triacontanol in improving the water status in drought-stressed rice seedlings, which may play a beneficial role in horticultural crop management to tolerate climatic fluctuations. The obtained results showed that TRIA alleviated the adverse effects caused by drought stress through molecular and physiological strategies, which contribute to improving the water status. Moreover, TRIA via increasing the content of free amino acids and soluble sugars under drought stress may increase the efficiency of stressed plants to retain water under its scarcity. Such an effect of TRIA was evident by the increase in the RWC and decrease in the MDA content and EL. In addition, TRIA highly induced the expression of aquaporin-related genes (*PIP1,1, PIP1,2, PIP2,4,* and *PIP2,5*) that might be involved in the regulation of water transport.

Author Contributions: H.H., B.M.A., and A.M.A. conceived and planned the experiments. H.H. performed the experiments. B.M.A. and A.M.A. contributed to sample preparation and formal analysis. H.H. contributed to the interpretation of the results and writing—review and editing of the manuscript. All authors provided critical feedback and helped shape the research, analysis, and manuscript. All authors have read and agreed to the published version of the manuscript.

Funding: This research received no external funding.

Institutional Review Board Statement: Not applicable.

Informed Consent Statement: Not applicable.

Data Availability Statement: The data presented in this study are available in this manuscript.

Conflicts of Interest: The authors declare no conflict of interest.

Abbreviations

Chl a	chlorophyll a
Chl b	chlorophyll b
EL	electrolyte leakage
Fv/Fm	the maximal photochemical efficiency
ABA	Abscisic acid
AQPs	Aquaporins
PIP1,1, PIP1,2, PIP2,4 and *PIP2,5*	Plasma membrane intrinsic proteins 1,1; 1,2; 2,4; and 2,5 genes
TRIA	triacontanol
RWC	relative water content

References

1. Kour, D.; Rana, K.L.; Yadav, A.N.; Sheikh, I.; Kumar, V.; Dhaliwal, H.S.; Saxena, A.K. Amelioration of drought stress in Foxtail millet (*Setaria italica* L.) by P-solubilizing drought-tolerant microbes with multifarious plant growth promoting attributes. *Environ. Sustain.* **2020**, *3*, 23–34. [CrossRef]
2. Sharma, M.; Gupta, S.K.; Majumder, B.; Maurya, V.K.; Deeba, F.; Alam, A.; Pandey, V. Salicylic acid mediated growth, physiological and proteomic responses in two wheat varieties under drought stress. *J. Proteom.* **2017**, *163*, 28–51. [CrossRef]
3. Sharma, S.; Mujumdar, P. Increasing frequency and spatial extent of concurrent meteorological droughts and heat waves in India. *Sci. Rep.* **2017**, *7*, 15582. [CrossRef] [PubMed]
4. Mckersie, B.D.; Ya'acov, Y.L. Oxidative stress. In *Stress and Stress Coping in Cultivated Plants*; Springer Science & Business Media: Berlin/Heidelberg, Germany, 1994; pp. 15–54.

5. Bray, E.A. Molecular responses to water deficit. *Plant Physiol.* **1993**, *103*, 1035. [CrossRef]
6. Okçu, G.; Kaya, M.D.; Atak, M. Effects of salt and drought stresses on germination and seedling growth of pea (*Pisum sativum* L.). *Turk. J. Agric. For.* **2005**, *29*, 237–242.
7. Farooq, M.; Wahid, A.; Kobayashi, N.; Fujita, D.; Basra, S.M.A. Plant drought stress: Effects, mechanisms and management. *Agron. Sustain. Dev.* **2009**, 185–212. [CrossRef]
8. Chong, P.; Li, H.; Li, Y. Physiological responses of seedling roots of the desert plant *Reaumuria soongorica* to drought stress. *Acta Pratacult. Sin.* **2015**, *24*, 72–80.
9. Zeid, I.M.; Shedeed, Z.A. Response of alfalfa to putrescine treatment under drought stress. *Biol. Plant.* **2006**, *50*, 635–640. [CrossRef]
10. Manickavelu, A.; Nadarajan, N.; Ganesh, S.K.; Gnanamalar, R.P.; Babu, R.C. Drought tolerance in rice: Morphological and molecular genetic consideration. *Plant Growth Regul.* **2006**, *50*, 121–138. [CrossRef]
11. Shinozaki, K.; Yamaguchi-Shinozaki, K. Gene networks involved in drought stress response and tolerance. *J. Exp. Bot.* **2007**, *58*, 221–227. [CrossRef]
12. Yamaguchi-Shinozaki, K.; Shinozaki, K. Transcriptional regulatory networks in cellular responses and tolerance to dehydration and cold stresses. *Annu Rev. Plant Biol.* **2006**, *57*, 781–803. [CrossRef]
13. Jacobsen, S.E.; Liu, F.; Jensen, C.R. Does root-sourced ABA play a role for regulation of stomata under drought in quinoa (*Chenopodium quinoa* Willd.). *Sci. Hortic.* **2009**, *122*, 281–287. [CrossRef]
14. Xu, W.; Jia, L.; Shi, W.; Liang, J.; Zhou, F.; Li, Q.; Zhang, J. Abscisic acid accumulation modulates auxin transport in the root tip to enhance proton secretion for maintaining root growth under moderate water stress. *New Phytol.* **2013**, *197*, 139–150. [CrossRef]
15. Puertolas, J.; Alcobendas, R.; Alarcón, J.J.; Dodd, I.C. Long-distance abscisic acid signalling under different vertical soil moisture gradients depends on bulk root water potential and average soil water content in the root zone. *Plant Cell Environ.* **2013**, *36*, 1465–1475. [CrossRef] [PubMed]
16. Puértolas, J.; Conesa, M.R.; Ballester, C.; Dodd, I.C. Local root abscisic acid (ABA) accumulation depends on the spatial distribution of soil moisture in potato: Implications for ABA signalling under heterogeneous soil drying. *J. Exp. Bot.* **2015**, *66*, 2325–2334. [CrossRef]
17. Khattab, H.I.; Emam, M.A.; Emam, M.M.; Helal, N.M.; Mohamed, M.R. Effect of selenium and silicon on transcription factors NAC5 and DREB2A involved in drought-responsive gene expression in rice. *Biol. Plant.* **2014**, *58*, 265–273. [CrossRef]
18. Luu, D.T.; Maurel, C. Aquaporins in a challenging environment: Molecular gears for adjusting plant water status. *Plant Cell Environ.* **2005**, *28*, 85–96. [CrossRef]
19. Hassan, M.U.; Aamer, M.; Chattha, M.U.; Ullah, M.A.; Sulaman, S.; Nawaz, M.; Zhiqiang, W.; Yanqin, M.; Guoqin, H. The role of potassium in plants under drought stress: Mini review. *J. Basic Appl. Sci.* **2017**, *13*, 268–271.
20. Maurel, C.; Verdoucq, L.; Luu, D.T.; Santoni, V. Plant aquaporins: Membrane channels with multiple integrated functions. *Annu. Rev. Plant Biol.* **2008**, *59*, 595–624. [CrossRef] [PubMed]
21. Uehlein, N.; Lovisolo, C.; Siefritz, F.; Kaldenhoff, R. The tobacco aquaporin NtAQP1 is a membrane CO_2 pore with physiological functions. *Nature* **2003**, *425*, 734–737. [CrossRef]
22. Hachez, C.; Moshelion, M.; Zelazny, E.; Cavez, D.; Chaumont, F. Localization and quantification of plasma membrane aquaporin expression in maize primary root: A clue to understanding their role as cellular plumbers. *Plant Mol. Biol.* **2006**, *62*, 305–323. [CrossRef]
23. Singh, M.; Khan, M.M.; Moinuddin, N.M. Augmentation of nutraceuticals, productivity and quality of ginger (*Zingiber officinale* Rosc.) through triacontanol application. *Plant Biosyst. Int. J. Deal. Asp. Plant Biol.* **2012**, *146*, 106–113.
24. Karam, E.A.; Keramat, B. Foliar spray of triacontanol improves growth by alleviating oxidative damage in coriander under salinity. *Indian J. Plant Physiol.* **2017**, *22*, 120–124. [CrossRef]
25. Naeem, M.; Khan, M.M.A.; Moinuddin. Triacontanol: A potent plant growth regulator in agriculture. *J. Plant Interact.* **2012**, *7*, 129–142. [CrossRef]
26. Ahmad, H.F.S.; Hassan, H.M.; El-Shafey, A.S. Effect of cadmium on growth, flowering and fruiting of *Zea mays* L. and possible roles of triacontanol in alleviating cadmium toxicity. *Egypt. J. Bot.* **2013**, *53*, 23–44. [CrossRef]
27. Zaid, A.; Mohammad, F.; Fariduddin, Q. Plant growth regulators improve growth, photosynthesis, mineral nutrient and antioxidant system under cadmium stress in menthol mint (*Mentha arvensis* L.). *Physiol. Mol. Biol. Plants* **2020**, *26*, 25–39. [CrossRef] [PubMed]
28. Islam, S.; Zaid, A.; Mohammad, F. Role of triacontanol in counteracting the ill effects of salinity in plants: A review. *J. Plant Growth Regul.* **2021**, *40*, 1–10. [CrossRef]
29. Borowski, E.; Blamowski, Z.K. The effect of triacontanol 'TRIA' and Asahi-SL on the development and metabolic activity of sweet basil (*Ocimum basilicum* L.) plants treated with chilling. *Folia Hortic.* **2009**, *21*, 39–48. [CrossRef]
30. Perveen, S.; Shahbaz, M.; Ashraf, M. Influence of foliar-applied triacontanol on growth, gas exchange characteristics, and chlorophyll fluorescence at different growth stages in wheat under saline conditions. *Photosynthetica* **2013**, *51*, 541–551. [CrossRef]
31. Perveen, S.; Iqbal, M.; Nawaz, A.; Parveen, A.; Mahmood, S. Induction of drought tolerance in *Zea mays* L. by foliar application of triacontanol. *Pak. J. Bot.* **2016**, *48*, 907–915.

32. Perveen, S.; Iqbal, M.; Parveen, A.; Akram, M.S.; Shahbaz, M.; Akber, S.; Mehboob, A. Exogenous triacontanol-mediated increase in phenolics, proline, activity of nitrate reductase, and shoot k+ confers salt tolerance in maize (*Zea mays* L.). *Braz. J. Bot.* **2017**, *40*, 1–11. [CrossRef]
33. Maresca, V.; Sorbo, S.; Keramat, B.; Basile, A. Effects of triacontanol on ascorbate-glutathione cycle in *Brassica napus* L. exposed to cadmium-induced oxidative stress. *Ecotoxicol. Environ. Saf.* **2017**, *144*, 268–274.
34. El-Shehaby, O.A. The effect of water stress-induced by polyethylene glycol on stomata and photosynthesis in Lupinus termis plants. *Egypt. J. Physiol. Sci.* **1994**, *18*, 381–392.
35. Strasser, R.J.; Tsimilli-Michael, M.; Srivastava, A. Analysis of the fluorescence transient. In *Chlorophyll Fluorescence: A Signature of Photosynthesis, Advances in Photosynthesis and Respiration Series*; George, C., Papageorgiou, C., Govindjee, G., Eds.; Springer: Dordrecht, The Netherlands, 2004; pp. 321–362.
36. Metzner, H.; Rau, H.; Senger, H. Untersuchungen zur synchronisierbarkeit einzelner pigmentmangel-mutanten von Chlorella. *Planta* **1965**, *65*, 186–194. [CrossRef]
37. Fairbairn, N.J. A modified anthrone reagent. *Chem. Ind.* **1953**, *4*, 285–313.
38. Muting, D.; Kaiser, E. Spectrophotometric method of determining of amino-N in biological material by means of the ninhydrin reaction. *Hoppe-Seylers Z. Physiol. Chem.* **1963**, *323*, 276–332.
39. Bates, L.S.; Waldren, R.P.; Teare, I.D. Rapid determination of free prolin for water-stress studies. *Plant Soil* **1973**, *39*, 205–207. [CrossRef]
40. Valentovič, P.; Luxová, M.; Kolarovič, L.; Gašparíková, O. Effect of osmotic stress on compatible solutes content, membrane stability and water relations in maize cultivars. *Plant Soil Environ.* **2006**, *52*, 186–191. [CrossRef]
41. Heath, R.L.; Packer, L. Photoperoxidation in isolated chloroplasts. I. kinetics and stoichiometry of fatty acid peroxidation. *Arch. Biochem. Biophys.* **1968**, *125*, 189–198. [CrossRef]
42. Pieczynski, M.; Marczewski, W.; Hennig, J.; Dolata, J.; Bielewicz, D.; Piontek, P.; Wyrzykowska, A.; Krusiewicz, D.; Strzelczyk-Zyta, D.; Konopka-Postupolska, D.; et al. Downregulation of CBP80 gene expression as a strategy to engineer a drought-tolerant potato. *Plant Biotechnol. J.* **2013**, *11*, 459–469. [CrossRef]
43. Wasfi, W.S.; Orrin, E.S. Identification of plant hormones from cotton ovules. *Plant Physiol.* **1975**, *55*, 550–554.
44. Tenea, G.N.; Bota, A.P.; Raposo, F.C.; Maquet, A. Reference genes for gene expression studies in wheat flag leaves grown under different farming conditions. *BMC Res. Notes* **2011**, *4*, 1–13. [CrossRef]
45. Livak, K.J.; Schmittgen, T.D. Analysis of relative gene expression data using real-time quantitative PCR and the $2-\Delta\Delta CT$ method. *Methods* **2001**, *25*, 402–408. [CrossRef]
46. Kabiri, R.; Hatami, A.; Oloumi, H.; Naghizadeh, M.; Nasibi, F.; Tahmasebi, Z. Foliar application of melatonin induces tolerance to drought stress in Moldavian balm plants (*Dracocephalum moldavica*) through regulating the antioxidant system. *Folia Hortic.* **2018**, *30*, 155–167. [CrossRef]
47. Delfine, S.; Tognetti, R.; Loreto, F.; Alvino, A. Physiological and growth responses to water stress in field-grown bell pepper (*Capsicum annuum* L.). *J. Hortic. Sci. Biotechnol.* **2002**, *77*, 697–704. [CrossRef]
48. Kabiri, R.; Nasibi, F.; Farahbakhsh, H. Effect of exogenous salicylic acid on some physiological parameters and alleviation of drought stress in Nigella sativa plant under hydroponic culture. *Plant Prot. Sci.* **2014**, *50*, 43–45. [CrossRef]
49. Hu, W.E.; Huang, C.; Deng, X.; Zhou, S.; Chen, L.; Li, Y.I.; Wang, C.; Ma, Z.; Yuan, Q.; Wang, Y.A.; et al. TaASR1, a transcription factor gene in wheat, confers drought stress tolerance in transgenic tobacco. *Plant Cell Environ.* **2013**, *36*, 1449–1464. [CrossRef]
50. Shahbaz, M.; Noreen, N.; Perveen, S. Triacontanol modulates photosynthesis and osmoprotectants in canola (*Brassica napus* L.) under saline stress. *J. Plant Interact.* **2013**, *8*, 350–359. [CrossRef]
51. Ramos-Zambrano, E.; Juárez-Yáñez, T.E.; Tapia-Maruri, D.; Camacho-Díaz, B.H.; Jiménez-Aparicio, A.R.; Martínez-Ayala, A.L. Effects of triacontanol and light on stomatal and photochemical responses in *Solanum lycopersicum* L. *J. Plant Growth Regul.* **2020**, 1–13. [CrossRef]
52. Khan, M.M.; Bhardwaj, G.; Naeem, M.; Mohammad, F.; Singh, M.; Nasir, S.; Idrees, M. Response of tomato (*Solanum lycopersicum* L.) to application of potassium and triacontanol. *XI Int. Symp. Process. Tomato* **2008**, *823*, 199–208. [CrossRef]
53. Pandey, N.; Upadhyay, S.K.; Tripathi, R.S. Effect of plant growth regulators and fertiuty levels on growth and yield of transplanted rice. *Indian J. Agric. Res.* **2001**, *35*, 205–207.
54. Moorthy, P.; Kathiresan, K. Physiological responses of mangrove seedling to triacontanol. *Biol. Plant.* **1993**, *35*, 577. [CrossRef]
55. Srivastava, N.K.; Sharma, S. Effect of triacontanol on photosynthesis, alkaloid content and growth in opium poppy (*Papaver somniferum* L.). *Plant Growth Regul.* **1990**, *9*, 65–71. [CrossRef]
56. Kumaravelu, G.; Livingstone, V.D.; Ramanujam, M.P. Triacontanol-induced changes in the growth, photosynthetic pigments, cell metabolites, flowering and yield of green gram. *Biol. Plant.* **2000**, *43*, 287–290. [CrossRef]
57. Naeem, M.; Khan, M.M.; Siddiqui, M.H. Triacontanol stimulates nitrogen-fixation, enzyme activities, photosynthesis, crop productivity and quality of hyacinth bean (*Lablab purpureus* L.). *Sci. Hortic.* **2009**, *121*, 389–396. [CrossRef]
58. Borowski, E.; Blamowski, Z.K.; Michałek, W. Effects of Tomatex/Triacontanol/on chlorophyll fluorescence and tomato (*Lycopersicon esculentum* Mill.) yields. *Acta Physiol. Plant* **2000**, *22*, 271–274. [CrossRef]
59. Chen, X.; Yuan, H.; Chen, R.; Zhu, L.; He, G. Biochemical and photochemical changes in response to triacontanol in rice (*Oryza sativa* L.). *Plant Growth Regul.* **2003**, *40*, 249–256. [CrossRef]

60. Muthuchelian, K.; Velayutham, M.; Nedunchezhian, N. Ameliorating effect of triacontanol on acidic mist-treated *Erythrina variegata* seedlings: Changes in growth and photosynthetic activities. *Plant Sci.* **2003**, *165*, 1253–1259. [CrossRef]
61. Chen, X.; Yuan, H.; Chen, R.; Zhu, L.; Du, B.; Weng, Q.; He, G. Isolation and characterization of triacontanol-regulated genes in rice (*Oryza sativa* L.): Possible role of triacontanol as a plant growth stimulator. *Plant Cell Physiol.* **2002**, *43*, 869–876. [CrossRef]
62. Trewavas, A.J.; Gilory, S. Signal transduction in plant cells. *Trends Genet.* **1991**, *7*, 356–361. [CrossRef]
63. Waqas, M.; Shahzad, R.; Khan, A.L.; Asaf, S.; Kim, Y.H.; Kang, S.M.; Bilal, S.; Hamayun, M.; Lee, I.J. Salvaging effect of triacontanol on plant growth, thermotolerance, macro-nutrient content, amino acid concentration and modulation of defense hormonal levels under heat stress. *Plant Physiol. Biochem.* **2016**, *99*, 118–125. [CrossRef]
64. Sanadhya, D.; Kathuria, E.; Malik, C.P. Effect of drought stress and its interaction with two phytohormones on *Vigna radiata* seed germination and seedling growth. *LS Int. J. Life Sci.* **2012**, *1*, 201–207. [CrossRef]
65. Suman, K.; Kondamudi, R.; Rao, Y.V.; Kiran, T.V.; Swamy, K.N.; Rao, P.R.; Subramanyam, D.; Voleti, S.R. Effect of triacontanol on seed germination, seedling growth and antioxidant enzyme in rice under poly ethylene glycol induced drought stress. *Agric. J.* **2013**, *60*, 132–137.
66. Franks, P.J. Passive and active stomatal control: Either or both? *New Phytol.* **2013**, *198*, 325–327. [CrossRef]
67. Kolattukudy, P.E.; Walton, T.J. The biochemistry of plant cuticular lipids. *Prog. Chem. Fats Other Lipids* **1973**, *13*, 119–175. [CrossRef]
68. Taiz, L.; Zeiger, E. *Plant Physiology*, 4th ed.; Sinauer Associates Inc. Publishers: Sunderland, MA, USA, 2006.
69. Dodd, I.C. Hormonal interactions and stomatal responses. *J. Plant Growth Regul.* **2003**, *22*, 32–46. [CrossRef]
70. Thakur, A.; Thakur, P.S.; Singh, R.P. Influence of paclobutrazol and triacontanol on growth and water relations in olive varieties under water stress. *Indian J. Plant. Physiol.* **1998**, *3*, 116–120.
71. Rahdari, P.; Hosseini, S.M.; Tavakoli, S. The studying effect of drought stress on germination, proline, sugar, lipid, protein and chlorophyll content in purslane (*Portulaca oleracea* L.) leaves. *J. Med. Plants Res.* **2012**, *6*, 1539–1547.
72. Demiral, T.; Türkan, I. Does exogenous glycinebetaine affect antioxidative system of rice seedlings under NaCl treatment? *Plant Physiol.* **2004**, *161*, 1089–1100. [CrossRef] [PubMed]
73. Tripathi, B.N.; Gaur, J.P. Relationship between copper-and zinc-induced oxidative stress and proline accumulation in Scenedesmus sp. *Planta* **2004**, *219*, 397–404. [CrossRef] [PubMed]
74. Sarwar, M.B.; Sadique, S.; Hassan, S.; Sania, R.I.; Rashid, B.; Mohamed, B.B.; Husnain, T. Physio-biochemical and molecular responses in transgenic cotton under drought stress. *J. Agric. Sci.* **2017**, *23*, 157–166.
75. Tahkokorpi, M.; Taulavuori, K.; Laine, K.; Taulavuori, E. After-effects of drought-related winter stress in previous and current year stems of *Vaccinium myrtillus* L. *Environ. Exp. Bot.* **2007**, *61*, 85–93. [CrossRef]
76. Sandhya, V.S.; Ali, S.Z.; Grover, M.; Reddy, G.; Venkateswarlu, B. Effect of plant growth promoting *Pseudomonas* spp. on compatible solutes, antioxidant status and plant growth of maize under drought stress. *Plant Growth Regul.* **2010**, *62*, 21–30. [CrossRef]
77. Silva, E.N.; Vieira, S.A.; Ribeiro, R.V.; Ponte, L.F.; Ferreira-Silva, S.L.; Silveira, J.A. Contrasting physiological responses of *Jatropha curcas* plants to single and combined stresses of salinity and heat. *J. Plant Growth Regul.* **2013**, *32*, 159–169. [CrossRef]
78. Mehta, P.; Jajoo, A.; Mathur, S.; Bharti, S. Chlorophyll a fluorescence study revealing effects of high salt stress on Photosystem II in wheat leaves. *Plant Physiol. Biochem.* **2010**, *48*, 16–20. [CrossRef]
79. Lane, B.G. Cellular desiccation and hydration: Developmentally regulated proteins, and the maturation and germination of seed embryos. *FASEB J.* **1991**, *5*, 2893–2901. [CrossRef]
80. Hoekstra, F.A.; Golovina, E.A.; Buitink, J. Mechanisms of plant desiccation tolerance. *Trends Plant Sci.* **2001**, *6*, 431–438. [CrossRef]
81. Rajasekaran, L.R.; Blake, T.J. New plant growth regulators protect photosynthesis and enhance growth under drought of jack pine seedlings. *J. Plant Growth Regul.* **1999**, *18*, 175–181. [CrossRef] [PubMed]
82. Khanam, D.; Mohammad, F. Plant growth regulators ameliorate the ill effect of salt stress through improved growth, photosynthesis, antioxidant system, yield and quality attributes in *Mentha piperita* L. *Acta Physiol. Plant* **2018**, *40*, 1–3. [CrossRef]
83. Martre, P.; Morillon, R.; Barrieu, F.; North, G.B.; Nobel, P.S.; Chrispeels, M.J. Plasma membrane aquaporins play a significant role during recovery from water deficit. *Plant Physiol.* **2002**, *130*, 2101–2110. [CrossRef]
84. Siefritz, F.; Tyree, M.T.; Lovisolo, C.; Schubert, A.; Kaldenhoff, R. PIP1 plasma membrane aquaporins in tobacco: From cellular effects to function in plants. *Plant Cell* **2002**, *14*, 869–876. [CrossRef]
85. Aharon, R.; Shahak, Y.; Wininger, S.; Bendov, R.; Kapulnik, Y.; Galili, G. Overexpression of a plasma membrane aquaporin in transgenic tobacco improves plant vigor under favorable growth conditions but not under drought or salt stress. *Plant Cell* **2003**, *15*, 439–447. [CrossRef] [PubMed]
86. Lian, H.L.; Yu, X.; Ye, Q.; Ding, X.S.; Kitagawa, Y.; Kwak, S.S.; Su, W.A.; Tang, Z.C. The role of aquaporin RWC3 in drought avoidance in rice. *Plant Cell Physiol.* **2004**, *45*, 481–489. [CrossRef]
87. Kaldenhoff, R.; Ribas-Carbo, M.I.; Sans, J.F.; Lovisolo, C.; Heckwolf, M.; Uehlein, N. Aquaporins and plant water balance. *Plant Cell Environ.* **2008**, *31*, 658–666. [CrossRef] [PubMed]
88. Parent, B.; Hachez, C.; Redondo, E.; Simonneau, T.; Chaumont, F.; Tardieu, F. Drought and abscisic acid effects on aquaporin content translate into changes in hydraulic conductivity and leaf growth rate: A trans-scale approach. *Plant Physiol.* **2009**, *149*, 2000–2012. [CrossRef] [PubMed]

89. Jang, J.Y.; Kim, D.G.; Kim, Y.O.; Kim, J.S.; Kang, H. An expression analysis of a gene family encoding plasma membrane aquaporins in response to abiotic stresses in Arabidopsis thaliana. *Plant Mol. Biol.* **2004**, *54*, 713–725. [CrossRef]
90. Zhu, C.; Schraut, D.; Hartung, W.; Schäffner, A.R. Differential responses of maize MIP genes to salt stress and ABA. *J. Exp. Bot.* **2005**, *56*, 2971–2981. [CrossRef]
91. Lian, H.L.; Yu, X.; Lane, D.; Sun, W.N.; Tang, Z.C.; Su, W.A. Upland rice and lowland rice exhibited different *PIP* expression under water deficit and ABA treatment. *Cell Res.* **2006**, *16*, 651–660. [CrossRef]

genes

MDPI

Article

The Seed Development Factors *TT2* and *MYB5* Regulate Heat Stress Response in *Arabidopsis*

Pierre Jacob [1], Gwilherm Brisou [1], Marion Dalmais [1], Johanne Thévenin [2], Froukje van der Wal [3], David Latrasse [1], Ravi Suresh Devani [1], Moussa Benhamed [1], Bertrand Dubreucq [2], Adnane Boualem [1], Loic Lepiniec [2], Richard G. H. Immink [3], Heribert Hirt [4,5] and Abdelhafid Bendahmane [1,*]

1 Institute of Plant Sciences Paris-Saclay, Université Paris-Saclay, Univ. Evry, INRAE, CNRS, 91405 Orsay, France; jpierre@email.unc.edu (P.J.); gwilherm.brisou@universite-paris-saclay.fr (G.B.); marion.dalmais@inrae.fr (M.D.); david.latrasse@ips2.universite-paris-saclay.fr (D.L.); ravi-sureshbhai.devani@universite-paris-saclay.fr (R.S.D.); moussa.benhamed@ips2.universite-paris-saclay.fr (M.B.); adnane.boualem@inrae.fr (A.B.)
2 Institut Jean-Pierre Bourgin, INRAE, AgroParisTech, Université Paris-Saclay, 78000 Versailles, France; Johanne.Thevenin@inrae.fr (J.T.); bertrand.dubreucq@inrae.fr (B.D.); loic.lepiniec@inrae.fr (L.L.)
3 Bioscience and Laboratory of Molecular Biology, Wageningen University and Research, 6708PB Wageningen, The Netherlands; froukje.vanderwal@wur.nl (F.v.d.W.); Richard.Immink@wur.nl (R.G.H.I.)
4 Darwin21, Biological and Environmental Science and Engineering Division (BESE), King Abdullah University of Science and Technology, Thuwal 23955-6900, Saudi Arabia; heribert.hirt@kaust.edu.sa
5 Max Perutz Laboratories, University of Vienna, 1030 Vienna, Austria
* Correspondence: abdelhafid.bendahmane@inrae.fr

Citation: Jacob, P.; Brisou, G.; Dalmais, M.; Thévenin, J.; van der Wal, F.; Latrasse, D.; Suresh Devani, R.; Benhamed, M.; Dubreucq, B.; Boualem, A.; et al. The Seed Development Factors *TT2* and *MYB5* Regulate Heat Stress Response in *Arabidopsis*. *Genes* **2021**, *12*, 746. https://doi.org/10.3390/genes12050746

Academic Editor: Patrizia Galeffi

Received: 6 April 2021
Accepted: 10 May 2021
Published: 15 May 2021

Publisher's Note: MDPI stays neutral with regard to jurisdictional claims in published maps and institutional affiliations.

Abstract: *HEAT SHOCK FACTOR A2* (*HSFA2*) is a regulator of multiple environmental stress responses required for stress acclimation. We analyzed *HSFA2* co-regulated genes and identified 43 genes strongly co-regulated with *HSFA2* during multiple stresses. Motif enrichment analysis revealed an over-representation of the site II element (SIIE) in the promoters of these genes. In a yeast 1-hybrid screen with the SIIE, we identified the closely related R2R3-MYB transcription factors TT2 and MYB5. We found overexpression of *MYB5* or *TT2* rendered plants heat stress tolerant. In contrast, *tt2*, *myb5*, and *tt2/myb5* loss of function mutants showed heat stress hypersensitivity. Transient expression assays confirmed that MYB5 and TT2 can regulate the *HSFA2* promoter together with the other members of the MBW complex, TT8 and TRANSPARENT TESTA GLABRA 1 (TTG1) and that the SIIE was involved in this regulation. Transcriptomic analysis revealed that *TT2/MYB5* target promoters were enriched in SIIE. Overall, we report a new function of TT2 and MYB5 in stress resistance and a role in SIIE-mediated *HSFA2* regulation.

Keywords: environmental stress; seed development; site II element; HSFA2; TT2/ MYB5-MBW complex

1. Introduction

Heat shock factors (HSFs) represent a widely conserved class of transcription factors involved in stress response and development [1,2]. Although they were first discovered in the context of the heat shock response, most biotic and abiotic stress responses require the concerted action of HSFs to regulate stress response and acclimation [3,4]. Elucidating the molecular mechanisms responsible for the regulation of *HSFs* is critical to enhance stress tolerance of plants.

Plant genomes contain a large number of *HSFs*; 21 in *Arabidopsis*, 25 in rice or 38 in soybean, compared to a single *HSF1* in *Saccharomyces cerevisiae* and seven in humans [5]. HSFs are grouped into three classes depending on the presence of specific protein domains/motifs. A class HSFs positively regulate gene expression as they exhibit the transcription activator motif AHA (aromatic and large hydrophobic residues in an acidic context), whereas B and C class HSFs are considered to function as transcriptional inhibitors or co-activators.

In *Arabidopsis thaliana*, *HSFA2* seems to be of particular importance [3]. Whereas the master regulators of the HSF pathway, the 4 *HSFA1s* are specifically modulated by different environmental cues, *HSFA2* is induced systemically. *HSFA2* overexpression (OE) is sufficient to rescue most of the *hsfa1s* quadruple mutant defects [6,7]. Accordingly, *HSFA2* OE leads to resistance against multiple environmental stresses [8–10].

Among *Arabidopsis* A-class *HSFs*, HSFA1d and HSFA1e have been found to regulate *HSFA2*. However, *HSFA2* is still highly induced in response to heat and/or high light stress in the double *hsfa1d/hsfa1e* mutant [11]. Consequently, at least one other positive regulator of *HSFA2* expression must exist.

In this study, we compared genes co-regulated with *HSFA2* under different stress conditions to define a cluster of *HSFA2* coregulated genes. We identified 43 genes strongly coregulated with *AtHSFA2* during cold, salt, heat, and hypoxia stress. Promoter analysis revealed the site II element (SIIE) to be enriched in the promoters of these genes. A yeast one hybrid (Y1H) screen, using *HSFA2* promoter as a bait, led to the identification of two close paralogs, the R2R3-MYB transcription factors TRANSPARENT TESTA 2 (TT2) and MYB5 as putative SIIE-binding proteins (SIIEBP).

TT2 was identified as a seed coat-specific factor responsible for proanthocyanidin (PA) accumulation, giving *Arabidopsis* seeds their characteristic brown color [12]. The function of TT2 in PA accumulation requires its interaction with a bHLH (TT8) and a WDR protein (TRANSPARENT TESTA GLABRA1 [TTG1]) to form an MBW (MYB/bHLH/WDR) complex that regulates several late anthocyanin/PA biosynthetic genes (LBGs). Multiple MBW complexes exist, with TT2 or the closely related MYB5 protein that differentially regulate LBGs including *BANYULS* (*BAN*) [12–17]. Even though MYB5 and TT2 are acting in similar protein complexes, they are spatially separated and not functionally redundant. *Myb5-1* mutants do not exhibit the transparent testa phenotype and most notably are defective in seed coat mucilage [18]. In MBW complexes, TT2 and MYB5 have been shown to interact with MYB core [C/T]NGTTR and/or and AC-rich elements, [A/C]CC[A/T]A[A/C], whereas bHLH partners bind E-box motifs (CANNTG/CACGTG) [17].

While *TT2* was previously described as a seed-specific factor, we found *TT2* along with its targets is induced in vegetative organs by heat stress. We show that TT2/MYB5-mediated *HFSA2* regulation involves the SIIE cis-element, in yeast and in planta. We also showed that TT2/MYB5-mediated *HFSA2* expression is significantly enhanced in the presence of TT8 and TTG1, two other members of the MBW complex. Consistent with this result, *Arabidopsis* plants overexpressing *TT2* and, to a lesser extent *MYB5*, exhibit enhanced resistance to heat stress and *tt2*, *myb5*, *tt8* and *tt2/myb5* loss of function mutants are more sensitive to heat stress. Gene expression analyses further confirm that *TT2* and *MYB5* upregulate genes related to multiple stress responses. Promoter enrichment analysis of differentially regulated gene networks revealed two different modes of transcriptional regulation, one depending on the AC-rich element and the E-box (secondary metabolic process related genes), and the other involving the SIIE, the E-box, and the HSE (genes involved in stress response). Overall, we report a new function of *TT2* and *MYB5* in stress response.

2. Materials and Methods

2.1. Plant Material

Experiments were performed on *Arabidopsis thaliana* accession *Columbia 0*, unless specified otherwise. Plants were grown in growth chamber with a 100 μmol.m^{-2}.s^{-1} light intensity with 8 h to 16 h of illumination per day for short and long day photoperiods, respectively. T-DNA insertion mutants *tt8-6* (N2105594), *tt2-5* (N2105593), *tt2-1* (NW83), and *myb5-1* (N2106725) were obtained from the NASC. The *tt2/myb5* double mutant was obtained from a cross between *tt2-1* and *myb5-1* and a homozygote individual was backcrossed three times in Col 0. Sequence data from this article can be found in The

Arabidopsis Information Resource under the following accession numbers: AT2G26150 (*HSFA2*), AT3G35550 (*TT2*), AT3G13540 (*MYB5*), AT5G24520 (*TTG1*), AT4G09820 (*TT8*), AT1G61720 (*BAN*), AT4G22880 (*LDOX*), AT5G42800 (*DFR*), AT1G56650 (*PAP1*).

2.2. Bio-Informatic Analyses

Genevestigator software was used for publicly available microarray data analyses [19]. Coregulated genes were determined using the coregulations tool with a 0.8 Pearson's coregulation coefficient cutoff. Genes' descriptions and promoter sequence consensus were the ones of TAIR10 (www.tair.com, accessed on February 2016). Promoter motif enrichment was analyzed using the MEME suite V4.12.0 [20]. MEME was used for de novo motif prediction and AME was used for detecting the enrichment of known motif.

2.3. Yeast One Hybrid

The REGIA collection was used for the screening, as previously described [21]. Site II element was de novo synthesized with cohesive ends corresponding to HindIII recognition site (Eurofins, Hamburg, Germany). Vector was digested accordingly and the following sequences were inserted in the bait vector by ligation: forward 5′ agc ttT CGT TAG AAA TAT ATT TAA GTA AAG TAT ATT ATG ATA TAT Ac 3′ and reverse 5′ tcg agT ATA TAT CAT AAT ATA CTT TAC TTA AAT ATA TTT CTA ACG Aa 3′. Cohesive ends are in lower case characters. The identity of the prey was confirmed by PCR followed by SANGER sequencing using target specific primers (Supplementary Table S1).

2.4. Plant Transformation

Coding sequences were amplified using the primers in Supplementary Table S10 (lower case indicates attB recombination sites). Amplified fragments were cloned into pDNR207 using the Gateway BP Clonase II enzyme mix (Thermo Fisher scientific n°11789020, Waltham, MA, USA) and then into a modified pGREENII0229 containing a gateway cassette using the Gateway LR Clonase II Enzyme mix (Thermo Fisher Scientific n°11791100, Waltham, MA, USA). Vectors were then transformed into *Agrobacterium tumefaciens* pMP90. Plant transformations were performed following the floral dip method.

2.5. Transient Protoplast Transformation and Flow Cytometry

TT2 and *MYB5* coding sequences were amplified and cloned in pDNR207 entry vector with the BP clonase II mix (Thermo Fisher Scientific n°11789020). They were subsequently cloned in pBluescript-derived, *Physcomitrella patens* expression vectors [22], using the Gateway LR clonase II enzyme mix (Thermo Fisher Scientific n°11791100). The *HSFA2* promoter was amplified from Col 0 genomic DNA and cloned in pDNR207 entry vector with the BP clonase II mix (Thermo Fisher Scientific n°11789020). The promoter was then introduced into the destination GFP expression vector with the Gateway LR clonase II enzyme mix (Thermo Fisher Scientific n°11791100). The mutated HSFA2 promoter was synthesized by Eurofins genomics. WT promoter containing vector and the mutated promoter were digested by AscI and EcoRV and the mutated promoter was introduced in the GFP expression vector by ligation. Protoplast were prepared as previously described [22], and transformed with 5 µg of each intended plasmid. Protoplast fluorescence was determined 48 h after transformation using flow cytometry. Protoplast suspensions were filtered through a 30-µm mesh. Flow cytometry was performed on a Partec CyFlow®Space instrument (Sysmex France, 93420 Villepinte, France), with a 488 nm solid sapphire 20 mW laser for excitation and using a FloMax®data acquisition and analysis software (Sysmex France, 93420 Villepinte, France). Green fluorescence was detected with a FITC 527nm/30nm band-pass filter (FL1 channel). Red chlorophyll-based fluorescence from living protoplasts was detected with a 610nm/30nm band-pass filter in the FL2 channel. The side light scatter (SSC) detector high voltage was set to 161.5 V. The photomultiplier tube voltages were adjusted to 275 V for FL1 and 475 V for FL2 (logarithmic amplification mode, four decades range, speed 4). For each sample, the GFP fluorescence per population of cells was calculated as the product of the average fluorescence intensity by the number of cells in the positive gate, normalized by the total number of living protoplasts in the transformation. The gate was drawn along a line of maximum GFP intensities for positive samples, when compared with protoplasts that were only transfected with *pBAN:GFP* as negative controls.

2.6. Heat Stress Resistance Assays

Seeds were gas-phased sterilized from 4 to 8 h. They were placed in an open tube inside a hermetic box containing a beaker with 50 mL of 9.6% bleach topped with a basket containing 3 mL of 37% HCl (Sigma-Aldrich, 258148, St. Louis, MO, USA). The box was sealed and shacked, and the HCl was poured into the bleach. After 4 to 8 h of sterilization, the seeds were sowed in 90 mm petri dishes on half strength Murashige and Skoog (MS) basal salt mixture (Sigma-Aldrich, M5524, St. Louis, MO, USA). The seeds were stratified at 4 °C for 48 h and seedlings were grown for 6 days in short day conditions before applying stress. Heat challenge consisted of 80 min at 44 °C. Plants were put back to control conditions immediately after stress and allowed to recover in control conditions for 10 days before survival rate was determined. Plants were considered dead if completely bleached, collapsed, or presenting a translucent, necrotic aspect.

2.7. RNA Extraction and Q-PCR

In vitro grown plantlets were used in all Q-PCR experiments. Plants were gathered and RNA was extracted with the Qiagen RNeasy plant mini kit (Qiagen, cat 74903, Hilden, Germany). The "user-developed protocol" for plant tissue was used without modifications. 1 µg of total RNAs were reverse transcribed using the Superscript II kit (Invitrogen, 18064, Carlsbad, CA, USA), according to the manufacturer's instructions. mRNAs were quantified by Q-PCR using the MESA GREEN qPCR MasterMix (Eurogentec, RT-SY2X-03+WOU, Liege, Belgium). Actin and 26S proteasome mRNA were both used as reference genes in all experiments. Runs were performed on the CFX384 Touch™ Real-Time PCR Detection System and relative mRNA levels were analysed using the software Bio-Rad CFX manager (http://www.bio-rad.com, accessed on February 2016). Primers used for Q-PCR are presented in Supplementary Table S2.

2.8. RNA-Sequencing Transcriptomic Analysis

Libraries were built from 2 µg of total RNA with the NEBNext®Ultra™ Directional RNA Library Prep Kit for Illumina (ref#E7420S) and sequenced on the Illumina NextSeq 500. Reads mapping and statistical analysis was performed with CLC Genomics Workbench 10 RNA-seq analysis suite. Gene ontology analyses were performed with AgriGO (http://bioinfo.cau.edu.cn/agriGO, accessed on August 2017). Gene regulatory network were modeled with Genemania (https://genemania.org/, accessed on August 2017). De novo motif discovery and motif enrichment analyses were performed with the MEME suite V4.12.0.

3. Results

3.1. Promoters of HSFA2 Coregulated Genes Are Enriched in the Site II Element Motif

To gain insight into the regulation of *HSFA2*, we used publicly available gene expression data (www.genevestigator.com, accessed on October 2015 [19]) and searched for *HSFA2* co-regulated genes. Genes showing a Pearson correlation coefficient above 0.8 were extracted. As expected, *HSFA2* was part of expression clusters strongly correlated during heat treatment (101 genes), cold stress (75 genes), hypoxia (178 genes), and salt stress (67 genes). One of the clusters, consisting of 43 genes, is systematically co-regulated with *HSFA2* and therefore called "*HSFA2* common stress cluster" (Figure 1a, Supplementary Table S3). The *HSFA2* common stress cluster is enriched in gene ontology terms "response to stress" and "protein folding" (Panther™ V12) and comprises several known HSFA2 targets (Figure 1b, Supplementary Table S1).

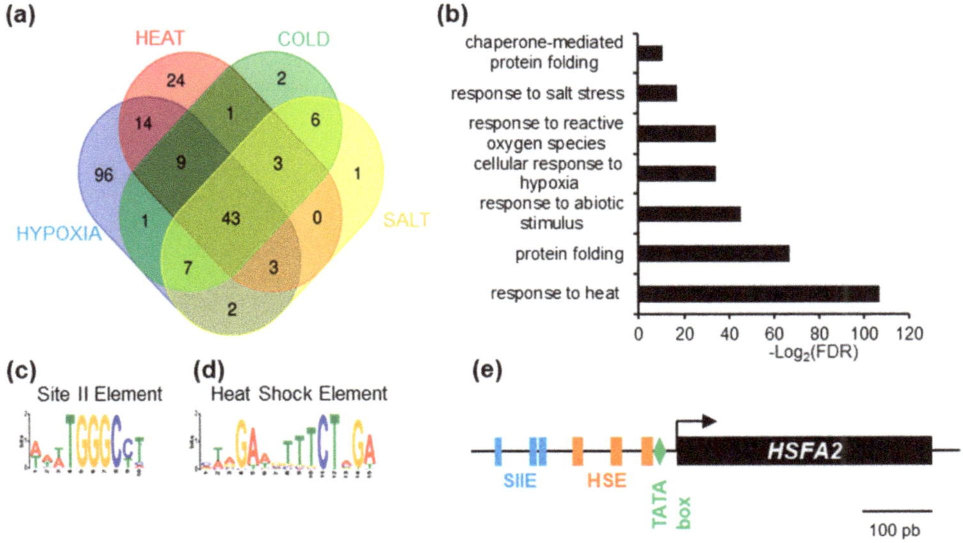

Figure 1. Identification of enriched cis-element in the promoter of HEAT SHOCK FACTOR A2 (HSFA2)-coregulated genes. (**a**) Venn diagram representing the intersection between genes identified as positively regulated with HSFA2 during cold, salt, heat, and hypoxia stress. (**b**) GO term enrichment of the 43 genes in the HSFA2 common stress cluster. (**c**) SII element (WWWTGGGCCT) and (**d**) Heat Shock Element (MWRGAWGTTTCTAGA) DNA motifs enriched in the promoter of the 43 *HSFA2* common stress cluster. (**e**) Schematic representation of *HSFA2* promoter featuring TATA box (−23), HSEs (−33; −77; −133) and SII (−186; −201; −250).

Search for cis-elements in the promoters of the *HSFA2* common stress cluster, using the MEME suite software [20] revealed two significantly over-represented motifs, the Heat Shock Element and the Site II element, hereafter called HSE and SIIE, respectively (Figure 1c,d). The HSE (5′-nGAAn-3′ repeats) was previously shown to be enriched in the promoters of HSFA2 and its targets [7,23], thereby validating the method (Figure 1d). Unexpectedly, we found that the majority (65.91%) of promoters ($p = 4.4 \times 10^{-12}$, Figure 1c) were also enriched in sequences corresponding to the SIIE motif (5′-(A/T)TGGGC(C/T) -3′ [24]. Analysis with PLMdetect further pinpointed an enrichment of the SIIE motif in a 200 bp window, between −136 bp and −332 bp upstream of the TSS [25]. Consistent with motifs involvement in transcription regulation, SIIEs were often found several times in the same promoters, in both forward and reverse orientation (Supplementary Figure S1). In the *HSFA2* promoter, 3 SIIEs were found in close proximity to the TSS (Figure 1e).

The SIIE was previously linked to cell cycle regulation as well as biotic and abiotic stress responses [24,26]. However, the identity of SIIE binding factors is still under debate. It was suggested that SIIEs interact with the TEOSINTE BRANCHED-1/CYCLOIDEA/PCF 20 (TCP20) factor [27]. However, in vivo experiments showed that TCP20 was not present in the region of SIIE motifs [28].

3.2. TT2 and MYB5, Two Related R2R3-MYBs TFs, Bind to HSFA2 Site II Element in Yeast

To identify upstream transcriptional regulators that could drive HSFA2 expression through binding of SIIE, we carried out a yeast one-hybrid (Y1H) screen. A 40 bp region of the *HSFA2* promoter from 175 to 216 bp before the TSS and containing two SIIEs, in sense and antisense orientation, was used as bait (Figure 2a). A similar sequence mutated on the SIIEs was used as a negative control to ensure the specificity of the interaction (Figure 2a). The REGIA collection, containing 1357 *Arabidopsis* transcription factors cloned in *Saccharomyces cerevisiae*, provided the set of different preys in the Y1H screen [21]. Clones harboring the mutated bait were subtracted from the list of positive clones possessing the wild type (WT) bait. Two TFs, TT2 (TRANSPARENT TESTA 2) and MYB5, fused to the GAL4 activation domain (GAL4AD), were identified to specifically activate transcription through SIIEs. To get an estimate of the strength of the interaction, yeast suspensions were diluted 10, 100, and 1000 fold and spotted on increasing concentrations of the antibiotic aureobasidin A (AurA), the reporter used for the screen. Growth inhibition was observed for the negative control (yeast with the bait but not the prey) at the standard concentration of 150 ng.µL^{-1}, and was complete at 200 ng.µL^{-1} AurA (Figure 2b). Yeast harboring SIIE as bait and the prey TT2-GAL4AD or MYB5-GAL4AD showed no growth inhibition at 200 ng.µL^{-1} AurA. Dilution experiments in the yeast cellular context further indicated that TT2 has a higher affinity for SIIEs than MYB5 (Figure 2b). Interestingly, TT2 and MYB5 are closely related transcription factors belonging to R2R3-MYBs.

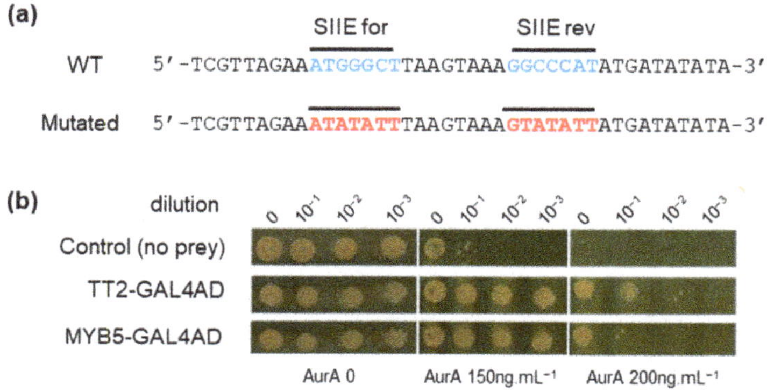

Figure 2. TRANSPARENT TESTA 2 (TT2) and MYB5 bind to the HEAT SHOCK FACTOR A2 (HSFA2) promoter. (**a**) Wild type (WT) and mutated sequences of the HSFA2 promoter used for the screening. (**b**) Yeast 1 hybrid interaction strength of TT2 and MYB5 with the HSFA2 promoter at different aureobasidin A, AurA, concentrations.

3.3. *TT2 or MYB5 Overexpression Induces HSFA2 Expression and Provides Stress Resistance in Planta*

To investigate the function of *TT2* and *MYB5* in heat stress, we constitutively expressed the two TFs under the control of CaMV35S promoter in *Arabidopsis*. Three and four independent transgenic lines, exhibiting strong *TT2* and *MYB5* transgene expression, respectively, were recovered (Supplementary Figure S2a,b). Analysis of *HSFA2* transcript accumulation in six-day-old seedlings revealed the constitutive expression of *HSFA2* in both *TT2* or *MYB5* overexpressor lines (Figure 3b). In accordance with previous reports describing *MYB5* as a 'weak' transcriptional activator [29], 35S:*TT2* plants accumulated more *HSFA2* transcripts than 35S:*MYB5* plants in control conditions (Figure 3b). As constitutive expression of *HSFA2* leads to thermotolerance, we assessed 35S:*TT2* and 35S:*MYB5* lines for heat stress resistance. Heat stress was applied on six-day-old seedlings and consisted of incubation of the seedlings for 80 min at 44 °C. To minimize experimental variability, plants were sown twice per plate, symmetrically, and in three randomized technical replicates (Figure 3a). Both 35S:*TT2* and 35S:*MYB5* lines showed enhanced resistance to heat stress in three independent biological replicates (Figure 3c,d, Supplementary Figure S2d). The heat stress resistance phenotype was stronger for 35S:*TT2* than for 35S:*MYB5* lines. Interestingly, no obvious developmental defects were observed in neither *TT2* nor *MYB5* overexpression plants in control conditions (Figure 3d, Supplementary Figure S2c). From this, we concluded that *TT2* and *MYB5* regulate *HSFA2* expression and lead to heat stress resistance.

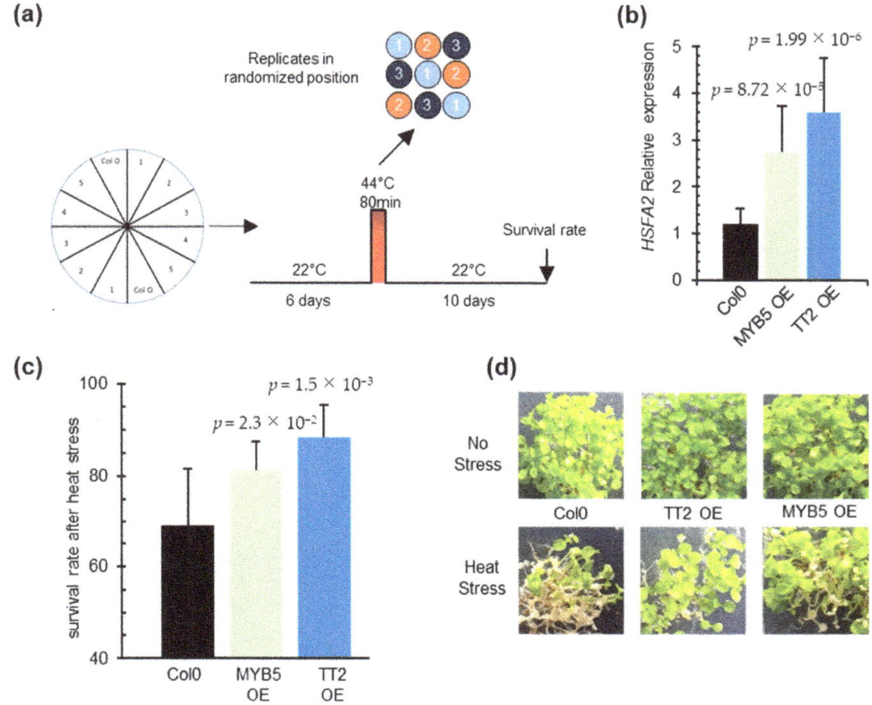

Figure 3. TRANSPARENT TESTA 2 (TT2) and MYB5 overexpression triggers upregulation of HEAT SHOCK FACTOR A2 (HSFA2) in planta and provides heat resistance. (**a**) Schematic representation of the resistance assay protocol. Each stress is applied on three identical petri dishes placed in randomized positions in the growth chamber. Lines were sowed symmetrically in vitro and grown in control conditions for six days. Heat stress consisted of 80 min at 44 °C. Survival rates were observed after 10 days. (**b**) HSFA2 expression in transgenic *TT2* or *MYB5* overexpressors (OEs) plants. Results are the mean of three independent measurements. Errors bars represent standard deviations (SD). Results are significant with *p*-values < 0.05 (Wilcoxon test). (**c**) Survival rates of TT2 OEs and MYB5 OEs plants after heat stress. Histograms present the mean of three independent measurements. Each measurement is the average of six randomized technical replicates. Error bars represent SD. *p*-values were obtained from Student's *t*-test. (**d**) Representative pictures of two-week-old TT2 OEs and MYB5 OEs plants grown in control and heat stress conditions.

3.4. TT2, MYB5, TT8, and TTG1 Activate HSFA2 Expression via SIIE Motifs

It has been shown that *TT2* and *MYB5* control outer seed coat development via a ternary MYB–BHLH–WDR (MBW) protein complex involving TT2/AtMYB123 or MYB5, TT8/AtBHLH042 and TTG1(TRANSPARENT TESTA GLABRA 1: WD-repeat protein) [29,30]. To test whether *TT2* and *MYB5* mediated *HSFA2* expression via SIIE involves the MBW complex, we exploited the *Physcomitrella patens* protoplast transient expression system. This consists of co-transformation of moss protoplasts with the investigated promoter and the test transcription factors. For quantification of the strength of the interaction, this method combines the advantages of GFP as a marker of promoter activity as a fast and reliable method for fluorescence measurements in cells with flow cytometry [22].

We co-transformed *P. patens* protoplasts with the *HSFA2* promoter GFP reporter construct p*HSFA2*:*GFP* in combinations with vectors expressing *TT2*, *MYB5*, *TT8*, or *TTG1*. As control, we used the *pHSFA2mut*:*GFP* reporter construct mutated in the SIIEs. The *HSFA2* promoter alone exhibited significant autoactivity that was reduced three-fold when the SIIEs were mutated (Figure 4a). This result indicated that homologous moss transcription factors are able to bind the SIIEs. Co-expression of *TT2* with *pHSFA2:GFP* increased the GFP signal 2.5-fold (Figure 4a). A similar increase was observed when *MYB5* was co-expressed with *pHSFA2:GFP* (Figure 4a). Both signals were significantly decreased when *TT2* or *MYB5* were co-expressed with the mutated promoter construct *pHSFA2mut:GFP* (Figure 4a). These results indicate that the SIIEs are necessary for the correct regulation of the *HSFA2* promoter by *TT2* or *MYB5*. Co-expression of *TT2* and *MYB5* further increased the activity of the *HSFA2* promoter, leading to a 3.9 fold increase of the GFP signal (Figure 4a). Consistent with previous reports describing MYB5-driven transcription as relatively weak [29], *MYB5/TT8/TTG1* co-expression triggered a 3.9-fold increase in GFP, whereas *TT2/TT8/TTG1* induced an 8-fold increase. The strongest signal was obtained by the *TT2/MYB5/TT8/TTG1* combination that yielded a 10-fold increase in GFP signal (Figure 4a). These results further confirmed that TT2, TT8, and TTG1 and to a lesser extent MYB5, cooperatively regulate the *HSFA2* promoter. Furthermore, GFP signals were weak when the different combinations of transcription factors were co-transformed with the promoter that lacks the SIIEs, indicating that the MBW complex activates *HSFA2* expression via SIIEs (Figure 4a).

3.5. The TT2/MYB5-MBW Complex Is Active in Vegetative Organs and Is Required for Heat Stress Resistance

We found that the TT2/MYB5-TT8-TTG1 complex can modulate the activity of the *HSFA2* promoter in transient assays and in transgenic overexpression lines (Supplementary Figures S3 and S4a). However, this complex was previously described as seed coat-specific [17]. To check whether the MBW complex could be active outside the seed context and responsible for the heat stress resistance phenotype, we looked for *BAN* transcript accumulation, a target of the MBW complex, in *TT2* OE lines by Q-PCR. We reverse transcribed RNA extracted from six-day-old in vitro grown plants. Consistent with previous reports describing the effect of *TT2* ectopic expression in roots and seedlings [12], we found *TT2*-overexpressing lines constitutively accumulate *BAN* transcripts in the absence of stress (Figure 4b). This confirmed that all the requirements for TT2-containing MBW complex activity were also met in vegetative organs.

To test if TT2/MYB5-mediated *HSFA2* expression has a biological role in heat stress resistance and is not an artifact of overexpression, we subjected *tt2-5*, *myb5-1* and *tt2-1/myb5* loss of function mutants to heat stress. As *tt2-1* and *myb5* derive from *Ler* and Col-0 ecotypes, respectively, the double mutant was backcrossed three times in Col-0 to generate *tt2/myb5* double mutants in a genetic background approaching *Col-0*. Both the simple, *tt2-5*, *myb5-1* and the double *tt2-1/myb5* mutants were compromised in basal heat stress resistance when compared to the wild type control Col-0 (Figure 4c). The heat stress sensitivity was also observed for the *tt2-1* mutant in the *Ler* background (Supplementary Figure S2e).

To investigate if *TT2* and *MYB5* function in heat resistance required the full MBW complex, we phenotyped *TT8* loss of function mutant, *tt8-6*, for heat resistance. TT8 is a central protein in the MBW ternary protein complex that is expressed in both seeds and vegetative tissues. Consistent with the *HSFA2* promoter-MBW interaction in *P. patens* protoplast transient expression assays, we found that basal heat stress resistance was compromised in *tt8-6* mutant plants (Figure 4c). As both *tt2/myb5* and *tt8* loss of function mutants are compromised in basal heat stress resistance, we concluded that the MBW ternary complex likely controls expression of genes required for basal heat stress resistance in *Arabidopsis*.

Previously only MBW complexes containing Production of Anthocyanin Pigments 1 protein, PAP1, were described to be active in vegetative organs [17]. To further confirm that the TT2-containing MBW complex is active in leaves during heat stress, we investigated *TT2* and *PAP1* expression before and 80 minutes and 24 hours after heat stress treatments. We detected a very low level expression of *TT2* in plants before stress, but a four-fold upregulation upon heat stress. Inversely, we observed downregulation of *PAP1* during heat stress (Figure 5). To quantify the activity of TT2 and PAP1 during stress, we analyzed the expression of their direct targets, using Q-PCR. PAP1 induces the expressions of *LDOX*, *DFR*, and *TT8* but not *BAN*, whereas the expression of *LDOX*, *DFR*, *BAN*, and *TT8* is induced by TT2 [17]. Only *BAN* and *TT8* transcripts showed a marked increase after heat stress. On the contrary, the non-specific *DFR* and *LDOX* were not found modulated by heat stress (Figure 5). Taken together, these results show the TT2/MYB5-MBW complex is active in vegetative organs under heat stress.

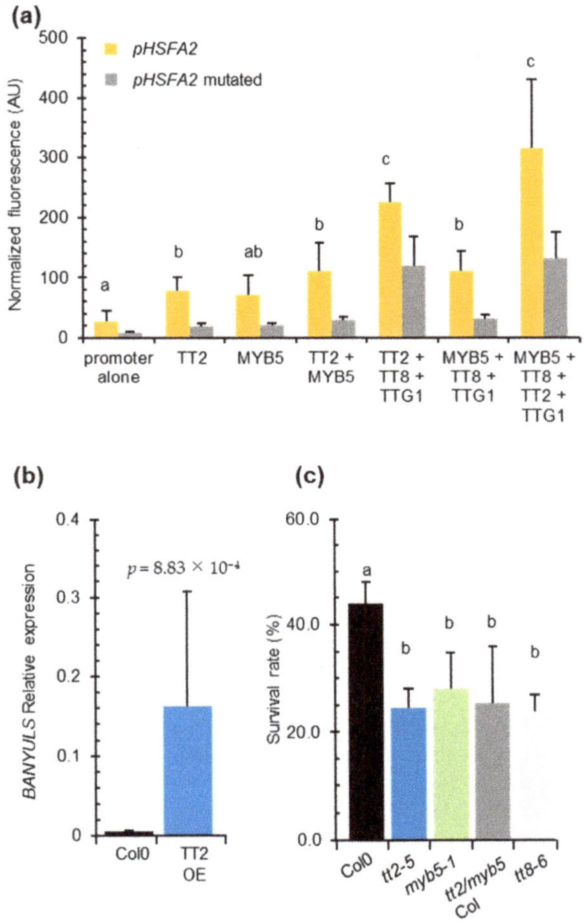

Figure 4. The entire MYB–BHLH–WDR (MBW) complex is required for *HEAT SHOCK FACTOR A2* (*HSFA2*) regulation and heat stress resistance in vegetative organs. (**a**) *Physcomitrella patens* protoplast transient transformation assays. *HSFA2* promoter or the mutated *HSFA2* promoter version lacking the three SIIelement were co-transformed in moss protoplasts with different transcription factors composing the MBW complex. Total fluorescence was normalized by the number of protoplasts per transformation. Differences between wild type and mutated promoter activities are all significantly different with *p*-values < 0.03. (**b**) *BANYULS* (*BAN*) expression in five independent *TRANSPARENT TESTA 2* (*TT2*) overexpressors. (**c**) Heat stress resistance potential of Col-0 compared to *tt2-5* and *myb5-1* and the double mutant *tt2/myb5* backcrossed three times in Col-0 (*tt2/myb5 Col*). Histograms are the mean of three independent experiments, each consisting of the average of six randomized technical replicates. *p*-values were obtained by the Student's *t*-test and error bars represent SDs. Letters indicate statistical significance with *p*-values < 0.05.

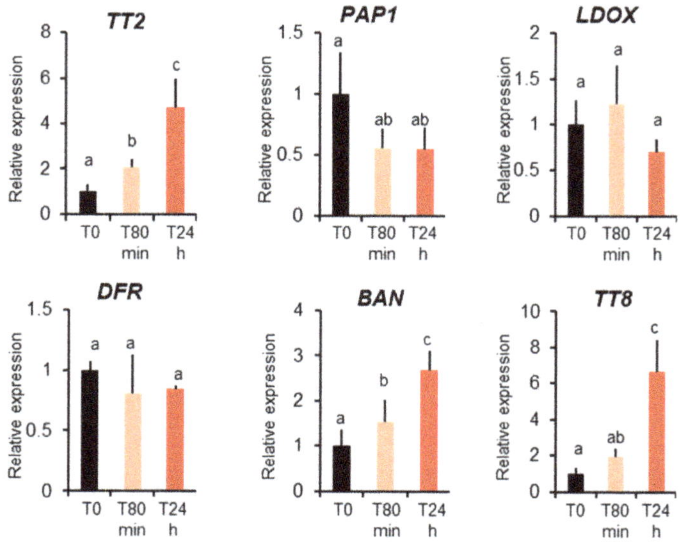

Figure 5. MYB–BHLH–WDR component and target expression during heat stress. Col 0-plants were collected before heat stress (T0), after 80 minutes of heat stress (T80min) and the next day (T24h). Gene expression was measured using real-time quantitative PCR. Results are the means of six independent experiments. Bars represent standard errors. Different letters indicate the statistical significance with p-values < 0.05 (Student's t-test).

3.6. *TT2 and MYB5 Redundantly Regulate Stress Response Genes*

We further investigated the impact of *TT2* and *MYB5* overexpression on gene expression through RNA-sequencing. We subjected unstressed, six-day-old Col-0, *TT2* OE and *MYB5* OE plants from three independent biological replicates to total RNA extraction and sequencing. We obtained approximately 40 million paired-end reads from each library. Reads were mapped and used to estimate gene expression using the RNAseq analysis suite of CLC genomics workbench (Qiagen). Compared to Col-0, *TT2* OE and *MYB5* OE showed 684 and 354 differentially expressed genes (DEGs), respectively, with FDR adjusted p-values under 0.05 (Supplementary Table S4). The overexpressed *TT2* or *MYB5* transcripts were excluded from subsequent analyses. Most DEGs (62.7%), in *MYB5* OE were similarly regulated in *TT2* OE (Figure 6a,b). Eighty-one genes were upregulated in both *TT2* OE and *MYB5* OE. Gene ontology (GO) analysis of those upregulated genes indicated enrichment in the "response to stress" or related categories (response to stimulus, abiotic stress, oxidative stress) as well as the "cellular development" and "cell differentiation" categories (Figure 6c). Interestingly, among the flavonoid biosynthesis genes, *LDOX, TTG2* were upregulated and *FLAVONOL SYNTHASE 1 (FLS1)* was downregulated in *TT2* and *MYB5* OE. *TT4, TT8, BAN,* and *DFR* were found upregulated in *TT2* OE only.

Downregulated genes in *TT2* OE and *MYB5* OE were enriched in "secondary metabolic process", "phenylpropanoid metabolic process" and "cellular amino acid derivative metabolic process" although less than 10% of the 138 genes belonged to these categories (Figure 6b,d). Overall, we found that *TT2* and *MYB5* redundantly control stress response genes while only *TT2* regulates late flavonoid biosynthesis genes.

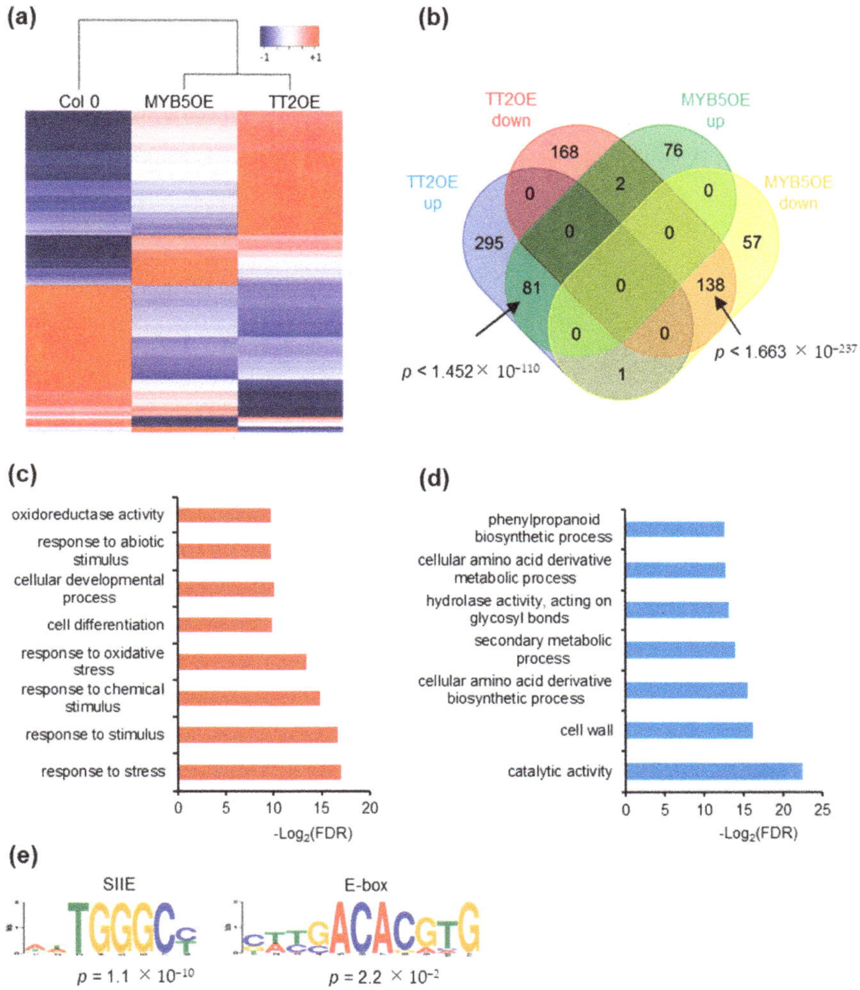

Figure 6. TRANSPARENT TESTA 2 (TT2) overexpressor (OE) and MYB5 OE redundantly regulate stress response genes. (**a**) Heatmap representation of *TT2* OE and *MYB5* OE up- (red) and downregulated (blue) genes. Heatmap was generated from mean RPKM values. (**b**) Venn diagram showing the overlap between *TT2* OE and *MYB5* OE differentially expressed genes, DEGs. 51.3% of the *MYB5* OE upregulated genes are also upregulated in *TT2* OE and 70.4% of the *MYB5* OE downregulated genes are also downregulated in *TT2* OE. The overlaps are statistically significant, hypergeometric test. (**c**) and (**d**) Gene ontology (GO) term enrichment analysis of common upregulated (**c**) or downregulated (**d**) genes in *TT2* and *MYB5* OEs. (**e**) DNA motif enriched in the promoter of genes upregulated in TT2 OEs plants.

We then looked for motif enrichment in the promoters of DEGs. De novo motif discovery was performed using 500 bp upstream of all genes upregulated by TT2 (Figure 6e). Notably, SIIE motif was found to be enriched together with motifs similar to the E-box (Figure 6e). Similar results were obtained from MYB5 upregulated genes. We confirmed this result by looking for enrichment of the motifs in the promoters of genes upregulated in *TT2* OE or *MYB5* OE and found the SIIE was enriched in both cases ($p = 1.32 \times 10^{-17}$ and $p = 6.91 \times 10^{-8}$, respectively). Promoters of *TT2* OE downregulated genes were also found to be enriched in SIIE ($p = 1.09 \times 10^{-2}$). These results further confirm the importance of the SIIE and E-box motif for *TT2/MYB5* regulation of transcription.

3.7. *TT2 Differentially Regulates Multiple Stress Response and Flavonoid Metabolism Genes*

As *TT2* is a more potent transcriptional regulator than *MYB5*, we used the DEG list of *TT2* to perform gene network analyses. We first performed a GO term enrichment analysis (agriGO; http://bioinfo.cau.edu.cn/agriGO, accessed on August 2017). Enriched GO terms in up- and downregulated genes are presented in Supplementary Tables S5 and S6. Similarly, to the GO term enrichment analysis performed on the *TT2/MYB5* overlap, we found *TT2* OE upregulated genes were strongly enriched in GO categories related to stress and secondary metabolic process. In particular, *HSFA3*, *HSFA7A*, and *MBF1C* (MULTIPROTEIN BINDING 1 C), three major regulators of multiple stress responses, were significantly upregulated (Table 1). Downregulated genes were mostly enriched in phenylpropanoid metabolic process but also showed enrichment in stress response-related terms (Table 2).

Table 1. Selection of *TRANSPARENT TESTA 2* overexpressor upregulated genes from the two major gene ontology categories, response to stress and secondary metabolic process.

Identifier	Fold Change	Gene Description
		Response to Stress
AT5G03720	1.57	HEAT SHOCK TRANSCRIPTION FACTOR A3 (HSFA3)
AT3G51910	1.33	HEAT SHOCK TRANSCRIPTION FACTOR A7A (HSFA7A)
AT3G24500	1.39	MULTIPROTEIN BRIDGING FACTOR 1C (MBF1C)
AT3G16770	1.52	ETHYLENE RESPONSE FACTOR 72 (ERF72)
AT5G59780	1.28	MYB DOMAIN PROTEIN 59 (MYB59)
AT4G31800	1.46	WRKY DNA-BINDING PROTEIN 18 (WRKY18)
AT5G52640	1.29	HEAT SHOCK PROTEIN 90.1 (HSP90.1)
AT2G19310	1.27	HSP20-like chaperones superfamily protein
AT5G51440	1.57	HSP20-like chaperones superfamily protein
AT1G63750	1.55	Disease resistance protein (TIR-NBS-LRR class) family
AT1G63860	1.42	Disease resistance protein (TIR-NBS-LRR class) family
AT5G41740	1.45	Disease resistance protein (TIR-NBS-LRR class) family
AT1G75830	3.04	Plant defensin 1.1 (PDF1.1)
AT4G22212	1.41	Encodes a defensin-like (DEFL)
AT2G43535	1.39	Encodes a defensin-like (DEFL)
AT2G21490	2.24	dehydrin LEA
AT2G47180	1.29	Galactinol synthase 1 (GolS1)
AT4G11650	1.57	OSMOTIN 34 (OSM34)
AT5G66400	1.32	RESPONSIVE TO ABA 18 (RAB18)
		Secondary Metabolic Process
AT1G71030	1.30	MYB-LIKE 2 (MYBL2)
AT5G11260	1.39	ELONGATED HYPOCOTYL 5 (HY5)
AT5G13930	2.15	TRANSPARENT TESTA 4 (TT4)
AT5G42800	34.65	DIHYDROFLAVONOL 4-REDUCTASE (DFR)
AT4G22880	6.34	LEUCOANTHOCYANIDIN DIOXYGENASE (LDOX)
AT1G61720	451.56	BANYULS (BAN)

Table 2. Selection of *TRANSPARENT TESTA 2* overexpressors downregulated genes from the two major gene ontology categories, response to stress and secondary metabolic process.

Identifier	Fold Change	Gene Description
		Response to Stress
AT5G08790	−1.48	ARABIDOPSIS NAC DOMAIN CONTAINING PROTEIN 81 (ANAC81)
AT5G06960	−1.30	TGACG MOTIF-BINDING FACTOR 5 (TGA5)
AT2G14610	−30.09	PATHOGENESIS-RELATED GENE 1 (PR 1)
AT2G17060	−1.97	Disease resistance protein (TIR-NBS-LRR class)
AT5G07390	−1.64	RESPIRATORY BURST OXIDASE HOMOLOG A (RBOHA)
AT5G08590	−1.39	SNF1-RELATED PROTEIN KINASE 2.1 (SNRK2.1)
AT1G55020	−2.42	LIPOXYGENASE 1 (LOX1)
AT2G37040	−1.44	PHE AMMONIA LYASE 1 (PAL1)
AT1G45145	−1.49	THIOREDOXIN H-TYPE 5 (TRX5)
AT1G05250	−13.72	PEROXIDASE 2 (PRX2)
AT5G51890	−1.30	PEROXIDASE 66 (PRX66)
AT2G38380	−1.47	Peroxidase superfamily protein
AT1G68850	−1.80	Peroxidase superfamily protein
AT5G64110	−1.30	Peroxidase superfamily protein
AT5G08670	−1.35	Encodes the mitochondrial ATP synthase beta-subunit
AT5G08680	−1.35	Encodes the mitochondrial ATP synthase beta-subunit
		Secondary Metabolic Process
AT5G08640	−1.72	FLAVONOL SYNTHASE 1 (FLS1)
AT5G25980	−2.88	THIOGLUCOSIDE GLUCOHYDROLASE 2 (TGG2)
AT5G26000	−1.35	THIOGLUCOSIDE GLUCOHYDROLASE 1 (TGG1)
AT1G17170	−1.47	GLUTATHIONE S-TRANSFERASE TAU 24 (GSTU24)
AT1G17180	−2.02	GLUTATHIONE S-TRANSFERASE TAU 25 (GSTU25)
AT1G78340	−1.54	GLUTATHIONE S-TRANSFERASE TAU 22 (GSTU22)
AT1G51680	−1.28	4-COUMARATE:COA LIGASE 1 (4CL1)
AT2G37040	−1.44	PHENYLALANINE AMMONIA LYASE 1 (PAL1)
AT3G10340	−1.44	PHENYLALANINE AMMONIA-LYASE 4 (PAL4)

To further characterize the TT2 DEGs, we used Genemania (https://genemania.org, accessed on August 2017) to build gene regulatory networks (GRN) from these categories. Two major networks of highly co-regulated genes could be defined from the upregulated genes, GRNs "response to stress" and "secondary metabolic process" (Figure 7, Supplementary Table S7). The stress response GRN contained genes involved in response to biotic stress (defense response to bacterium) and multiple abiotic stresses (oxidative, salt, heat, high light, and cold stress), together with genes involved in multiple hormones response (Table 1, Supplementary Tables S7 and S8). Seventeen genes of the *HSFA2* common stress cluster were found in the stress response GRN (Supplementary Table S7). Secondary metabolic process GRN was found mostly involved in the production of PA (Table 1, Supplementary Table S7 and S9). Two GRNs of highly co-regulated genes were found downregulated in *TT2* OE (Figure 7, Supplementary Table S7). The downregulated GRN "response to stress" contained genes related to fungus response together with genes involved in cold, osmotic, and wounding stress response (Supplementary Table S7 and S10). The downregulated GRN "secondary metabolic process" contained genes involved in flavonols biosynthesis, lignin biosynthesis, and early steps of the general phenylpropanoid metabolic pathway (Supplementary Table S7 and S11).

To screen for shared features between the promoters of the different GRNs, we used the MEME suite V4.12.0 tool to perform de novo motif enrichment analysis. Promoters of upregulated genes from the stress response GRN were found enriched in motifs similar to the SIIE, the E-box, and in HSE (Figure 7). Interestingly, promoters of genes belonging to the secondary metabolism GRN were enriched in E-box motif and in AC-rich element although with high E-values (Figure 7). Promoters of downregulated GRNs were only enriched in motifs corresponding to AC-rich element.

Overall, *TT2* performs a complex modulation of both stress responses and anthocyanin/PA biosynthesis genes and this modulation seems to involve different cis-elements.

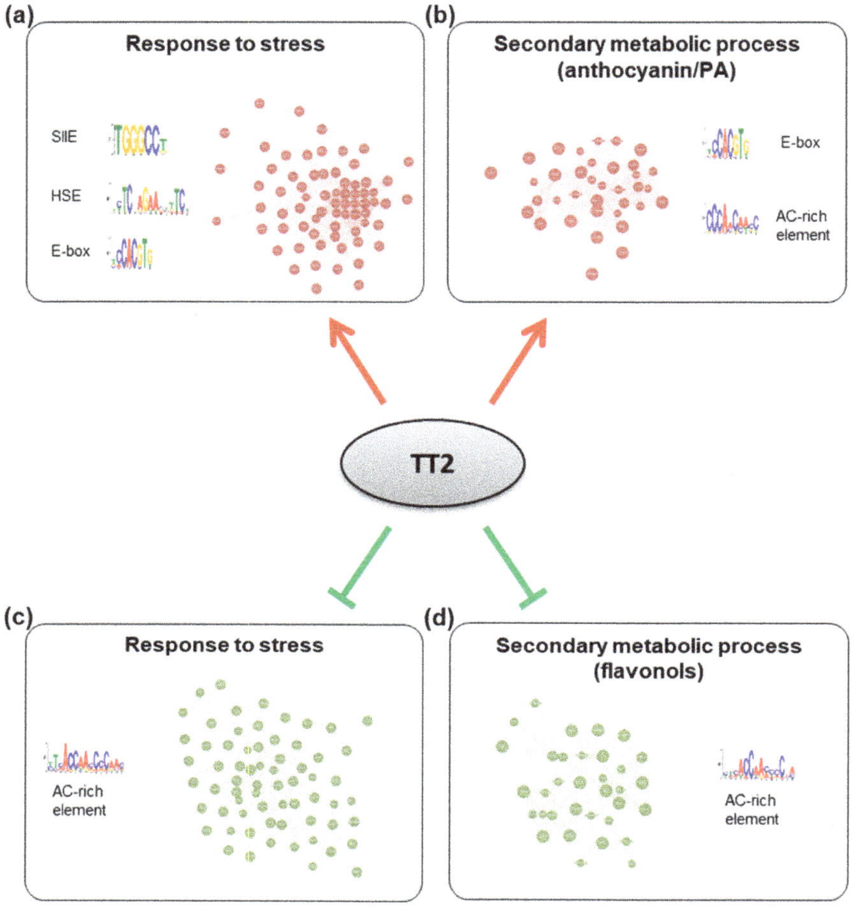

Figure 7. Gene regulatory networks (GRNs) regulated by TRANSPARENT TESTA 2 (TT2) overexpressors (OE). GRNs were constructed from differentially expressed genes in *TT2* OEs. (**a**) Upregulated genes belonging to the "response to stress" gene ontology (GO) category belong to a tightly coregulated set of genes including *HEAT SHOCK FACTOR A2, A3, A7A, MULTIPROTEIN BRIDGING FACTOR 1C,* and many chaperones. Promoters of these genes were found to be enriched in SII element, heat shock element and E-box motifs. (**b**) Upregulated genes belonging to the "secondary metabolic process" GO category formed another network including multiple flavonoids biosynthesis genes (*BANYULS, LEUCOANTHOCYANIDIN DIOXYGENASE, TRANSPARENT TESTA 5* and *6*, etc …). Promoters of these genes were enriched in E-box and a motif similar to the AC-rich element. A comparable AC-rich element motif was enriched in downregulated GRNs (**c**,**d**).

4. Discussion

4.1. Role of the TT2/MYB5 in the Plant Heat Stress Response

In this report, we provide strong evidence for the involvement of TT2 and MYB5 in the heat stress response. We show in transient and stable transformation experiments that TT2 mediates *HSFA2* regulation directly or indirectly via SIIE motifs (Figures 2 and 4). We also show that *TT2* and *MYB5* overexpressing plantlets are heat stress resistant, whereas *tt2*, *myb5*, and *tt2/myb5* loss of function mutants are impaired in heat stress resistance (Figures 3 and 4, Supplementary Figure S2). These results extend previous reports of *tt2* mutant seeds, being more sensitive to salt, sucrose, and ABA that were attributed to an increased permeability of the seed coat to stressors [29]. However, as *myb5*-1 does not exhibit the yellow seed phenotype but show the same sensitivity to heat stress, the hypothesis of the increased permeability of the seed coat to stressors is unlikely (Figure 4). In addition, the significant modulation of stress-related genes in OE plants further establishes a role of *TT2* and *MYB5* as regulators of stress response (Figures 6 and 7).

TT2 and MYB5 were previously thought to be seed coat-specific factors whereas PAP1 was found to be present in vegetative organs. Here, we show that *TT2* is expressed at low levels in vegetative parts of the plant and that the expression of *TT2*, *BAN*, and *TT8* is modulated by heat stress (Figure 5). Consistent with previous reports [12], *BAN*, *DFR*, *LDOX*, *TT4*, and *TT8* were also induced constitutively in *TT2* OE lines (Figure 4, Table 1 and Supplementary Table S4). These results confirm that the full MBW complex is present and active in vegetative organs. It would be very interesting to localize the tissues where *TT2* and *TT8* are induced during stress to better understand their precise functions. Previously, the PAP1-4-containing MBW complex was thought to be the only MBW complex present in vegetative organs. In our study, *PAP1* was found to be downregulated by heat stress in wild type plants, while *LDOX* and *DFR* were not induced, suggesting that a PAP1-containing MBW is not involved in the heat response (Figure 5). Overall, our results indicate that *TT2* and *MYB5* are genuinely involved in heat stress resistance in vegetative organs, warranting further investigations into the role and mechanisms of MBW complex formation and functioning in different stress conditions.

4.2. *TT2/MYB5 in Stress Resistance Is Independent of Anthocyanin/PA Metabolism*

The MBW complex was demonstrated to regulate anthocyanin and PA accumulation. Flavonoids have been described as important components of multiple biotic and abiotic stress responses [31]. However, we do not believe that these molecules play a major role in the resistance phenotypes associated with *TT2/MYB5* OEs. *TT2* OE alone is not sufficient to trigger ectopic anthocyanin/PA accumulation [32], but is sufficient to increase basal heat stress resistance (Figure 3). In addition, *myb5-1* mutants are not impacted in flavonoid metabolism and do not exhibit the yellow seed phenotype. Still, they show the same defect in thermotolerance as *tt2* mutants.

Other groups reported 35S:*TT2* plants were inducing strong *BAN* expression without exhibiting PA and PA precursor accumulation in vegetative organs [12]. In our study, we found the PA biosynthesis genes *DFR*, *LDOX*, *BAN* to be specifically upregulated in *TT2* OE plants, while flavonol (*FLS1*) and anthocyanin (*GSTs*) biosynthesis genes were downregulated (Tables 1 and 2). At the same time, several enzymes of the very early steps of the general phenylpropanoid metabolic pathway were downregulated (*4CL1*, *PAL1* and *PAL4*; Table 2 and Supplementary Table S4). These results most likely explain the absence of PA accumulation in *TT2* OE plants and show that PA accumulation requires other seed-specific factors.

Gene expression analysis further supports the view of *TT2* functioning in stress resistance and in flavonoid biosynthesis through independent mechanisms. RNA-seq analysis performed in this study suggests that *MYB5* and *TT2* redundantly regulate stress response genes but not flavonoid biosynthesis genes, which were found to be differentially regulated in *TT2* OE plants only. This suggests stress resistance is not the consequence of anthocyanin/PA accumulation. Previous studies of developing *tt2-5* seeds showed that 33.6% and 32.1% of up- and down-regulated genes, respectively, were involved in stress/defense response, whereas only 0.2% and 5.4% of up- and down-regulated genes were involved in flavonoid metabolism [33]. Our findings that *TT2* OE upregulates the expression of many stress-related genes rule out the possibility that this modulation is due to seed coat defects or an overall lack of anthocyanins/PA (Figure 6). Motif enrichment analyses also suggest the regulation of stress response and secondary metabolic process is performed directly, or indirectly, by different protein complexes. SIIE motifs were found to be enriched in the promoters of stress response genes but not in the promoters of secondary metabolic process-related genes (Figure 7). Consequently, we attribute enhanced stress resistance to the modulation of SIIE-regulated genes in general.

4.3. TT2 Dual Modes of Transcriptional Regulation

TT2 and MYB5 binding sites have been previously identified as the MYB core and AC-rich elements [34]. Contrary to what we observe for *HSFA2* promoter regulation, the *BAN* promoter, which does not contain SIIEs, could not be activated by TT2 or MYB5 alone but required the full MBW complex for proper activation [22]. This indicates that the regulation of *HSFA2* and *BAN* promoters by TT2/MYB5-TT8-TTG1 involves different mechanisms. Examining the *HSFA2* promoter, we found several putative binding sites for TT8 (E-boxes), and MYBs (AC-rich elements, Figure 6), which may explain the residual activation of the *HSFA2* promoter mutated in the SIIEs (Figure 4). Indeed, a combination of TT2-TT8-TTG1 but not MYB5-TT8-TTG1 retained a strong ability to activate the mutated *HSFA2* promoter. This tends to indicate that TT2 and MYB5 alone regulate transcription through SIIEs while TT2-TT8-TTG1 but not MYB5-TT8-TTG1 can activate transcription through other means. Accordingly, GRNs upregulated by *TT2* OE were linked with different cis-regulatory elements (Figure 7). Stress response GRN was linked with HSE, SIIE, and E-box motifs, whereas secondary metabolic process GRN was linked to the E-box and AC-rich element.

Overall, we propose that the *HSFA2* promoter is regulated by a combination of TT2/MYB5 through SIIEs and the MBW complex through AC-rich and E-box motifs. Interestingly, the *HSFA2* promoter has two tandem repeats of AC-rich elements and E-box motifs separated by exactly 21 nucleotides. Our proposed model of TT2 and MYB5 involvement in *HSFA2* promoter regulation is presented in Figure 8.

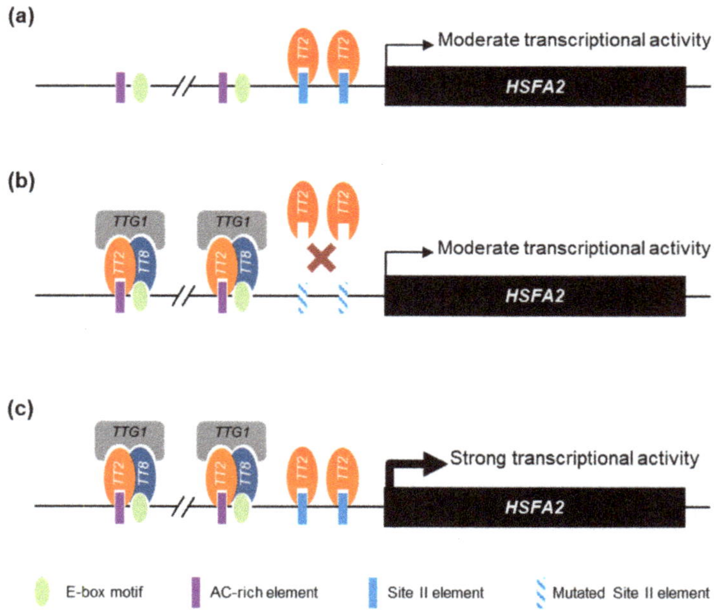

Figure 8. Proposed model of *HEAT SHOCK FACTOR A2* (*HSFA2*) promoter regulation by TRANSPARENT TESTA 2 (TT2) and MYB5. (**a**) When expressed alone, TT2 and MYB5 are able to drive a moderate transcriptional activity mostly dependent on the SII element (SIIE). (**b**) Together with the other members of the MYB–BHLH–WDR (MBW) complex, TRANSPARENT TESTA 8 (TT8) TRANSPARENT TESTA GLABRA 1 (TTG1), TT2 acquires new DNA binding specificities. When the SIIE is mutated, the MBW complex is still able to induce a moderate transcriptional activity through the E-box and AC-rich element motifs. (**c**) The concerted action of TT2-TT8-TTG1 and TT2 or MYB5 monomers trigger a strong activation of the *HSFA2* promoter. For simplicity, we represent the different proteins as regulating transcription through direct DNA binding.

Supplementary Materials: The following are available online at https://www.mdpi.com/article/10.3390/genes12050746/s1, Figure S1: Graphical representation of the promoter of the 43 coregulated genes showing enriched DNA motifs, Figure S2: *TT2* and *MYB5* are involved in plant stress resistance, Table S1. Genes strongly coregulated with HSFA2 during heat, cold, hypoxia, and salt stresses, Table S2. Genes differentially expressed in TT2 OE or MYB5 OE compared to wild-type plants, Table S3. GO terms enriched in TT2OE upregulated genes, Table S4. GO terms enriched in TT2OE downregulated genes. Table S5. GRNs controlled by TT2, Table S6. GO terms enrichment in upregulated GRN "response to stress", Table S7. GO terms enrichment in upregulated GRN "secondary metabolic process", Table S8. GO terms enrichment in downregulated GRN "response to stress", Table S9. GO terms enrichment in downregulated GRN "secondary metabolic process", Table S10. Primer used for Y2H prey cloning, Table S11. Primer used for expression analysis.

Author Contributions: Conceptualization, A.B. (Abdelhafid Bendahmane), H.H., L.L., R.G.H.I., A.B. (Adnane Boualem) and P.J.; investigation, P.J., G.B., M.D., J.T., F.v.d.W., B.D., R.S.D., D.L., and M.B.; resources, R.G.H.I. and L.L.; writing—original draft preparation, P.J.; writing—review and editing, A.B. (Abdelhafid Bendahmane), H.H., L.L., R.G.H.I., A.B. (Adnane Boualem), and P.J.; supervision, project administration and funding acquisition, A.B. (Abdelhafid Bendahmane) and H.H. All authors have read and agreed to the published version of the manuscript.

Funding: This work was supported by Gautier Semences SA and the ANRT (CIFR2012/1125). IPS2 and IJPB benefit from the support of the Labex Saclay Plant Sciences-SPS (ANR-10-LABX-0040-SPS).

Data Availability Statement: All the data supporting the results of this paper are present in the paper and/or the supplementary materials. The RNA-Seq data analysed in this article have been deposited in NCBI's Gene Expression Omnibus (GEO) and are accessible through GEO Series accession number GSE171922 (https://www.ncbi.nlm.nih.gov/geo/query/acc.cgi?acc=GSE171922). All the materials and relevant data are available from the corresponding author on request.

Acknowledgments: We thank P. Audigier for plant handling, and are grateful for the cell biology imaging platform and research facilities provided by the Institute of Plant-Science Paris-Saclay (IPS2).

Conflicts of Interest: The authors declare no competing interests.

References

1. Fragkostefanakis, S.; Simm, S.; Paul, P.; Bublak, D.; Scharf, K.-D.; Schleiff, E. Chaperone network composition in *Solanum lycopersicum* explored by transcriptome profiling and microarray meta-analysis. *Plant Cell Environ.* **2014**, *38*, 693–709. [CrossRef]
2. Fragkostefanakis, S.; Mesihovic, A.; Simm, S.; Paupière, M.J.; Hu, Y.; Paul, P.; Mishra, S.K.; Tschiersch, B.; Theres, K.; Bovy, A.; et al. HsfA2 Controls the Activity of Developmentally and Stress-Regulated Heat Stress Protection Mechanisms in Tomato Male Reproductive Tissues. *Plant Physiol.* **2016**, *170*, 2461–2477. [CrossRef] [PubMed]
3. Jacob, P.; Hirt, H.; Bendahmane, A. The heat-shock protein/chaperone network and multiple stress resistance. *Plant Biotechnol. J.* **2017**, *15*, 405–414. [CrossRef]
4. Swindell, W.R.; Huebner, M.; Weber, A.P. Transcriptional profiling of Arabidopsis heat shock proteins and transcription factors reveals extensive overlap between heat and non-heat stress response pathways. *BMC Genom.* **2007**, *8*, 125. [CrossRef] [PubMed]
5. Scharf, K.-D.; Berberich, T.; Ebersberger, I.; Nover, L. The plant heat stress transcription factor (Hsf) family: Structure, function and evolution. *Biochim. Biophys. Acta Bioenerg.* **2012**, *1819*, 104–119. [CrossRef]
6. Hahn, A.; Bublak, D.; Schleiff, E.; Scharf, K.-D. Crosstalk between Hsp90 and Hsp70 Chaperones and Heat Stress Transcription Factors in Tomato. *Plant Cell* **2011**, *23*, 741–755. [CrossRef]
7. Liu, H.-C.; Charng, Y.-Y. Common and Distinct Functions of Arabidopsis Class A1 and A2 Heat Shock Factors in Diverse Abiotic Stress Responses and Development. *Plant Physiol.* **2013**, *163*, 276–290. [CrossRef] [PubMed]
8. Nishizawa, A.; Yabuta, Y.; Yoshida, E.; Maruta, T.; Yoshimura, K.; Shigeoka, S. Arabidopsis heat shock transcription factor A2 as a key regulator in response to several types of environmental stress. *Plant J.* **2006**, *48*, 535–547. [CrossRef]
9. Ogawa, D.; Yamaguchi, K.; Nishiuchi, T. High-level overexpression of the Arabidopsis HsfA2 gene confers not only increased themotolerance but also salt/osmotic stress tolerance and enhanced callus growth. *J. Exp. Bot.* **2007**, *58*, 3373–3383. [CrossRef] [PubMed]
10. Banti, V.; Mafessoni, F.; Loreti, E.; Alpi, A.; Perata, P. The Heat-Inducible Transcription Factor HsfA2 Enhances Anoxia Tolerance in Arabidopsis. *Plant Physiol.* **2010**, *152*, 1471–1483. [CrossRef]
11. Nishizawa-Yokoi, A.; Nosaka, R.; Hayashi, H.; Tainaka, H.; Maruta, T.; Tamoi, M.; Ikeda, M.; Ohme-Takagi, M.; Yoshimura, K.; Yabuta, Y.; et al. HsfA1d and HsfA1e Involved in the Transcriptional Regulation of HsfA2 Function as Key Regulators for the Hsf Signaling Network in Response to Environmental Stress. *Plant Cell Physiol.* **2011**, *52*, 933–945. [CrossRef]

12. Nesi, N.; Jond, C.; Debeaujon, I.; Caboche, M.; Lepiniec, L. The Arabidopsis TT2 Gene Encodes an R2R3 MYB Domain Protein That Acts as a Key Determinant for Proanthocyanidin Accumulation in Developing Seed. *Plant Cell* **2001**, *13*, 2099–2114. [CrossRef] [PubMed]
13. Baudry, A.; Caboche, M.; Lepiniec, L. TT8 controls its own expression in a feedback regulation involving TTG1 and homologous MYB and bHLH factors, allowing a strong and cell-specific accumulation of flavonoids in *Arabidopsis thaliana*. *Plant J.* **2006**, *46*, 768–779. [CrossRef] [PubMed]
14. Lepiniec, L.; Debeaujon, I.; Routaboul, J.M.; Baudry, A.; Pourcel, L.; Nesi, N.; Caboche, M. Genetics and biochemistry of seed flavonoids. *Annu. Rev. Plant Biol.* **2006**, *57*, 405–430. [CrossRef] [PubMed]
15. Feyissa, D.N.; Løvdal, T.; Olsen, K.M.; Slimestad, R.; Lillo, C. The endogenous GL3, but not EGL3, gene is necessary for anthocyanin accumulation as induced by nitrogen depletion in Arabidopsis rosette stage leaves. *Planta* **2009**, *230*, 747–754. [CrossRef]
16. Appelhagen, I.; Jahns, O.; Bartelniewoehner, L.; Sagasser, M.; Weisshaar, B.; Stracke, R. Leucoanthocyanidin Dioxygenase in Arabidopsis thaliana: Characterization of mutant alleles and regulation by MYB–BHLH–TTG1 transcription factor complexes. *Gene* **2011**, *484*, 61–68. [CrossRef]
17. Xu, W.; Dubos, C.; Lepiniec, L. Transcriptional control of flavonoid biosynthesis by MYB–bHLH–WDR complexes. *Trends Plant Sci.* **2015**, *20*, 176–185. [CrossRef]
18. Giraud, E.; Ng, S.; Carrie, C.; Duncan, O.; Low, J.; Lee, C.P.; Van Aken, O.; Millar, A.H.; Murcha, M.; Whelan, J. TCP Transcription Factors Link the Regulation of Genes Encoding Mitochondrial Proteins with the Circadian Clock in Arabidopsis thaliana. *Plant Cell* **2011**, *22*, 3921–3934. [CrossRef]
19. Hruz, T.; Laule, O.; Szabo, G.; Wessendorp, F.; Bleuler, S.; Oertle, L.; Widmayer, P.; Gruissem, W.; Zimmermann, P. Genevestigator V3: A Reference Expression Database for the Meta-Analysis of Transcriptomes. *Adv. Bioinform.* **2008**, *2008*, 1–5. [CrossRef]
20. Bailey, T.L.; Boden, M.; Buske, F.A.; Frith, M.; Grant, C.E.; Clementi, L.; Ren, J.; Li, W.W.; Noble, W.S. MEME SUITE: Tools for motif discovery and searching. *Nucleic Acids Res.* **2009**, *37*, w202–w208. [CrossRef]
21. Castrillo, G.; Turck, F.; Leveugle, M.; Lecharny, A.; Carbonero, P.; Coupland, G.; Paz-Ares, J.; Oñate-Sánchez, L. Speeding Cis-Trans Regulation Discovery by Phylogenomic Analyses Coupled with Screenings of an Arrayed Library of Arabidopsis Transcription Factors. *PLoS ONE* **2011**, *6*, e21524. [CrossRef] [PubMed]
22. Thevenin, J.; Dubos, C.; Xu, W.; Le Gourrierec, J.; Kelemen, Z.; Charlot, F.; Nogué, F.; Lepiniec, L.; Dubreucq, B. A new system for fast and quantitative analysis of heterologous gene expression in plants. *New Phytol.* **2011**, *193*, 504–512. [CrossRef] [PubMed]
23. Kim, S.-J.; Tsukiyama, T.; Lewis, M.S.; Wu, C. Interaction of the DNA-binding domain of Drosophila heat shock factor with its cognate DNA site: A thermodynamic analysis using analytical ultracentrifugation. *Protein Sci.* **1994**, *3*, 1040–1051. [CrossRef] [PubMed]
24. Ma, S.; Shah, S.; Bohnert, H.J.; Snyder, M.; Dinesh-Kumar, S.P. Incorporating Motif Analysis into Gene Co-expression Networks Reveals Novel Modular Expression Pattern and New Signaling Pathways. *PLoS Genet.* **2013**, *9*, e1003840. [CrossRef] [PubMed]
25. Bernard, V.; Lecharny, A.; Brunaud, V. Improved detection of motifs with preferential location in promoters. *Genome* **2010**, *53*, 739–752. [CrossRef] [PubMed]
26. Evrard, A.; Ndatimana, T.; Eulgem, T. FORCA, a promoter element that responds to crosstalk between defense and light signaling. *BMC Plant Biol.* **2009**, *9*, 2. [CrossRef] [PubMed]
27. Trémousaygue, D.; Garnier, L.; Bardet, C.; Dabos, P.; Hervé, C.; Lescure, B. Internal telomeric repeats and 'TCP domain' protein-binding sites co-operate to regulate gene expression in *Arabidopsis thaliana* cycling cells. *Plant J.* **2003**, *33*, 957–966. [CrossRef]
28. Hervé, C.; Dabos, P.; Bardet, C.; Jauneau, A.; Auriac, M.C.; Ramboer, A.; Lacout, F.; Tremousaygue, D. In Vivo Interference with AtTCP20 Function Induces Severe Plant Growth Alterations and Deregulates the Expression of Many Genes Important for Development. *Plant Physiol.* **2009**, *149*, 1462–1477. [CrossRef]
29. Xu, W.; Grain, D.; Bobet, S.; Le Gourrierec, J.; Thévenin, J.; Kelemen, Z.; Lepiniec, L.; Dubos, C. Complexity and robustness of the flavonoid transcriptional regulatory network revealed by comprehensive analyses of MYB–b HLH–WDR complexes and their targets in A rabidopsis seed. *New Phytol.* **2014**, *202*, 132–144. [CrossRef]
30. Gonzalez, A.; Mendenhall, J.; Huo, Y.; Lloyd, A. TTG1 complex MYBs, MYB5 and TT2, control outer seed coat differentiation. *Dev. Biol.* **2009**, *325*, 412–421. [CrossRef]
31. Nakabayashi, R.; Yonekura-Sakakibara, K.; Urano, K.; Suzuki, M.; Yamada, Y.; Nishizawa, T.; Matsuda, F.; Kojima, M.; Sakakibara, H.; Shinozaki, K.; et al. Enhancement of oxidative and drought tolerance in Arabidopsis by overaccumulation of antioxidant flavonoids. *Plant J.* **2014**, *77*, 367–379. [CrossRef]
32. Sharma, S.B.; Dixon, R.A. Metabolic engineering of proanthocyanidins by ectopic expression of transcription factors in *Arabidopsis thaliana*. *Plant J.* **2005**, *44*, 62–75. [CrossRef] [PubMed]
33. Chen, M.; Wang, Z.; Zhu, Y.; Li, Z.; Hussain, N.; Xuan, L.; Guo, W.; Zhang, G.; Jiang, L. The Effect of TRANSPARENT TESTA2 on Seed Fatty Acid Biosynthesis and Tolerance to Environmental Stresses during Young Seedling Establishment in Arabidopsis. *Plant Physiol.* **2012**, *160*, 1023–1036. [CrossRef] [PubMed]
34. Kelemen, Z.; Sebastian, A.; Xu, W.; Grain, D.; Salsac, F.; Avon, A.; Berger, N.; Tran, J.; Dubreucq, B.; Lurin, C.; et al. Analysis of the DNA-Binding Activities of the Arabidopsis R2R3-MYB Transcription Factor Family by One-Hybrid Experiments in Yeast. *PLoS ONE* **2015**, *10*, e0141044. [CrossRef] [PubMed]

MDPI
St. Alban-Anlage 66
4052 Basel
Switzerland
Tel. +41 61 683 77 34
Fax +41 61 302 89 18
www.mdpi.com

Genes Editorial Office
E-mail: genes@mdpi.com
www.mdpi.com/journal/genes

www.ingramcontent.com/pod-product-compliance
Lightning Source LLC
Chambersburg PA
CBHW040749250926
51850CB00007BB/1453
9783036574097